THE STORY OF THE HUMAN BODY

'In thoroughly enjoyable and edifying prose, Lieberman . . . leads a fascinating journey through human evolution. He comprehensively explains how evolutionary forces have shaped the human species as we know it . . . . He balances a historical perspective with a contemporary one . . . while asking how we might control the destiny of our species. He argues persuasively that "cultural evolution is now the dominant force of evolutionary change acting on the human body"' *Publishers Weekly*

'This truly absorbing book explores why our bodies are not designed for our current day lifestyles . . . reading this book, you are taken on a fascinating journey through evolution and learn how our diets have changed . . . an empowering read that we just couldn't put down' *Optimum Nutrition*

'In thoughtful, lucid prose backed up by a hard-to-fathom amount of research, Lieberman gives us the language to understand the history of our ancestors – the history that lives on in our minds and in our bodies . . . *The Story of the Human Body*, expertly researched and told in an original voice, will make you look at your own body more critically – and perhaps treat it with a little more respect. After all, we sit at the edge of millions of years of small refinements that stretched this part and shortened that piece. Lieberman shows how it all fits together and that it was no accident' *Everyday eBook*

'Eloquent and precise . . . Lieberman is the first to point out that modern living and technology have made our lives better in many ways. Still, a look back at where we came from can tell us a lot about where we're headed, he says – and how we might alter that course for the better' *Grist.org*

'Lieberman holds nothing back . . . He cleverly and comprehensively points out the perils of possessing Paleolithic anatomy and physiology in a modern world and bemoans "just how out of touch we have become with our bodies" . . . If we want to continue our phenomenal run as a species, it is essential to understand (and embrace) our evolutionary legacy' *Booklist*

'A massive review of where we came from and what ails us now . . . Would that industry and governments take heed' *Kirkus Reviews*

ABOUT THE AUTHOR

Daniel Lieberman is the Chair of the Department of Human Evolutionary Biology at Harvard and a leader in the field. He has written nearly 100 articles, many appearing in the journals *Nature* and *Science*, and his cover story on barefoot running in *Nature* was picked up by major media the world over. His research and discoveries have been highlighted in newspapers and magazines, including *The New York Times*, the *Boston Globe*, *Discover* and *National Geographic*.

The Story of the Human Body

Evolution, Health and Disease

DANIEL LIEBERMAN

PENGUIN BOOKS

PENGUIN BOOKS

Published by the Penguin Group
Penguin Books Ltd, 80 Strand, London WC2R ORL, England
Penguin Group (USA) Inc., 375 Hudson Street, New York, New York 10014, USA
Penguin Group (Canada), 90 Eglinton Avenue East, Suite 700, Toronto, Ontario, Canada M4P 2Y3
(a division of Pearson Penguin Canada Inc.)
Penguin Ireland, 25 St Stephen's Green, Dublin 2, Ireland (a division of Penguin Books Ltd)
Penguin Group (Australia), 707 Collins Street, Melbourne, Victoria 3008, Australia
(a division of Pearson Australia Group Pty Ltd)
Penguin Books India Pvt Ltd, 11 Community Centre, Panchsheel Park, New Delhi – 110 017, India
Penguin Group (NZ), 67 Apollo Drive, Rosedale, Auckland 0632, New Zealand
(a division of Pearson New Zealand Ltd)
Penguin Books (South Africa) (Pty) Ltd, Block D, Rosebank Office Park,
181 Jan Smuts Avenue, Parktown North, Gauteng 2193, South Africa

Penguin Books Ltd, Registered Offices: 80 Strand, London WC2R ORL, England

www.penguin.com

First published in the United States of America by Pantheon Books,
a Division of Random House, LLC 2013
First published in Great Britain by Allen Lane 2013
Published in Penguin Books 2014
012

Copyright © Daniel E. Lieberman, 2013

The moral right of the author has been asserted

All rights reserved

Printed and bound in Great Britain by Clays Ltd, Elcograf S.p.A.

Except in the United States of America, this book is sold subject
to the condition that it shall not, by way of trade or otherwise, be lent,
re-sold, hired out, or otherwise circulated without the publisher's
prior consent in any form of binding or cover other than that in
which it is published and without a similar condition including this
condition being imposed on the subsequent purchaser

978-0-141-39995-9

www.greenpenguin.co.uk

To my parents

To my family

Contents

Preface

Like most people, I am fascinated by the human body, but unlike most folks, who sensibly relegate their interest in people's bodies to evenings and weekends, I have made the human body the focus of my career. In fact, I am extremely lucky to be a professor at Harvard University, where I teach and study how and why the human body is the way it is. My job and my interests allow me to be a jack-of-all trades. In addition to working with students, I study fossils, I travel to interesting corners of the earth to see how people use their bodies, and I do experiments in the lab on how human and animal bodies work.

Like most professors, I also love to talk, and I enjoy people's questions. But of all the questions I am commonly asked, the one I used to dread the most was "What will human beings look like in the future?" I hated this question! I am a professor of human *evolutionary* biology, which means I study the past, not what lies ahead. I am not a soothsayer, and the question made me think of tawdry science fiction movies that depict humans of the distant future as having enormous brains, pale and tiny bodies, and shiny clothing. My reflexive answer was always something along the lines of: "Human beings aren't evolving very much because of culture." This response is a variant of the standard answer that many of my colleagues give when asked the same question.

I have since changed my mind about this question and now consider the human body's future to be one of the most important issues we can think about. We live in paradoxical times for our bodies. On the one hand, this era is probably the healthiest in human history. If you live in a developed country, you can reasonably expect all your offspring to survive childhood, to live to their dotage, and to become parents and grandparents. We have conquered or quelled many diseases that used to kill people in droves: smallpox, measles, polio, and the plague. People are taller, and formerly life-threatening conditions like appendicitis, dysentery, a broken leg, or anemia are easily remedied. To be sure, there is still too much malnutrition and disease in some countries, but these evils are often the result of bad government and social inequality, not a lack of food or medical know-how.

On the other hand, we could be doing better, much better. A wave of obesity and chronic, preventable illnesses and disabilities is sweeping across the globe. These preventable diseases include certain cancers, type 2 diabetes, osteoporosis, heart disease, strokes, kidney disease, some allergies, dementia, depression, anxiety, insomnia, and other illnesses. Billions of people are also suffering from ailments like lower back pain, fallen arches, plantar fasciitis, myopia, arthritis, constipation, acid reflux, and irritable bowel syndrome. Some of these troubles are ancient, but many are novel or have recently exploded in prevalence and intensity. To some extent, these diseases are on the rise because people are living longer, but most of them are showing up in middle-aged people. This epidemiological transition is causing not just misery but also economic woe. As baby boomers retire, their chronic illnesses are straining health-care systems and stifling economies. Moreover, the image in the crystal ball looks bad because these diseases are also growing in prevalence as development spreads across the planet.

The health challenges we face are causing an intense worldwide conversation among parents, doctors, patients, politicians, journalists, researchers, and others. Much of the focus has been on obesity. Why are people getting fatter? How do we lose weight and change our diets? How do we prevent our children from becoming overweight? How can we encourage them to exercise? Because of the urgent necessity to help people who are sick, there is also an intense focus on devising new cures for increasingly common noninfectious diseases. How do we treat and cure cancer, heart disease, diabetes, osteoporosis, and the other illnesses most likely to kill us and the people we love?

As doctors, patients, researchers, and parents debate and investigate these questions, I suspect that few of them cast their thoughts back to the ancient forests of Africa, where our ancestors diverged from the apes and stood upright. They rarely think about Lucy or Neanderthals, and if they do consider evolution it is usually to acknowledge the obvious fact that we used to be cavemen (whatever that means), which perhaps implies that our bodies are not well adapted to modern lifestyles. A patient with a heart attack needs immediate medical care, not a lesson in human evolution.

If I ever suffer a heart attack, I too want my doctor to focus on the exigencies of my care rather than on human evolution. This book, however, argues that our society's general failure to think about human evolution is a major reason we fail to prevent preventable diseases. Our bodies have a story—an evolutionary story—that matters intensely. For one, evolution explains why our bodies are the way they are, and thus yields clues on how to avoid getting sick. Why are we so liable to become fat? Why do we sometimes choke on our food? Why do we have arches in our feet that flatten? Why do we have backs that ache? A related reason to consider the human body's evolutionary story is to help understand what our bodies are and are not adapted for. The answers to this question are tricky and unintuitive but have profound implications for making sense of what promotes health and disease and for comprehending why our bodies sometimes naturally make us sick. Finally, I think the most pressing reason to study the human body's story is that it isn't over. We are still evolving. Right now, however, the most potent form of evolution is not biological evolution of the sort described by Darwin, but cultural evolution, in which we develop and pass on new ideas and behaviors to our children, friends, and others. Some of these novel behaviors, especially the foods we eat and the activities we do (or don't do), make us sick.

Human evolution is fun, interesting, and illuminating, and much of this book explores the amazing journey that created our bodies. I also try to highlight the progress achieved by farming, industrialization, medical science, and other professions that have made this era the best of all times *so far* to be a human. But I am no Pangloss, and since our challenge is to do better, the last few chapters focus on how and why we get sick. If Tolstoy were writing this book, perhaps he might write that "all healthy bodies are alike; each unhealthy body is unhealthy in its own way."

The core subjects of this book—human evolution, health, and disease—are enormous and complex. I have done my best to try to keep the facts, explanations, and arguments simple and clear without dumbing them down or avoiding essential issues, especially for serious diseases such as breast cancer and diabetes. I have also included many references, including websites, where you can investigate further. Another struggle was to find the right balance between breadth and depth. Why our bodies are the way they are is simply too large a topic to cover because bodies are so complex. I have therefore focused on just a few aspects of our bodies' evolution that relate to diet and physical activity, and for every topic I cover, there are at least ten I don't. The same caveat applies to the final chapters, which focus on just a few diseases that I chose as exemplars of larger problems. Moreover, research in these fields is changing fast. Inevitably some of what I include will become out of date. I apologize.

Finally, I have rashly concluded the book with my thoughts about how to apply the lessons of the human body's past story to its future. I'll spill the beans right now and summarize the core of my argument. We didn't evolve to be healthy, but instead we were selected to have as many offspring as possible under diverse, challenging conditions. As a consequence, we never evolved to make rational choices about what to eat or how to exercise in conditions of abundance and comfort. What's more, interactions between the bodies we inherited, the environments we create, and the decisions we sometimes make have set in motion an insidious feedback loop. We get sick from chronic diseases by doing what we evolved to do but under conditions for which our bodies are poorly adapted, and we then pass on those same conditions to our children, who also then get sick. If we wish to halt this vicious circle then we need to figure out how to respectfully and sensibly nudge, push, and sometimes oblige ourselves to eat foods that promote health and to be more physically active. That, too, is what we evolved to do.

The Story of
the Human Body

1

Introduction

What Are Humans Adapted For?

> If we open a quarrel between the past and the
> present, we shall find that we have lost the future.

—WINSTON CHURCHILL

Have you ever heard of the "Mystery Monkey," which provided a sideshow to the 2012 Republican National Convention in Tampa, Florida? The monkey in question, an escaped rhesus macaque, had been living for more than three years on the city's streets scavenging food from Dumpsters and trash cans, dodging cars, and cleverly evading capture by frustrated wildlife officials. It became a local legend. Then, as hordes of politicians and journalists descended on the city for the convention, the Mystery Monkey gained sudden international fame. Politicians were quick to use the monkey's story as an opportunity to promote their views. Libertarians and liberals hailed the monkey's persistent evasion of capture as symbolic of the instinct to be free from unjust intrusions on people's (and monkeys') freedom. Conservatives interpreted the years of failed efforts to capture the monkey as symbolic of inept, wasteful government. Journalists could not resist telling the story of the Mystery Monkey and its would-be captors as a metaphor for the political circus going on elsewhere in town. Most folks simply wondered what a solitary macaque was doing in suburban Florida, where it obviously didn't belong.

As a biologist and anthropologist, I viewed the Mystery Monkey along with the reactions it inspired through a different lens—as emblematic of the evolutionarily naïve and inconsistent way that humans view our place in nature. On the face of it, the monkey epitomizes how some animals survive superbly in conditions for which they were not originally adapted. Rhesus macaques evolved in southern Asia, where their ability to forage for many different foods enables them to inhabit grasslands, woodlands, and even mountainous regions. They also thrive in villages, towns, and cities, and they are commonly used in laboratories. In this regard, the Mystery Monkey's talent for surviving off trash in Tampa is unsurprising. However, the general conviction that a free-range macaque didn't belong in a Florida city reveals how poorly we apply the same line of reasoning to ourselves. When considered from an evolutionary perspective, the monkey's presence in Tampa was no more incongruous than the presence of the vast majority of humans living in cities, suburbs, and other modern environments.

You and I exist about as far removed from our natural environment as the Mystery Monkey. More than six hundred generations ago, everybody everywhere was a hunter-gatherer. Until relatively recently—the blink of an eye in evolutionary time—your ancestors lived in small bands of fewer than fifty people. They moved regularly from one camp to the next, and they survived by foraging for plants as well as hunting and fishing. Even after agriculture was invented starting about 10,000 years ago, most farmers still lived in small villages, labored daily to produce enough food for themselves, and never imagined an existence now common in places like Tampa, Florida, where people take for granted cars, toilets, air-conditioning, cell phones, and an abundance of highly processed, calorie-rich food.

I am sorry to report that the Mystery Monkey was finally captured in October 2012, but how much should we be concerned that the vast majority of humans today still exist, as the Mystery Monkey once did, in novel conditions for which our bodies were not originally adapted? In many respects, the answer is "very little," because life at the start of the twenty-first century is pretty good for the average human being, and, overall, our species is thriving, in large part due to social, medical, and technological progress made over the last few generations. There are more than seven billion people alive, a large percentage of whom expect their children and grandchildren to live, as they will, into their seventies or above. Even countries with widespread poverty have achieved great progress: average life expectancy in India was less than fifty in 1970, but today is more than sixty-five.[1] Billions of people will live longer, grow taller, and enjoy more comfort than most kings and queens of the past.

Yet as good as things are, they could be much better, and there are plenty of reasons to worry about the human body's future. Apart from potential threats posed by climate change, we are also confronting a massive population boom combined with an epidemiological transition. As more people are living longer and fewer are dying young from diseases caused by infections or insufficient food, exponentially more middle-aged and elderly people are suffering from chronic noninfectious diseases that used to be rare or unknown.[2] Cosseted by an embarrassment of riches, a majority of adults in developed countries such as the United States and the United Kingdom are unfit and overweight, and the prevalence of childhood obesity is skyrocketing globally, presaging billions more unfit and obese people in the decades to come. Poor fitness and excess weight, in turn, are accompanied by heart disease, strokes, and various cancers, as well as a multitude of costly, chronic illnesses such as type 2 diabetes and osteoporosis. Patterns of disability are also changing in disturbing ways as more people around the globe suffer from allergies, asthma, myopia, insomnia, flat feet, and other problems. Stated succinctly, lower mortality is being replaced by higher morbidity (ill health). To some extent, this shift is occurring because fewer people are dying when they are young from communicable diseases, but we mustn't confuse diseases that become more common in older people with diseases that are actually caused by normal aging.[3] Morbidity and mortality at every age are significantly affected by lifestyle. Men and women aged forty-five to seventy-nine who are physically active, eat plenty of fruits and vegetables, do not smoke, and consume alcohol moderately have on average one-fourth the risk of death during a given year than people with unhealthy habits.[4]

The soaring incidence of so many people with chronic diseases not only portends an escalation of suffering but also gargantuan medical bills. More than eight thousand dollars is spent per person each year on health care in the United States, adding up to nearly 18 percent of the nation's gross domestic product (GDP).[5] A large percentage of this money is spent on treating preventable illnesses like type 2 diabetes and heart disease. Other countries spend less on health care, but their costs are rising at worrying rates as chronic diseases mount (France, for example, now spends about 12 percent of its GDP on health care). As China, India, and other developing countries become wealthier, how will they cope with these illnesses and their costs? Clearly, we need to lower the cost of health care and to develop new, inexpensive treatments for the billions of current and future sick people. Yet wouldn't it be better to prevent these illnesses in the first place? But how?

Which brings us back to the story of the Mystery Monkey. If people deemed it necessary to remove the monkey from the suburbs of Tampa, where it doesn't belong, then maybe we should also return its former human neighbors to a more biologically normal state of nature. Even though humans, like rhesus macaques, can survive and multiply in a wide range of environments (including suburbs and laboratories), wouldn't we enjoy better health if we ate the foods we were adapted to consume and exercised as our ancestors used to? The logic that evolution primarily adapted humans to survive and reproduce as hunter-gatherers rather than as farmers, factory laborers, or white-collar workers is inspiring a growing movement of modern-day cavemen. Followers of this approach to health contend that you would be healthier and happier if you ate and exercised more like your Stone Age ancestors. You can start by adopting a "paleodiet." Eat plenty of meat (grass-fed, of course), as well as nuts, fruits, seeds, and leafy plants, and shun all processed foods with sugar and simple starches. If you are really serious, supplement your diet with worms, and never eat grains, dairy products, or anything fried. You can also incorporate more Paleolithic activities into your daily routine. Walk or run 10 kilometers (6.2 miles) a day (barefoot, of course), climb a few trees, chase squirrels in the park, throw rocks, eschew chairs, and sleep on a board instead of a mattress. To be fair, advocates of primal lifestyles are not advocating that you quit your job, move to the Kalahari Desert, and abandon all the best conveniences of modern life such as toilets, cars, and the Internet (which is essential to blog about your Stone Age experiences to other similarly minded folks). They *are* suggesting that you rethink how you use your body, especially what you eat and how you exercise.

But are they right? If a more Paleolithic lifestyle is obviously healthier, why don't more people live this way? What are the drawbacks? Which foods and activities should we drop or adopt? Although it is obvious that humans are poorly adapted to gorging on too much junk food and lounging about in chairs all day long, our forebears also did not evolve to eat domesticated plants and animals, read books, take antibiotics, drink coffee, and run barefoot on glass-strewn streets.

These and other issues beg the fundamental question at the heart of this book: *What are human bodies adapted for?*

This is a profoundly challenging question to answer and it demands multiple approaches, one of which is to explore the evolutionary story of the human body. How and why did our bodies evolve to be the way they are? What foods did we evolve to eat? What activities did we evolve to do? Why do we have big brains, no fur, arched feet, and other distinctive features? As we shall see, the answers to these questions are fascinating, often hypothetical, and sometimes counterintuitive. A first order of business, however, is to consider the deeper, thornier question of what "adaptation" means. In truth, the concept of adaptation is notoriously tricky to define and apply. Just because we evolved to eat certain foods or do certain activities doesn't mean they are good for us, or that other foods and activities aren't better. Thus, before we tackle the story of the human body, let's consider how the concept of adaptation derives from the theory of natural selection, what the term really means, and how it might be relevant to our bodies today.

How Natural Selection Works

Like sex, evolution elicits equally strong opinions from those who study it professionally and those who consider it so wrong and dangerous that they believe the subject shouldn't be taught to children. Yet, despite much controversy and passionate ignorance, the idea that evolution occurs should not be contentious. Evolution is simply change over time. Even die-hard creationists recognize that the earth and its species have not always been the same. When Darwin published *On the Origin of Species* in 1859, scientists were already aware that former portions of the ocean floor, replete with shells and marine fossils, somehow had been thrust up into mountainous highlands. Discoveries of fossil mammoths and other extinct creatures testified that the world had altered profoundly. What was radical about Darwin's theory was its breathtakingly comprehensive explanation for how evolution occurs through natural selection without any agency.[6]

Natural selection is a remarkably simple process that is essentially the outcome of three common phenomena. The first is *variation*: every organism differs from other members of its species. Your family, your neighbors, and other humans vary widely in weight, leg length, nose shape, personality, and so on. The second phenomenon is *genetic heritability*: some of the variations present in every population are inherited because parents pass their genes on to their offspring. Your height is much more heritable than your personality, and which language you speak has no genetically heritable basis at all. The third and final phenomenon is *differential reproductive success*: all organisms, including humans, differ in how many offspring they produce who, themselves, survive to reproduce. Often, differences in reproductive success seem small and inconsequential (my brother has one more child than I do), but these differences can be dramatic and significant when individuals have to struggle or compete to survive and reproduce. Every winter, about 30 to 40 percent of the squirrels in my neighborhood perish, as did similar proportions of humans during great famines and plagues. The Black Death wiped out at least a third of Europe's population between 1348 and 1350.

If you agree that variation, heritability, and differential reproductive success occur, then you must accept that natural selection occurs, because the inevitable outcome of these combined phenomena is natural selection. Like it or not, natural selection just happens. Stated formally, natural selection occurs whenever individuals with heritable variations differ in the number of surviving offspring they have compared to other individuals in the population (in other words, they differ in their *relative fitness*)[7]. Natural selection occurs most commonly and strongly when organisms inherit rare, harmful variations, like hemophilia (the inability to form blood clots), that impair an individual's ability to survive and reproduce. Such traits are less likely to be passed on to the next generation, thus reducing or eliminating them from the population. This sort of filter is called negative selection and often leads to a lack of change over time within a population, maintaining the status quo. Occasionally, however, positive selection occurs when an organism inherits by chance an *adaptation,* a new, heritable feature that helps it survive and reproduce better than its competitors. Adaptive features, by their very nature, tend to increase in frequency from generation to generation, causing change over time.

On the face of it, adaptation appears to be a straightforward concept that should be similarly straightforward to apply to humans, Mystery Monkeys, and other living beings. If a species evolved— and hence is presumably "adapted" to a particular diet or habitat— then members of that species should be most successful eating those foods and living in those circumstances. We have little difficulty accepting that lions, for example, are adapted for the African savanna rather than temperate forests, desert islands, or zoos. By the same logic, if lions are adapted, hence best suited, to the Serengeti, aren't humans adapted, hence optimally suited, to living as hunter-gatherers? For many reasons, the answer is "not necessarily," and considering how and why this is the case has profound ramifications for thinking about how the evolutionary story of the human body is relevant to its present and future.

The Thorny Concept of Adaptation

Your body has many thousands of obvious adaptations. Your sweat glands help you stay cool, your brain helps you think, and your gut's enzymes help you digest. These attributes are adaptations because they are useful, inherited features that were shaped by natural selection and that promote survival and reproduction. You normally take these adaptations for granted, and their adaptive value often becomes evident only when they fail to function properly. For example, you might consider earwax a useless annoyance, but these secretions are actually beneficial because they help prevent ear infections. However, not all our bodies' features are adaptations (I can think of nothing useful about my dimples, nostril hair, or tendency to yawn), and many adaptations function in counterintuitive or unpredictable ways. Appreciating what we are adapted for requires us to identify true adaptations and interpret their relevance. This, however, is easier said than done.

A first problem is to identify which features are adaptations and why. Consider your genome, which is a sequence of about three billion pairs of molecules (known as base pairs) that code for slightly more than twenty thousand genes. Every instant of your life, thousands of your body's cells are replicating these billions of base pairs, each time with nearly perfect accuracy. It would be logical to infer that these billions of lines of code are all vital adaptations, but it turns out that nearly a third of your genome has no apparent function but exists because it somehow got added or lost its function over eons.[8] Your phenotype (your observable traits, such as the color of your eyes or the size of your appendix) is also replete with features that perhaps once had a useful role but no longer do, or which are simply the by-products of the way you developed.[9] Your wisdom teeth (if you still have them) exist because you inherited them, and they affect your ability to survive and reproduce no more than many other features you may have, such as a double-jointed thumb, an ear whose lower lobe is attached to the skin of the cheek, or nipples if you are a male. It is therefore erroneous to assume that all features are adaptations. Further, while it is easy to make up "just-so" stories about each feature's adaptive value (an absurd example being that noses evolved to hold eyeglasses), careful science requires testing whether particular features are actually adaptations.[10]

Although adaptations are not as widespread and easy to identify as you might assume, your body is nonetheless loaded with them. However, what makes an adaptation truly *adaptive* (that is, it improves an individual's ability to survive and reproduce) is often dependent on context. This realization was, in fact, one of the key insights that Darwin gained from his celebrated trip around the world on the *Beagle*. Darwin inferred (after he returned to London) that variations in beak shape among the finches of the Galápagos Islands are adaptations for eating different foods. During the wet season, longer and thinner beaks help finches eat preferred foods such as cactus fruits and ticks, but during dry periods, shorter and thicker beaks help finches eat less desirable foods such as seeds, which are harder and less nutritious.[11] Beak shapes, which are genetically heritable and vary within populations, are thus subject to natural selection among the Galápagos finches. As rainfall patterns fluctuate seasonally and annually, finches with longer beaks have relatively fewer offspring during dry spells, and finches with shorter beaks have relatively fewer offspring during wet spells, causing the percentage of short and long beaks to change. The same processes apply to other species, including humans. Many human variations such as height, nose shape, and the ability to digest foods like milk are heritable and evolved among certain populations because of specific environmental circumstances. Fair skin, for example, does not protect against sunburns but is an adaptation to help cells below the skin's surface synthesize enough vitamin D in temperate habitats with low levels of ultraviolet radiation during the winter.[12]

If adaptations are context dependent, what contexts matter the most? Here things can get tricky in consequential ways. Since adaptations are, by definition, features that help you have more offspring than others in your population, it follows that selection for adaptations will be most potent when the number of surviving descendants you have is most likely to vary. Put crudely, adaptations evolve most strongly when the going gets tough. As an example, your ancestors from about 6 million years ago mostly consumed fruit, but that doesn't mean their teeth were just adapted to chew figs and grapes. If rare but serious droughts made fruit scarce, then individuals with bigger, thicker molars that helped them chew other, less preferred foods such as tough leaves, stems, and roots would have had a strong selective advantage. Along the same lines, the nearly universal tendency to crave rich food like cake and cheeseburgers and store the excess calories as fat is maladaptive under today's conditions of relentless abundance, but it must have been highly advantageous in the past when food was scarcer and less calorific.

Adaptations also have costs that balance their benefits. Every time you do something, you can't do something else. Further, as conditions inevitably change, the relative costs and benefits of variations inevitably change too, depending on context. Among the Galápagos finches, thick beaks are less effective for eating cactuses, thin beaks are less effective for eating hard seeds, and intermediate beaks are less effective for eating both kinds of foods. Among humans, having short legs is advantageous for conserving heat in cold climates but disadvantageous for walking or running long distances efficiently. One consequence of these and other compromises is that natural selection rarely, if ever, achieves perfection because environments are always changing. As rainfall, temperatures, foods, predators, prey, and other factors shift and vary seasonally, annually, and over longer time spans, the adaptive value of every feature also changes. Each individual's adaptations are thus the imperfect product of an endless series of constantly altering compromises. Natural selection constantly pushes organisms toward optimality, but optimality is almost always impossible to achieve.

Perfection may be unattainable, but bodies function remarkably well under a wide range of circumstances because of the way evolution accumulates adaptations in bodies much like the way you probably keep accumulating new kitchen utensils, books, or items of clothing. Your body is a jumble of adaptations that accrued over millions of years. An analogy for this hodgepodge effect is a palimpsest, an ancient manuscript page that was written on more than once and thus contains multiple layers of texts that begin to mix up over time as the more superficial texts rub away. Like a palimpsest, a body has multiple related adaptations that sometimes conflict with one another, but at other times work in combination to help you function effectively in a broad range of conditions. Consider your diet. Human teeth are superbly adapted for chewing fruit because we evolved from apes that mostly ate fruit, but our teeth are extremely ineffective for chewing raw meat, especially tough game. Later, we evolved other adaptations such as the ability to make stones into tools and cook that now allow us to chew meat, coconuts, nettles, and just about everything else that isn't poisonous. Multiple interacting adaptations, however, sometimes lead to compromises. As later chapters will explore, humans evolved adaptations to walk and run upright, but these limited our ability to sprint fast or climb with great agility.

The final and most important point about adaptation is really a crucial caveat: no organism is primarily adapted to be healthy, long-lived, happy, or to achieve many other goals for which people strive. As a reminder, adaptations are features shaped by natural selection that promote relative reproductive success (fitness). Consequently, adaptations evolve to promote health, longevity, and happiness *only insofar as these qualities benefit an individual's ability to have more surviving offspring*. To return to an earlier topic, humans evolved to be prone to obesity not because excess fat makes us healthy, but because it increases fertility. Along the same lines, our species' proclivities to be worried, anxious, and stressed cause much misery and unhappiness, but they are ancient adaptations to avoid or cope with danger. And we not only evolved to cooperate, innovate, communicate, and nurture, but also to cheat, steal, lie, and murder. The bottom line is that many human adaptations did not necessarily evolve to promote physical or mental well-being.

All in all, trying to answer the question "What are humans adapted for?" is paradoxically both simple and quixotic. On the one hand, the most fundamental answer is that humans are adapted to have as many children, grandchildren, and great-grandchildren as possible! On the other hand, how our bodies actually manage to pass themselves on to the next generation is anything but straightforward. Because of your complex evolutionary history, you are not adapted for any single diet, habitat, social environment, or exercise regime. From an evolutionary perspective, there is no such thing as optimal health. As a result, humans—like our friend the Mystery Monkey—not only survive but sometimes also thrive in novel conditions for which we did not evolve (like the suburbs of Florida).

If evolution provides no easy-to-follow guidelines for optimizing health or preventing illness, then why should anyone interested in his or her well-being think about what happened in human evolution? How are apes, Neanderthals, and early Neolithic farmers relevant to your body? I can think of two very important answers, one involving the evolutionary past, and the other involving the evolutionary present and future.

Why the Human Evolutionary Past Matters

Everybody and every body has a story. Your body in fact has several stories. One is the story of your life, your biography: who your parents are and how they met, where you grew up, and how your body was molded by life's vicissitudes. The other story is evolutionary: the long chain of events that transformed your ancestors' bodies from one generation to the next over millions of years, and which made your body different from that of a *Homo erectus,* a fish, and a fruit fly.[13] Both stories are worth knowing, and they share certain common elements: characters (including putative heroes and villains), settings, chance events, triumphs, and tribulations.[14] Both stories can also be approached using the scientific method by framing them as hypotheses whose facts and assumptions can be questioned and rejected.

The evolutionary history of the human body is an interesting yarn. One of its most valuable lessons is that we are not an inevitable species: had circumstances been different, even slightly so, we would be very different creatures (or in all probability we wouldn't exist at all). For many people, however, the chief reason to tell (and test) the story of the human body is to shed light on why we are the way we are. Why do we have big brains, long legs, especially visible belly buttons, and other peculiarities? Why do we walk on just two legs and communicate with language? Why do we cooperate so much and cook our food? A related, urgent, and practical reason to consider how the human body evolved is to help evaluate what we are and are not adapted for, hence why we get sick. In turn, evaluating why we get sick is essential for preventing and treating diseases.

To appreciate this logic, consider the example of type 2 diabetes, an almost entirely preventable disease whose incidence is soaring throughout the world. This disease arises when cells throughout your body cease to respond to insulin, a hormone that shuttles sugar out of the bloodstream and stores it as fat. When the inability to respond to insulin sets in, the body starts acting like a broken heating system that fails to deliver heat from the furnace to the rest of the house, causing the furnace to overheat while the house freezes. With diabetes, blood sugar levels keep rising, which in turn stimulates the pancreas to produce even more insulin, but with futile results. After several years, the fatigued pancreas cannot produce enough insulin, and blood sugar levels stay persistently high. Too much blood sugar is toxic and causes horrid health problems and eventually death. Fortunately, medical science has become adept at recognizing and treating the symptoms of diabetes early on, enabling millions of diabetics to survive for decades.

On the face of it, the evolutionary history of the human body seems irrelevant to treating patients with type 2 diabetes. Because these patients need urgent, costly care, thousands of scientists now study the disease's causal mechanisms, such as how obesity makes certain cells resistant to insulin, how overworked insulin-producing cells in the pancreas stop functioning, and how certain genes predispose some but not others to the disease. Such research is essential for better treatment. But what about preventing the disease in the first place? To prevent a disease or any other complex problem one not only needs to know about its proximate, causal mechanisms, but also its deeper underlying roots. Why does it occur? In the case of type 2 diabetes, *why* are humans so susceptible to this disease? *Why* do our bodies sometimes cope poorly with modern lifestyles in ways that lead to type 2 diabetes? *Why* are some people more at risk? *Why* aren't we better at encouraging people to eat healthier food and be more physically active in order to prevent the disease?

Efforts to answer these and other *why* questions impel us to consider the evolutionary history of the human body. No one has ever expressed this imperative better than the pioneering geneticist Theodosius Dobzhansky, who famously wrote, "Nothing in biology makes sense except in the light of evolution."[15] Why? Because life is most essentially the process by which living things use energy to make more living things. Therefore if you want to know why you look, function, and get sick differently from your grandparents, your neighbor, or the Mystery Monkey, you need to know the biological history—the long chain of processes—by which you, your neighbor, and the monkey came into being differently. The important details of this story, morever, go back many, many generations. Your body's various adaptations were selected to help your ancestors survive and reproduce in an untold number of distant incarnations, not just as hunter-gatherers, but also as fish, monkeys, apes, australopiths, and more recently as farmers. These adaptations explain and constrain how your body normally functions in terms of how you digest, think, reproduce, sleep, walk, run, and more. It follows that considering the body's long evolutionary history helps explain why you and others get sick or injured when you behave in ways for which you are poorly or insufficiently adapted.

Returning to the problem of why humans get type 2 diabetes: the answer does not lie solely in the cellular and genetic mechanisms that precipitate the disease. More deeply, diabetes is a growing problem because human bodies, like those of captive primates, were adapted primarily for very different conditions that render us inadequately adapted to cope with modern diets and physical inactivity.[16] Millions of years of evolution favored ancestors who craved energy-rich foods, including simple carbohydrates like sugar that used to be rare, and who efficiently stored excess calories as fat. In addition, few if any of your distant ancestors had the opportunity to become diabetic by being physically inactive and by eating lots of soda and donuts. Apparently, our ancestors also did not experience strong selection to adapt to the causes of other recent diseases and disabilities like hardening of the arteries, osteoporosis, and myopia. The fundamental answer to why so many humans are now getting sick from previously rare illnesses is that many of the body's features were adaptive in the environments for which we evolved but have become maladaptive in the modern environments we have now created. This idea, known as the mismatch hypothesis, is the core of the new, emerging field of evolutionary medicine, which applies evolutionary biology to health and disease.[17]

The mismatch hypothesis is the focus of the second part of this book, but to figure out which diseases are or are not caused by evolutionary mismatches requires more than a superficial consideration of human evolution. Some simplistic applications of the mismatch hypothesis propose that since humans evolved to be hunter-gatherers we are therefore optimally adapted to a hunter-gatherer way of life. This kind of thinking can lead to naïve prescriptions based on what Bushmen of the Kalahari or the Inuit of Alaska have been observed to eat and do. One problem is that hunter-gatherers themselves are not always healthy, and they are highly variable, in large part because they inhabit a wide range of environments including deserts, rain forests, woodlands, and the arctic tundra. There is no one ideal, quintessential hunter-gatherer way of life. More important, as discussed above, natural selection did not necessarily adapt hunter-gatherers (or any creature) to be healthy, but instead to have as many babies as possible who then survived to breed as well. It also bears repeating that human bodies (including those of hunter-gatherers) are palimpsest-like compilations of adaptations that were accumulated and modified over countless generations. Before our ancestors were hunter-gatherers they were apelike bipeds, and before then they were monkeys, small mammals, and so on. And since then, some populations have evolved new adaptations to being farmers. Consequently, there was no single environment for which the human body evolved, and hence is adapted. Thus, to answer the question "What are we adapted for?" requires that we consider not just hunter-gatherers realistically, but also look at the long chain of events that led to the evolution of hunting and gathering, as well as what happened since we started to farm our food. As an analogy, trying to understand what the human body is adapted for by focusing on just hunter-gatherers is like trying to understand the result of a football game from watching just part of the fourth quarter.

The bottom line is that we have much to gain from considering in more than superficial depth the story of how and why the human body evolved if we wish to comprehend what humans are (and are not) adapted for. Like every family story, our species' evolutionary history is rewarding to learn but confusingly messy and full of gaps. Trying to figure out the family tree of human ancestors can make keeping track of the characters in *War and Peace* seem like child's play. However, more than a century of intense research has yielded a coherent and generally accepted understanding of how our lineage evolved from being apes in an African forest into modern humans who inhabit most of the globe. Leaving aside the precise details of the family tree (essentially, who begat whom), the story of the human body can be boiled down to five major transformations. None of them were inevitable, but each altered our ancestors' bodies in different ways by adding new adaptations and by removing others.

TRANSITION ONE: *The very earliest human ancestors diverged from the apes and evolved to be upright bipeds.*

TRANSITION TWO: *The descendants of these first ancestors, the australopiths, evolved adaptations to forage for and eat a wide range of foods other than mostly fruit.*

TRANSITION THREE: *About 2 million years ago, the earliest members of the human genus evolved nearly (though not completely) modern human bodies and slightly bigger brains that enabled them to become the first hunter-gatherers.*

TRANSITION FOUR: *As archaic human hunter-gatherers flourished and spread across much of the Old World, they evolved even bigger brains and larger, more slowly growing bodies.*

TRANSITION FIVE: *Modern humans evolved special capacities for language, culture, and cooperation that allowed us to disperse rapidly across the globe and to become the sole surviving species of human on the planet.*

Why Evolution Matters for the Present and Future, Too

Do you think that evolution is just the study of the past? I used to, and so does my dictionary, which defines evolution as "the process by which different kinds of living organisms are thought to have developed and diversified from earlier forms during the history of the earth." I am dissatisfied with this definition because evolution (which I prefer to define as change over time) is also a dynamic process that is still occurring today. Contrary to what some people assume, the human body didn't stop evolving once the Paleolithic ended. Instead, natural selection is still relentlessly chugging along, and will carry on as long as people inherit variations that influence, even slightly, how many offspring they have who survive and then breed again. As a result, our bodies are not entirely the same as our ancestors' bodies were a few hundred generations ago. Along the same lines, our descendants hundreds of generations from now will also differ from us.

Evolution, in addition, isn't just about biological evolution. How genes and bodies change over time is incredibly important, but another momentous dynamic to grapple with is *cultural evolution,* now the most powerful force of change on the planet and one that is radically transforming our bodies. Culture is essentially what people learn, and so cultures evolve. Yet a crucial difference between cultural and biological evolution is that culture doesn't change solely through chance but also through intention, and the source of this change can come from anyone, not just your parents. Culture can therefore evolve with breathtaking rapidity and degree. Human cultural evolution got its start millions of years ago, but it accelerated dramatically after modern humans first evolved around 200,000 years ago, and it has now reached dizzying speeds. Looking back on the last few hundred generations, two cultural transformations have been of vital importance to the human body and need to be added to the list of evolutionary transformations above:

TRANSITION SIX: *The Agricultural Revolution, when people started to farm their food instead of hunt and gather.*

TRANSITION SEVEN: *The Industrial Revolution, which started as we began to use machines to replace human work.*

Although these last two transformations did not generate new species, it is difficult to exaggerate their importance for the story of the human body because they radically altered what we eat and how we work, sleep, regulate our body temperature, interact, and even defecate. Although these and other shifts in our bodies' environments have spurred some natural selection, they have mostly interacted with the bodies we inherited in ways we have yet to fathom. Some of these interactions have been beneficial, especially allowing us to have more children. Others, however, have been deleterious, including a host of novel mismatch diseases caused by contagion, malnutrition, and physical inactivity. Over the last few generations we have learned how to conquer or curb many of these diseases, but other chronic, noninfectious mismatch diseases—many linked to obesity—are now increasing rapidly in prevalence and intensity. By any standard, the human body's evolution is far from over thanks to rapid cultural change.

I would therefore argue that, when applied to humans, Dobzhansky's brilliant statement that "nothing in biology makes sense except in the light of evolution" applies not just to evolution by natural selection *but also to cultural evolution.* To go a step further, since cultural evolution is now the dominant force of evolutionary change acting on the human body, it follows that we can better understand why more people are getting chronic noninfectious mismatch diseases and how to prevent these illnesses by considering interactions between cultural evolution and our inherited, still-evolving bodies. These interactions sometimes set in motion an unfortunate dynamic that typically works as follows: First, we get sick from noninfectious mismatch diseases caused by our bodies being poorly or inadequately adapted to the novel environments we have created through culture. Then, for various reasons, we sometimes fail to prevent these mismatch diseases. In some cases, we don't understand a disease's causes well enough to prevent it. Often, efforts at prevention fail because it is difficult or impossible to change the novel environmental factors responsible for the mismatch. Occasionally, we even promote mismatch diseases by treating their symptoms so effectively that we unwittingly perpetuate their causes. In all cases, however, by not addressing the novel environmental causes of mismatch diseases we permit a vicious circle to occur that allows the disease to remain prevalent or sometimes to become more common or severe. This feedback loop is not a form of biological evolution because we don't pass on mismatch diseases directly to our children. Instead it is a form of cultural evolution because we pass on the environments and behaviors that cause them.

But I am getting ahead of myself, and the human body's story. Before we think about how biological and cultural evolution interact, we first need to consider the long trajectory of evolutionary history, how we evolved the capacity for culture, and what the human body is really adapted for. This exploration requires turning back the clock about 6 million years or so to a forest somewhere in Africa . . .

PART I

Apes and Humans

2

Upstanding Apes

How We Became Bipeds

Your hands than mine are quicker for a fray,
My legs are longer though, to run away.

—SHAKESPEARE, *A Midsummer Night's Dream*

The forest, as usual, is quiet apart from the muted sounds of
rustling leaves, buzzing insects, and a few chirping birds. Suddenly, pandemonium breaks out as three chimpanzees tear through
the trees high above the forest floor, leaping spectacularly from
branch to branch, hair bristling, screaming wildly as they chase a
group of colobus monkeys at breakneck speed. In less than a minute, an experienced older chimp makes a magnificent jump, catches
a terrified monkey that was heading his way, and dashes its brains
out against a tree. The hunt is over as suddenly as it started. As the
victor rips his prey into pieces and starts to consume the flesh, other
chimps hoot with excitement. Any humans watching, however, are
likely to be shocked. Observing chimps hunt can be disturbing, not
just because of the violence, but also because we prefer to think of
them as gentle, intelligent cousins. Sometimes they seem mirrors
of our better selves, but when hunting, chimps reflect humanity's

darker tendencies in their craving for flesh, their capacity for violence, and even their lethal use of teamwork and strategy.

The scene also highlights fundamental contrasts between human and chimp bodies. Apart from the obvious anatomical differences such as fur, snouts, and walking on all fours, chimps' spectacular hunting skills underscore how athletically pathetic humans are in many ways. Humans almost always hunt with weapons because no person alive could possibly match a chimp for speed, power, and agility, especially in the trees. Despite my desire to be like Tarzan, I climb trees clumsily, and even practiced tree climbers must ascend and descend gingerly and cautiously. The ability to scamper up a tree trunk as if it were a ladder, leap between precarious branches, and make a flying grab through the air at a fleeing monkey while landing safely on a bough or branch is far beyond the skill of the most highly trained human gymnast. Although watching a chimp hunt is disturbing, I find it impossible not to admire the inhuman acrobatic capabilities of these chimps with which we share more than 98 percent of our genetic code.

Humans are comparatively poor athletes on land as well. The world's speediest humans can sprint about 23 miles (37 kilometers) per hour for less than half a minute. For most of us plodders, such speeds seem superhuman, but numerous mammals, including chimps and goats, easily run at twice that speed for many minutes without the help of coaches or years of intense training. I can't even outrun a squirrel. Running humans are also unwieldy and unsteady, unable to make rapid turns. Even the slightest bump or nudge can cause a runner to tumble to the ground. Finally, we lack power. An adult male chimp weighs 15 to 20 kilograms (33 to 44 pounds) less than most human males, yet efforts to measure their strength indicate that a typical chimp can muster more than twice as much muscle force as the brawniest of elite human athletes.[1]

As we start our exploration of the human body's story in order to ask what humans are adapted for, a key first question is: why and how did humans become so ill adapted to life in trees, as well as feeble, slow, and awkward?

The answer begins with becoming upright, apparently the first major transformation in human evolution. If there was any one key initial adaptation, a spark that set the human lineage off on a separate evolutionary path from the other apes, it was likely bipedalism, the ability to stand and walk on two feet. In his typically prescient fashion, Darwin first suggested this idea in 1871. Lacking any fossil record, Darwin made his conjecture by reasoning that the earliest human ancestors evolved from apes; by becoming upright, they emancipated their hands from locomotion, freeing them for making and using tools, which then favored the evolution of larger brains, language, and other distinctive human features:

> Man alone has become a biped; and we can, I think, partly see how he has come to assume his erect attitude, which forms one of his most conspicuous characters. Man could not have attained his present dominant position in the world without the use of his hands, which are so admirably adapted to act in obedience to his will. . . . But the hands and arms could hardly have become perfect enough to have manufactured weapons, or to have hurled stones and spears with a true aim, as long as they were habitually used for locomotion and for supporting the whole weight of the body, or, as before remarked, so long as they were especially fitted for climbing trees. . . . If it be an advantage to man to stand firmly on his feet and to have his hands and arms free, of which, from his pre-eminent success in the battle of life, there can be no doubt, then I can see no reason why it should not have been advantageous to the progenitors of man to have become more and more erect or bipedal. They would thus have been better able to defend themselves with stones or clubs, to attack their prey, or otherwise to obtain food. The best built individuals would in the long run have succeeded best, and have survived in larger numbers.[2]

A century and a half later, we now have enough evidence to suggest that Darwin was probably right. Thanks to a peculiar set of contingent circumstances—many of them initiated by climate change—the oldest known members of the human lineage developed several adaptations to stand and walk on just two legs more easily and frequently than apes. Today, we are so thoroughly adapted to being habitually bipedal, we rarely give our unusual way of standing, walking, and running much thought. But look around you: how many other creatures, apart from birds (or kangaroos if you live in Australia), do you see tottering or hopping about on just two legs? The evidence suggests that of all the human body's major transformations over the last few million years, this adaptive shift was one of the most momentous, not only because of its advantages, but also because of its disadvantages. Therefore, learning about how our early ancestors became adapted to being upright is a principal starting point for recounting the human body's journey. As a first step, let's meet those primordial ancestors, beginning with the last ancestor we shared with apes.

The Elusive Missing Link

The term "missing link," which goes back to the Victorian era, is a frequently misused word that generally refers to key transitional species in the history of life. Although many fossils are glibly labeled missing links, there is one especially fundamental species in the record of human evolution that is well and truly missing: the last common ancestor (LCA) of humans and the other apes. To our great frustration, this important species so far remains entirely unknown. Like chimps and gorillas, the LCA most likely lived, as Darwin inferred, in an African rain forest, an environment inhospitable to the preservation of bones, and thus to the creation of a fossil record. Bones that fall to the forest floor quickly rot and then dissolve. For this reason, there are few informative fossil remnants of the chimpanzee and gorilla lineages, and the chances are slim of finding fossil remains of the LCA.[3]

Although absence of evidence is not evidence for absence, it sure does lead to rampant speculation. A dearth of fossils from the part of the family tree where the LCA belongs has occasioned much conjecture and debate regarding this elusive missing link. Even so, we can make some reasonable inferences about when and where the LCA lived and what it was like by making careful comparisons of the similarities and differences between humans and apes in conjunction with what we know about our evolutionary tree. This tree, illustrated in figure 1, shows that there are three living species of African apes, and that humans are more closely related

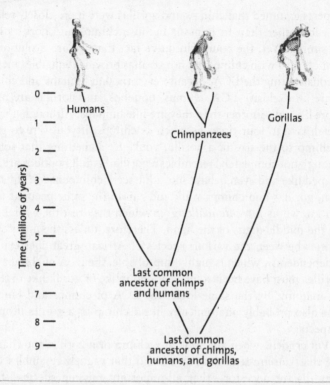

FIGURE 1. Evolutionary tree of humans, chimpanzees, and gorillas. This tree shows the two species of chimps (bonobos and common chimpanzees); some experts divide gorillas into more than one species.

to the two species of chimpanzees, common chimps and pygmy chimps (also known as bonobos), than to gorillas. Figure 1, which is based on extensive genetic data, also shows that the human and chimp lineages diverged about 8 to 5 million years ago (the exact date remains the subject of debate). Strictly speaking, humans are a special subset of the ape family termed *hominins,* defined as all species more closely related to living humans than to chimpanzees or other apes.[4]

Our especially close evolutionary relationship to chimps came as a surprise to scientists in the 1980s when the molecular evidence necessary to resolve this tree became available. Before then, most experts assumed that chimps and gorillas were more closely related to each other than to humans because chimps and gorillas look so similar. Yet, the counterintuitive fact that we are evolutionary first cousins with chimps but not gorillas provides valuable clues for reconstructing the LCA, because even though humans and chimps share an exclusive LCA, chimps, bonobos, and gorillas are much more like one another than they are like humans. Although gorillas weigh two to four times as much as chimps, if you were to grow a chimp to the size of a gorilla, you'd get something that sort of (though not completely) resembles a gorilla.[5] Adult bonobos are also shaped like and even behave like adolescent chimpanzees.[6] In addition, gorillas and chimps walk and run in the same peculiar fashion known as *knuckle walking,* in which they rest their forelimbs on the middle digits of the hand. Therefore, unless the many similarities between the various species of African great apes evolved independently, which is highly improbable, the LCA of chimps and gorillas must have been somewhat chimplike or gorillalike in terms of anatomy. By the same logic, the LCA of chimps and humans was also probably anatomically like a chimp or a gorilla in many respects.

Put crudely, when you look at a chimp or a gorilla, the chances are that you are regarding an animal that vaguely resembles your very distant ancestor—that all-important missing species—from several hundred thousand generations ago. I must emphasize, however, that this hypothesis is impossible to test definitively without direct fossil evidence, leaving plenty of room for differing opinions. Some paleoanthropologists think that the way humans stand and walk upright is reminiscent of the way that gibbons, a more distantly related ape, swing below and travel on top of branches. In fact, for more than one hundred years, when chimps and gorillas were thought to be first cousins, many scholars reasoned that humans evolved from an unknown species that was sort of gibbonish.[7] Alternatively, a few paleoanthropologists speculate that the LCA was a monkeylike creature that walked on top of branches and climbed trees using all four limbs.[8] These views notwithstanding, the balance of evidence suggests that the very first species in the human lineage evolved from an ancestor that wasn't considerably different from today's chimps and gorillas. This inference, it turns out, has major implications for understanding how and why the first hominins apparently evolved to be upright. Fortunately, unlike the still-missing LCA, we have tangible evidence of these very ancient ancestors.

Who Were the First Hominins?

When I was a student, there were no useful fossils to record what happened during the first few million years of human evolution. Lacking data, many experts had no choice but to assume (sometimes blithely) that the oldest fossils then known, such as Lucy, who lived about 3 million years ago, were good stand-ins for earlier, missing hominins. However, since the mid-1990s we have been blessed by the discovery of many fossils from the first few million years of the human lineage. These primordial hominins have abstruse, unmellifluous names, yet they have caused us to rethink what the LCA was like, and, more important, they reveal much about the origins of bipedalism and other features that made the first hominins different from the other apes. Currently, four species of early hominins, two of which are shown in figure 2, have been found. Before discussing what these species were like, what they were adapted for, and their relevance to later events in human evolution, here are some basic facts about who they were and where they came from.

The oldest known proposed species of hominin is *Sahelanthropus tchadensis,* discovered in Chad in 2001 by an intrepid French team under the leadership of Michel Brunet. Recovering fossils of this species required years of grueling, dangerous fieldwork because they had to be excavated from beneath the sands of the southern part of the Sahara Desert. Today, this area is a barren, inhospitable place, but millions of years ago it was a partly wooded habitat near a giant lake. *Sahelanthropus* is mostly known from a single, nearly complete cranium (nicknamed Toumaï, which means "hope of life" in the language of the region it was found) shown in figure 2, as well as some teeth, jaw fragments, and a few other bones.[9] According to Brunet and his colleagues, *Sahelanthropus* is at least 6 million years old and may be as old as 7.2 million years.[10]

Another proposed species of early hominin from Kenya, named

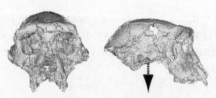

Sahelanthropus tchadensis

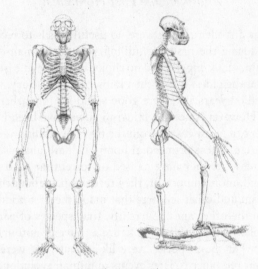

Ardipithecus ramidus

FIGURE 2. Two early hominins. Top, cranium of *Sahelanthropus tchadensis* (nick-named Toumaï); bottom, a reconstruction of *Ardipithecus ramidus* (nicknamed Ardi). The angle of the foramen magnum in Toumaï indicates a vertically oriented upper neck, a clear sign of bipedalism. The reconstruction of the partial *Ardipithecus* skeleton suggests that she was adapted for bipedal walking as well as climbing trees. Image of *Sahelanthropus* courtesy of Michel Brunet; drawing of *Ardipithecus* copyright © 2009 Jay Matternes.

Orrorin tugenensis, is about 6 million years old.[11] Unfortunately, there are only a few scraps of this enigmatic species: a single jaw fragment, some teeth, and some limb bone fragments. We still know little about *Orrorin,* in part because there is not much to study, and in part because the fossils have not yet been comprehensively analyzed.

The richest trove of early hominin fossils was discovered in Ethiopia by an international team led by Tim White and colleagues from the University of California, Berkeley. These fossils have been assigned to two different species from yet another genus, *Ardipithecus*. The older species, *Ardipithecus kadabba*, is dated to between 5.8 and 5.2 million years ago and is so far known from a handful of bones and teeth.[12] The younger species, *Ardipithecus ramidus*, dated to 4.5 to 4.3 million years ago, includes a much larger collection of fossils, including a remarkable partial skeleton of a female nicknamed Ardi, shown in figure 2.[13] This species is also represented by numerous fragments (mostly teeth) of more than a dozen other individuals. Ardi's skeleton is the focus of intense research because it gives us a rare, exciting opportunity to figure out how she and other early hominins stood, walked, and climbed.

You could fit all the fossils from *Ardipithecus, Sahelanthropus,* and *Orrorin* in a single shopping bag. Even so, they yield concrete glimpses of the earliest phases of human evolution during the first few million years after we split from the LCA. One unsurprising revelation is that these early hominins are generally apelike. As predicted by our close relationship to the African great apes, they bear many resemblances to chimps and gorillas in details of the teeth, crania, and jaws, as well as their arms, legs, hands, and feet.[14] For example, their skulls have small brains in the size range of chimps, a substantial browridge above the eyes, big front teeth, and long, projecting snouts. Many features of the feet, arms, hands, and legs of Ardi are also similar to what one sees in African apes, especially chimps. In fact, some experts have suggested these ancient species are too apelike to actually be hominins.[15] I think, however, that they are bona fide hominins for several reasons, the most important of which is that they bear indications that they were adapted to walking upright on two legs.

Will the First Hominin Please Stand Up?

Egocentric creatures that we are, humans often mistakenly consider our quintessential features to be special when in fact they are simply unusual. Bipedalism is no exception. Like many parents,

I fondly remember when my daughter took her first triumphant steps, which suddenly made her seem so much more human than our dog. A common belief (especially among proud parents) is that walking upright is particularly challenging and difficult, perhaps because it takes human children many years to learn to walk well, and because few other animals are habitual bipeds. In actual fact, the reason children don't toddle until they are about a year old and then walk and run awkwardly for a few more years is that many of their neuromuscular skills also require considerable time to mature.[16] Just as it takes years for our big-brained children to walk properly, it also takes them years to speak rather than babble, control their bowels, and manipulate tools with skill. In addition, although habitual bipedalism is rare, occasional bipedalism is unexceptional. Apes sometimes stand and walk on two legs, as do many other mammals (including my dog). Yet human bipedalism is different from what apes do in one key respect: we habitually stand and walk very efficiently because we gave up the ability to be quadrupeds. Whenever chimps and other apes walk upright, they lurch about with an awkward and energetically costly gait because they lack a few key adaptations, shown in figure 3, that enable you and me to walk well. What is especially exciting about the first hominins is that they, too, have some of these adaptations, indicating that they were also upright bipeds of some sort. However, if Ardi is generally representative of these hominins, they still retained many ancestral features useful for climbing trees. Although we are struggling to reconstruct precisely how Ardi and other early hominins walked when they weren't climbing, there is no question that they walked very differently from you and me in a much more apelike fashion. This type of early bipedalism was probably a critical intermediate form of upright locomotion that set the stage for later, more modern gaits, and it was made possible by several adaptations we still retain in our bodies today.

The first of these adaptations is the shape of the hips. If you watch a chimpanzee walk upright, observe that it keeps its legs far apart and its upper body sways from side to side like an unstable drunkard. Sober humans, in contrast, sway their torsos almost imperceptibly, which means we can spend most of our energy moving forward instead of stabilizing the upper body. Our steadier gait

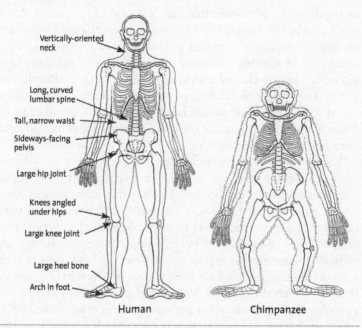

Vertically-oriented
neck

Long, curved
lumbar spine

Tall, narrow waist

Sideways-facing
pelvis

Large hip joint

Knees angled
under hips

Large knee joint

Large heel bone

Arch in foot

Human   **Chimpanzee**

FIGURE 3. Comparison of a human and chimpanzee highlighting some of the adaptations for upright standing and walking in humans. Figure adapted from D. M. Bramble and D. E. Lieberman (2004). Endurance running and the evolution of *Homo*. *Nature* 432: 345–52.

is largely attributable to a simple change in the shape of the pelvis. As figure 3 shows, the large, broad bone that forms the upper part of the pelvis (the ilium) is tall and faces backward in apes, but this part of the hip is short and faces sideways in humans. This sideways orientation is a crucial adaptation for bipedalism because it allows the muscles on the side of the hips (the small gluteals) to stabilize the upper body over each leg during walking when only one leg is on the ground. You can demonstrate this adaptation for yourself by standing on one leg as long as possible while keeping your trunk upright. (Go ahead and try!) After a minute or two, you'll feel these muscles tire. Chimps cannot stand or walk this way because their hips face backward, permitting the same muscles only to extend the leg behind them. The sole way a chimp can avoid falling sideways when one leg is on the ground is by markedly tilting its trunk to the side above that leg. Not so Ardi. Although Ardi's pelvis was badly distorted and had to be reconstructed extensively, she appears to have a shortened and sideways-facing ilium, just like a human.[17] In addition, the femur of *Orrorin* has an especially large hip joint, a long neck, and a wide upper shaft, features that allowed its hip muscles to stabilize the torso efficiently when walking and to withstand the high side-to-side bending forces this action causes.[18] These features inform us that the first hominins didn't have to lurch from side to side when walking.

Another important adaptation for being a biped is an S-shaped spine. Like other quadrupeds, apes have spines that curve gently (the front side is slightly concave), so when they stand upright, their trunks naturally tilt forward. As a result, the ape's torso is positioned unstably in front of its hips. In contrast, the human spine has two pairs of curves. The lower, lumbar curve is made possible by having more lumbar vertebrae (apes usually have three or four, whereas humans usually have five), several of which have a wedged shape in which the top and bottom surfaces are not parallel. Just as wedge-shaped stones allow architects to construct arched structures like bridges, wedged vertebrae curve the lower spine inward above the pelvis, positioning the torso stably above the hips. Human chest and neck vertebrae create another, gentler curve at the top of the spine, which orients the upper neck downward rather than backward from the skull. Although we have yet to find any early hominin lumbar vertebrae, the shape of Ardi's pelvis hints at a long lumbar region.[19] An even more telling clue of having an S-shaped spine adapted for bipedalism comes from the shape of the *Sahelanthropus* cranium. The necks of chimps and other apes emerge from near the backs of their skulls at a slightly horizontal angle, but Toumaï's cranium, shown in figure 2, is so complete we can deduce confidently that his upper neck was nearly vertical when he was standing or walking.[20] This configuration could be possible only if Toumaï's spine had a backward curve in the lower spine, the neck, or both.

Yet more crucial adaptations for upright locomotion that appear in early hominins are at the other end of the body, in the foot. Walking humans usually land first on the heel and then, as the rest of the foot makes contact with the ground, we stiffen the foot's arch, enabling us to push the body upward and forward at the end of stance, mostly with the big toe. The shape of the human arch is created by the shapes of the foot's bones, as well as by many ligaments and muscles that secure the bones in place like cables in a suspension bridge, and which become taut (to varying extents) when the heel comes off the ground. In addition, the surfaces of the joints between the toes and the rest of the foot in humans are very rounded and point slightly upward, helping us bend our toes at an extreme angle (hyperextend) when we push off. The feet of chimps and other apes lack an arch, preventing them from pushing off against a stiffened foot, and their toes are unable to extend as much as humans'.

Importantly, Ardi's foot (along with a younger partial foot that could belong to the same genus) bears some traces that the middle was partly stiffened, and it has toe joints that were capable of bending upward at the end of stance.[21] These features suggest that Ardi, like humans but unlike chimps, had feet capable of generating effective propulsion when walking upright.

The evidence I just summarized for bipedalism in the first hominins is electrifying but admittedly scant. There is a great deal we don't know about how these species stood, walked, and ran because we lack much of Ardi's skeleton, and we know almost nothing about the skeletons of *Sahelanthropus* and *Orrorin*. Nonetheless, there is sufficient evidence to indicate that these ancient species stood and walked differently than you and me to a large extent because they retained numerous ancient adaptations for climbing trees. Ardi's foot, for example, had a highly muscular and divergent big toe that was very capable of grasping around branches or tree trunks. Its other toes were long and fairly curved, and its ankle tilted slightly inward. These features, which are useful for climbing, caused her foot to function differently than modern feet. When walking, she probably used her feet more like a chimpanzee, keeping her weight along the outside of the foot rather than rolling it in (*pronating*) like a human.[22] Ardi also had short legs, and if she walked along the outside of her feet, then she might have walked with a wider stance than people today. Perhaps she had slightly bent knees as well. As you might expect, there is plenty of other evidence for tree-climbing abilities in Ardi's upper body, which had long, powerfully muscled forearms and long, curved fingers.[23]

Standing back from the details, the overall picture that emerges of the first hominins is that they were certainly not quadrupeds when they were on the ground but instead were occasional bipeds who stood and walked upright in a distinctively nonhuman manner when they were not climbing trees. They could not stride as efficiently as humans, but they were probably able to walk upright with more efficiency and stability than a chimp or a gorilla. However, these ancient ancestors were also adept climbers who likely spent a considerable portion of their time aloft. If we could observe them climbing, we'd probably marvel at their ability to scamper up boughs and jump from branch to branch, but they might have been less agile than a chimp. If we could observe them walking, we'd think their gait was slightly odd as they stepped on the sides of their long, inwardly angled feet, taking short strides. It is tempting to imagine them wobbling about unstably on two legs like upright chimps (or drunken humans), but this is unlikely. I suspect they were proficient at both walking and climbing, but they did so in a distinctive fashion unlike any creature alive today.

Dietary Differences

Animals move about for many reasons, including to escape predators and to fight, but a principal reason to walk or run is to get dinner. Accordingly, before we consider why bipedalism initially evolved we need to highlight one additional suite of features, all related to diet, which distinguishes the first hominins.

For the most part, the earliest hominins like Toumaï and Ardi have apelike faces and teeth, suggesting that they ate a rather apelike diet that was dominated by ripe fruit. For example, they have wide front teeth shaped like spatulas, which are well suited for biting into fruits just as you do when you sink your teeth into an apple. They also have cheek teeth with low cusps that are perfectly shaped for crushing the flesh of fibrous fruits. However, there are a few subtle hints that these early members of the human lineage were slightly better adapted than chimps to eating low-quality foods in addition to fruit. One difference is that their cheek teeth are moderately bigger and thicker than those of apes such as chimps and gorillas.[24] Larger, thicker molars would have been better able to break down harder, tougher items of food like stems and leaves. Second, Ardi and Toumaï are a little less snouty because of slightly more forward-placed cheekbones and more vertical faces.[25] This configuration positions the chewing muscles so they produce higher bite forces for breaking down tougher and harder foods. Finally, the canines (fangs) of early hominin males are smaller, shorter, and less dagger-shaped than those of male chimps.[26] Although some researchers believe that smaller male canines suggest that males fought less with one another, an alternative and more convincing explanation is that smaller canines were adaptations to help them chew tougher, more fibrous food.[27]

Putting the evidence together, we can conjecture with some confidence that the first hominins probably gorged as much as they could on fruit, but natural selection favored those better able to resort to eating less desirable, tough, fibrous foods, like the woody stems of plants, which require lots of hard chewing to break down. These diet-related differences are frankly subtle. However, when we consider them in combination with what we know about their locomotion and the environments in which they lived, we can begin to hypothesize why the first hominins became bipedal, thus setting the human lineage off on a very different evolutionary path from our ape cousins.

Why Be a Biped?

Plato once defined humans as featherless bipeds, but he didn't know about dinosaurs, kangaroos, and meerkats. In actual fact, we humans are the only striding, featherless, and tailless bipeds. Even so, tottering about on two legs has evolved only a few times, and there are no other bipeds that resemble humans, making it hard to evaluate the comparative advantages and disadvantages of being a habitually upright hominin. If hominin bipedalism is so exceptional, why did it evolve? And how did this strange manner of standing and walking influence subsequent evolutionary changes to the hominin body?

It is impossible to ever know for sure why natural selection favored adaptations for bipedalism, but I think the evidence most strongly supports the idea that regularly standing and walking upright was initially selected to help the first hominins forage and obtain food more effectively in the face of major climate change that was occurring when the human and chimpanzee lineages diverged.

Climate change is a topic of intense interest today because of evidence that humans are warming the earth by burning massive quantities of fossil fuel, but it has long been an influential factor in human evolution, including during the time when we split from the apes. Figure 4 graphs the temperature of the earth's oceans over the last 10 million years.[28] As you can see, between 10 and 5 million years ago, the entire earth's climate cooled considerably. Although this cooling happened over millions of years and with endless fluctuations between warmer and colder periods, the overall effect in Africa was to cause rain forests to shrink and woodland habitats to expand.[29] Now imagine yourself as the LCA—a large-bodied, fruit-eating ape—during this period. If you were living in the heart of the rain forest, you probably would not have noticed much of a difference. But if you had the misfortune to be living at the margins of the forest, then this change must have been stressful. As the forest around you shrinks and becomes woodland, the ripe fruits you hunger after become less abundant, more dispersed, and more seasonal. These changes would sometimes require you to travel farther to get the same amount of food, and you'd resort more frequently to eating fallback foods, which are more abundant but lower in quality than preferred foods such as ripe fruit. Typical fallback foods for chimpanzees include the fibrous stems and leaves of plants, as well as various herbs,[30] and the evidence for climate change suggests that the first hominins would have needed to find and eat such foods more often and more intensely than chimps do. Perhaps they were more like orangutans, whose habitats are not as continuously bountiful as those of chimps, requiring them to eat very tough stems and even bark when fruit is unavailable.[31]

Just as the tough get going when the going gets tough, natural selection acts most strongly not during times of plenty, but dur-

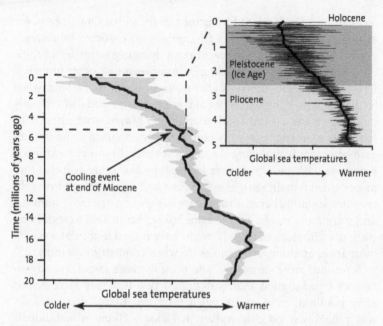

FIGURE 4. Climate change during human evolution. The left graph plots how global sea temperatures have fallen over the last 20 million years, with a major cooling event around the time the human and chimpanzee lineages diverged. This graph is expanded on the right to highlight the last 5 million years. The mean temperature indicated by the central line is an average of the many large, rapid fluctuations (shown by the zigs and zags). Note the major cooling at the beginning of the Ice Age. The graph is modified from J. Zachos et al. (2001). Trends, rhythms, and aberrations in global climate 65 Ma to present. *Science* 292: 686–93.

ing times of stress and scarcity. If, as we think, the LCA was a mostly fruit-eating ape that lived in a rain forest, then natural selection would have favored the two major transformations we see in very early hominins such as Toumaï and Ardi. The first shift is that hominins with bigger, thicker cheek teeth and the ability to chew more forcefully would have been better able to consume more tough, fibrous fallback foods. The second but more extensive shift, bipedalism, is a little harder to appreciate as an adaptation to climate change but was probably even more important in the long run for several reasons, one of which may be surprising.

One obvious advantage of bipedalism is that standing on two

feet can make it easier to forage for certain fruits. Orangutans, for example, sometimes stand nearly upright on branches when feeding in trees, reaching for precariously hanging foods by keeping their knees straight and holding on to at least one other branch.[32] Chimps and some monkeys also stand in a similar fashion when feeding on low-hanging berries and fruits.[33] So, bipedalism initially might have been a postural adaptation. Maybe competition for food was so intense that early hominins better able to stand upright gathered more fruit during seasons of scarcity. In this context, early hominins with more sideways facing hips and other features that helped them remain upright might have had an advantage over others when standing because they spent less energy, had more stamina, and were more stable. By the same token, being able to stand and walk upright more effectively might have helped hominins to carry more fruit, as chimps sometimes do when competition is intense.[34]

A second, more surprising, and possibly more important advantage of bipedalism is that walking on two legs may have helped early hominins save energy when traveling. Recall that the LCA was probably a knuckle walker. Knuckle walking is a decidedly peculiar way to walk on all fours, and it is also energetically costly. Laboratory studies that have enticed chimps to walk on treadmills while wearing oxygen masks have found that these apes spend four times more energy to walk (on either two or four limbs) a given distance than humans.[35] Four times! This extraordinary difference occurs because chimps have short legs, they sway from side to side, and they always walk with bent hips and knees. As a result, chimps constantly spend lots of energy contracting their back, hip, and thigh muscles to keep from toppling over and collapsing to the ground. Not surprisingly, chimps walk comparatively little, only about 2 or 3 kilometers a day (about 1 to 2 miles).[36] For the same amount of energy, a human can walk between 8 and 12 kilometers (5 to 7.5 miles). Therefore, if early hominins were able to walk bipedally with less lurching and with straighter hips and knees, they would have had a substantial energetic advantage over their knuckle-walking cousins. Being able to walk farther using the same amount of energy would have been a very beneficial adaptation as the rain forests shrank, fragmented, and opened up, causing preferred foods to become rarer and more dispersed. Keep in mind, however, that although the way humans walk on two legs is vastly more economical than the way chimps knuckle walk, the first hominins may have been only slightly more efficient than chimps and not as efficient as later hominins.

As one might expect, other selective pressures are hypothesized to have favored bipedalism in the first hominins. Additional suggested advantages of being upright include improved abilities to make and use tools, to see over tall grasses, to wade across streams, and even to swim. None of these hypotheses bear up under scrutiny. The oldest stone tools don't appear until millions of years after bipedalism evolved. In addition, apes can and do stand up just fine to wade and look about, and it takes considerable imagination to convince oneself that humans are well adapted for swimming either in terms of cost or speed. (Spending much time in some African lakes or rivers is also a surefire way to become a crocodile's meal.) Another longstanding idea is that bipedalism was initially selected to help hominins carry food, perhaps so males could provision females, just as hunter-gatherer men do today. In fact, one formulation of this idea is that bipedalism evolved to favor males who exchanged food for sex with females.[37] Titillating as the idea may seem—especially in light of the fact that human females, unlike their chimp counterparts, display no overt signals when they are ovulating—the hypothesis is unconvincing for several reasons, not the least of which is that human females often provision males. In addition, we don't yet know how much bigger early hominin males were than females, but in later species of hominins, males were about 50 percent bigger than females.[38] This kind of size difference between sexes is strongly associated with males vigorously competing with one another for sexual access to females rather than wooing females through cooperation and food sharing.[39]

In short, many lines of evidence suggest that climate change spurred selection for bipedalism in order to improve early hominins' ability to acquire the fallback foods they needed to eat when fruit was not available. More evidence is needed to test this scenario fully, but whatever its cause, the shift to standing and walking upright was the first major transformation in human evolution. But why was bipedalism such a big deal for what followed in human evolution? What makes it such a fundamentally important adaptation?

Why Bipedalism Matters

The tangible world around us usually appears so normal and so natural that it is tempting and sometimes comforting to assume that everything we perceive has a purpose, perhaps by design, and that things are as they should be. This way of thinking can lead one to believe that humans are as much a certainty as the moon in the sky and the laws of gravity. Although selection for bipedalism played an initial, fundamental role in the first stages of human evolution, the contingent circumstances by which it arose highlight the fallacy of its inevitability. Had early hominins not become bipeds, then humans would never have evolved as they did, and you would probably not be reading this. Further, bipedalism initially evolved because of an improbable series of events, all of which were contingent on earlier circumstances that were driven by chance shifts in the world's climate. Bipedal hominins probably neither could have nor would have evolved if knuckle-walking, fruit-eating apes hadn't previously evolved to live in the African rain forest. In addition, had the earth not cooled substantially those many millions of years ago, the conditions that favored the beginnings of bipedalism among these apes might never have existed. Our being here is the result of many rolls of the dice.

Whatever its causes, was habitually standing and walking on two legs the spark that ignited later developments in human evolution? In some ways, the kind of intermediate bipedalism we see in Ardi and company seems like an improbable trigger for what followed. As we have seen, the first hominins were like their African ape cousins in many respects, with the major exception of being upright on the ground. If a surviving relict population of very early hominins were to be discovered, we'd be more likely to send them to zoos than boarding schools because they had modest, chimp-size brains. In this respect, Darwin was prescient to speculate in 1871 that, of all the characteristics that make humans distinct, it was bipedalism rather than big brains, language, or tool use that first set the human lineage off on its separate path from the other apes. Darwin's reasoning was that bipedalism initially emancipated the hands from locomotion, allowing natural selection to subsequently favor additional capabilities such as making and using tools. In turn, these capabilities selected for bigger brains, language, and other cognitive skills that have made humans so exceptional in spite of our lack of speed, strength, and athletic prowess.

Darwin appears to have been right, but a major problem with his hypothesis was that he did not account for how or why natural selection favored bipedalism in the first place, and he could not explain why freeing the hands then selected for tool making, cognition, and language. After all, kangaroos and dinosaurs also have unencumbered hands, but they didn't evolve big brains and tool-making abilities. Such arguments led many of Darwin's successors to argue that it was big brains rather than bipedalism that led the way in human evolution.

More than one hundred years later, we now have a better idea of how and why bipedalism initially evolved and why it was such a monumental and consequential shift. As we have seen, the first bipeds didn't get up on two feet in order to free their hands; instead they probably became upright in order to forage more efficiently and to reduce the cost of walking (if the LCA was a knuckle walker). In this respect, bipedalism was probably an expedient adaptation for fruit-loving apes to survive better in more open habitats as Africa's climate cooled. Moreover, the evolution of habitual bipedalism did not require an immediate radical transformation of the body. Although few mammals habitually stand and walk on two legs, the anatomical features that make hominins effective bipeds are actually just modest shifts that were evidently subject to natural selection. Consider lumbar regions. In any population of chimps, you'll find that about half of them have three lumbar vertebrae, the other half have four, and a very tiny number have five, thanks to heritable genetic variations.[40] If having five lumbar vertebrae gave some apes a few million years ago a slight advantage when standing and walking, they would have been more likely to have passed that variation on to their offspring. The same selective processes must have applied to other features that improved the LCA's ability to be bipedal, such as how wedged its lumbar vertebrae were, the orientation of its hips, and the stiffness of its feet. How long it took for selection to transform a population of the LCA into the first bipedal hominins is unknown, but it could have occurred only if the initial intermediate stages had some benefit. Put differently, the first hominins must have had a slight reproductive advantage from being just partly better at standing or walking upright.

Change always generates new contingencies and new challenges. Once bipedalism evolved, it created new conditions for further evolutionary change to occur. Darwin, of course, understood this logic, but he mostly considered how bipedalism led to further evolutionary change by focusing on its *advantages* rather than the *disadvantages*. Yes, bipedalism did free the hands and set the stage for subsequent selection based on tool making. But these additional selective changes don't seem to have become important for millions of years, and they didn't inevitably follow from having a spare pair of limbs. What Darwin didn't give much consideration to was that bipedalism also posed new and substantive challenges for hominins. We are so used to being bipedal—it seems so normal—that we sometimes forget what a problematic mode of locomotion it can be. Ultimately, these challenges may have been just as important as its benefits for subsequent events in human evolution.

One major drawback with being bipedal is coping with pregnancy. Pregnant mammals, four-legged or two-legged, have to carry a lot of extra weight not only from the fetus but also from the placenta and extra fluids. By full term, a pregnant human mother's weight increases by as much as 7 kilograms (15 pounds). But unlike in quadrupedal mothers, this extra mass has a tendency to cause her to fall over because it shifts her center of gravity well in front of the hips and feet. As any pregnant mother-to-be will tell you, she becomes less stable and less comfortable as her pregnancy progresses, requiring her either to contract her back muscles more, which is tiring, or to lean backward, shifting her center of mass back over her hips. Although this characteristic pose saves energy, it places extra shearing stresses on the lumbar vertebrae of the lower back as they try to slide away from one another. Lower back pain is thus a common, debilitating problem for human mothers. Yet we can see that natural selection helped hominin mothers cope with this extra load by increasing the number of wedged vertebrae over which females arch their lower spines: three in females versus two in males.[41] This extra curving reduces shearing forces in the spine. Natural selection also favored females whose lumbar vertebrae have more reinforced joints to bear these stresses. And, as you would predict, these adaptations for coping with the unique problems of being a pregnant biped are very ancient and can be seen in the oldest vertebral columns of hominins so far discovered.

Another consequential disadvantage of bipedalism is loss of speed. When early hominins became bipeds they surrendered the ability to gallop. By any conservative estimate, not being able to gallop limited our early ancestors to being about half as fast as a typical ape when sprinting. In addition, two limbs are much less stable than four and make it harder to turn quickly when running. Predators such as lions, leopards, and saber-toothed tigers probably had a field day hunting hominins, making it especially perilous for our ancestors to venture into open habitats (and risk not being anyone's ancestor). Bipedalism probably also hampered the ability to climb trees with as much agility as a quadrupedal ape. It is hard to tell for sure, but early bipeds were probably unable to hunt the way chimpanzees do, by leaping through the trees. Giving up speed, power, and agility set the stage for natural selection to eventually (millions of years later) make our ancestors tool makers and endurance runners. Becoming bipedal also led to other quintessential human problems like sprained ankles, lower back pain, and knee troubles.

Yet in spite of the many disadvantages of being bipedal, the benefits of walking and standing upright must have outweighed the costs at every evolutionary stage. Early hominins apparently trudged about parts of Africa in search of fruits and other foods in spite of their lack of speed and agility on the ground. These hominins were also probably quite adept at climbing trees, and as far as we can tell, their overall way of life endured for at least 2 million years. But then another burst of evolution occurred around 4 million years ago that gave rise to a diverse group of hominins known collectively as the australopiths. The australopiths are important not only because they are a testament to the initial success and subsequent importance of bipedalism, but also because they set the stage for later, even more revolutionary shifts that further transformed the human body.

3

Much Depends on Dinner

How the Australopiths Partly Weaned Us Off Fruit

Since Eve ate apples, much depends on dinner.

—BYRON, *Don Juan*

Like me, you probably eat mostly soft and highly processed food, little of it fruit. If you added up the amount of time you actually spent chewing, it would total less than half an hour per day. This is odd for an ape. Every day, from dawn to dusk, a chimpanzee spends nearly half its wakeful hours chewing like a raw foodist.[1] Chimps typically eat forest fruits like wild figs, wild grapes, and palm fruits, none of which are as sweet and easy to chew as the domesticated bananas, apples, and oranges that you and I enjoy. Instead they are slightly bitter, less sweet than a carrot, extremely fibrous, and they have tough outer coverings. In order to get enough calories from eating such fruits all day long, a chimp consumes prodigious quantities, sometimes a kilogram (2.2 pounds) in an hour and then waits about two hours for its stomach to empty before gorging again.[2] Chimps and other apes must also resort sometimes to lower-quality foods such as leaves and gnarly stems when fruit is not abundant. When and why did we stop spending most of the day eating fruit? How did adaptations for eating different foods affect our bodies' evolution?

Adaptations for eating foods other than mostly fruit are at the heart of the second major transformation in the story of the human body. As we have seen, the first hominins probably needed to eat leaves and stems on occasion, but the trend toward increased dietary diversity accelerated dramatically about 4 million years ago in their descendants, a confusing group of species informally called the australopiths (so named because many of them belong to the genus *Australopithecus*). These diverse and fascinating ancestors occupy a special place in human evolution because their efforts to feed themselves changed what we are adapted for in ways still evident every time we look in the mirror. The most obvious of these shifts are adaptations in our teeth and face for chewing hard and tough foods. Even more important, the benefits of foraging far and wide favored further adaptations for more habitual and efficient long-distance walking than we see in Ardi and other earlier hominins. The combination of these adaptations, which were largely driven by the exigencies of climate change, had momentous implications, setting the stage for the evolution of the genus *Homo* a few million years later and for many important features of the human body. Were it not for the australopiths, your body would be very different, and you would probably be spending much more time in trees, mostly gorging on fruit.

Lucy's Gang: The Australopiths

The australopiths lived in Africa between about 4 and 1 million years ago, and we know much about them thanks to a rich fossil record of their remains. The most famous fossil of all is, of course, the glam girl herself, Lucy, a diminutive female who lived in Ethiopia 3.2 million years ago. Unfortunately for her (but luckily for us), Lucy died in a marsh, which quickly covered her up, preserving slightly more than a third of her skeleton. Lucy is just one among many hundreds of fossils belonging to a species known as *Australopithecus afarensis,* which lived in eastern Africa between 4 and 3 million years ago. *Au. afarensis,* in turn, is just one of more than half a dozen different species of australopiths. Unlike today, when there is only one living species of hominin, *Homo sapiens,* there used to be several species living at any one time, and the australopiths were an especially diverse bunch. In order to give you a quick who's who of these relatives, I've summarized their basic details in table 1. Keep in mind that some of these species are known from just a few fossil specimens, so paleontologists do not entirely agree on how to define them. Because of uncertainties and the differences among the species, a good way to make sense of the various australopiths is to divide them into two general groups: the smaller-toothed graciles and the bigger-toothed robusts. The best-known species of gracile australopiths are *Au. afarensis* (of Lucy fame), which comes from eastern Africa, and *Au. africanus* and *Au. sediba,* which come from southern Africa. The best-known robust australopiths are *Au. boisei* and *Au. robustus,* from eastern and southern Africa respectively. Figure 5 illustrates what a few of these species might have looked like.

Instead of focusing on the names and dates of these species, let's consider what they were generally like as well as some of the varia-

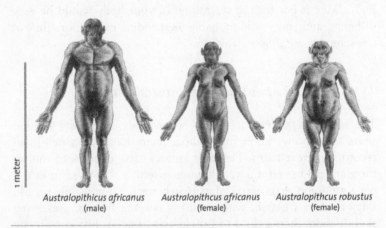

Australopithicus africanus (male)	*Australopithcus africanus* (female)	*Australopithcus robustus* (female)

FIGURE 5. Reconstructions of two species of australopiths. On the left, a male and female *Australopithecus africanus;* on the right, a female *Australopithecus robustus.* Note the relatively long arms, short legs, wide waists, and large faces. Reconstructions copyright © 2013 John Gurche.

tions they reveal. If you could observe a group of them, your first impression might be that they were upright apes. In terms of size, they were more like chimpanzees than humans: females averaged 1.1 meters in height (3 feet 7 inches) and weighed between 28 and 35 kilograms (62 to 77 pounds), while males averaged 1.4 meters in height (4 feet 7 inches) and weighed between 40 and 50 kilograms (88 to 100 pounds).[3] Lucy, for example, was just under 65 pounds (29 kilograms), but a partially complete skeleton of a male *from the same species* (nicknamed Kadanuumuu, which means "big man") weighed about 55 kilograms (121 pounds).[4] This means that male australopiths were about 50 percent larger than females, a size difference typical of species such as gorillas or baboons, in which males regularly fight with one another for access to females. Australopith heads were also generally apelike, with small brains only just a little larger than a chimpanzee's, and they retained long snouts and large browridges. Like chimps, their legs were relatively short and their arms were relatively long, but their toes and fingers were neither as long and curved as a chimp's nor as short and straight as a human's. Their arms and shoulders were powerful, well suited to climbing in trees. Finally, if you could be like Jane Goodall and observe them for years, you'd discover that the australopiths had an apelike rate of growth and reproduction: they took about twelve years to grow into adulthood and females probably had offspring every five to six years.[5]

In other respects, however, the australopiths were different not just from apes but also from the first hominins we previously discussed. One very noticeable and important contrast is what they ate. Although there is much variation, the australopiths as a whole probably ate much less fruit and instead relied more heavily on tubers, seeds, plant stems, and other foods that are hard and tough. The key evidence for this inference are their many adaptations for being prodigious chewers. Compared to presumed ancestors such as *Ardipithecus,* they had bigger teeth, more massive jaws, and faces that were wider and taller, with very forwardly placed cheekbones and large chewing muscles. These characteristics, however, vary among species, and are especially extreme in the three species of robust australopiths: *Au. boisei, Au. robustus,* and *Au. aethiopicus.* Put crudely, these robust species are the hominin equivalent of

TABLE 1. Early hominin species

Species	Date (millions of years ago)	Locations found	Brain size (cm³)	Body mass (kg)
Early hominins				
Sahelanthropus tchadensis	7.2–6.0	Chad	360	?
Orrorin tugenensis	6	Kenya	?	?
Ardipithecus kadabba	5.8–4.3	Ethiopia	?	?
Ardipithecus ramidus	4.4	Ethopia	280–350	30–50
Gracile Australopiths				
Australopithecus anamensis	4.2–3.9	Kenya, Ethiopia	?	?
Australopithecus afarensis	3.9–3.0	Tanzania, Kenya, Ethiopia	400–550	25–50
Australopithecus africanus	3.0–2.0	South Africa	400–560	30–40
Australopithecus sediba	2.0–1.8	South Africa	420–450	?
Australopithecus garhi	2.5	Ethiopia	450	?
Kenyanthropus platyops	3.5–3.2	Kenya	400–450	?
Robust Australopiths				
Australopithecus aethiopicus	2.7–2.3	Kenya, Ethiopia	410	?
Australopithecus boisei	2.3–1.3	Tanzania, Kenya, Ethiopia	400–550	34–50
Australopithecus robustus	2.0–1.5	South Africa	450–530	32–40

cows. The most specialized of the robust australopiths, *Au. boisei,* for example, had molars twice the size of yours, and its cheekbones were so wide, tall, and forwardly positioned that its face looks like a soup plate. Its chewing muscles were the size of small steaks. After Mary and Louis Leakey first discovered the species in 1959, people were so impressed with its heavy-duty jaws that it got nick-named "Nutcracker Man." In terms of the rest of their anatomy, the robust australopith species apparently differed little from their gracile cousins.[6]

The other distinctive but also variable characteristic of the australopiths to consider is how they walked. Like Ardi and the other first hominins, they were bipeds, but some species of australopiths

walked with a more humanlike striding gait thanks to many features they share with us, such as widely spaced hips, a stiff foot
with a partial arch, and a short big toe in line with the other toes.
Smoking-gun evidence for australopith bipedalism comes from the
Laetoli footprints, a trail made by several individuals—including a
male, a female, and a child—who walked across a wet ash plain in
northern Tanzania about 3.6 million years ago. These footprints
and other clues preserved in their skeletons suggest that australopith species such as *Au. afarensis* walked upright habitually and
efficiently. Other australopith species such as *Au. sediba,* however,
may have been better suited to climbing trees and walked with
shorter strides more along the outside of the foot.[7]

How did the australopiths come to be? Why were there so many
species and how did they differ? And, most important, what role
did these creatures play in the evolution of the human body? The
answers to these questions generally have to do with the continued
challenges of finding dinner as Africa's climate kept changing.

The First Junk Food Diet

You and I are unusual in many ways, not the least of which is
that when we ask the question "What's for dinner?" we have an
unprecedented choice of abundant, nutritious foods available to us.
Like other animals, however, our australopith ancestors ate only
what they could find, not in fruit-filled forests as their predecessors enjoyed, but in more open habitats with fewer trees. To make
matters worse, during the geological epoch in which they lived, the
Pliocene (5.3 to 2.6 million years ago), the earth became slightly
cooler and Africa continued to become drier. While these changes
occurred in fits and starts (as shown by the many zigs and zags of
figure 4), the overall trend in Africa during the australopith era
was the expansion of open woodland and savanna habitats, widely
diminishing and scattering the availability of fruit.[8] This fruit crisis
undoubtedly exerted strong selective pressures on the australopiths,
favoring individuals better able to gain access to other foods.

So it was that the australopiths (some species more than others)
were pushed to forage regularly for lower quality foods—so-called

fallback foods that one eats when preferred foods are unavailable. Humans still have to eat fallback foods on rare occasions. Acorns were a common food of last resort throughout Europe during the Middle Ages, and many Dutch people resorted to eating tulip bulbs to avoid starvation during the severe winter famine of 1944. As we have already seen, apes also have fallback foods; they consume leaves, plant stems, herbs, and even bark when ripe fruit is unavailable. An important point about fallback foods is that they can be the difference between life and death, so natural selection tends to act strongly on adaptations that help animals eat them.[9] We often say "you are what you eat," but evolutionary logic dictates that sometimes "you are what you'd rather not eat."

What were the fallback foods of Lucy and other australopiths? And what is the evidence that natural selection for such foods had any appreciable effect on their bodies' evolution? These questions are impossible to answer definitively, but we can make some reasonable inferences. First, there is evidence that the australopiths lived in habitats that had some fruit-bearing trees, so they probably ate fruits when they could get them, just as human foragers still do today in the tropics. It is therefore hardly surprising that their skeletons retain some adaptations for climbing trees like long arms with long, curved fingers, and their teeth have many of the features one typically sees in fruit-eating apes, including wide upper incisors that tilt forward slightly (helpful for peeling), and broad molars with low cusps (helpful for crushing pulp). However, habitats such as woodlands have lower densities of fruiting trees than rain forests, and the fruit tends to be more seasonal. It is almost certain that the australopiths faced shortages of fruit during certain times of year, and these shortages would have been extreme during drought years. Under such conditions, they probably did what the great apes do: fall back on other digestible but less desirable plants. Chimps, for example, will eat leaves (think grape leaves), plant stems (think uncooked asparagus), and herbs (think fresh bay leaves).

Studies of australopith teeth and ecological analyses of their habitats suggest that the australopiths had diverse and complex diets that included not only fruits but also edible leaves, stems, and seeds,[10] but it is highly likely that some of them also started to dig for food, thus adding new, very important, and highly nutritious fallback foods to their diet. Although most plants store carbohydrates aboveground in seeds, fruits, or in the pithy center of stems, some plants like potatoes and ginger store their energy reserves underground as roots, tubers, or bulbs, thus hiding them from herbivores like birds and monkeys and preventing them from being desiccated by the sun. These plant parts are known collectively as underground storage organs, or USOs. USOs are hard to find and they require some effort and skill to extract, but they are rich sources of food and water, and they tend to be available year-round, including dry seasons. In the tropics, one finds USOs in marshes (sedges like papyrus have edible tubers), but also in open habitats such as woodlands and savannas.[11] Many hunter-gatherers rely heavily on USOs, which sometimes make up a third or more of their diet. We now eat domesticated USOs, such as potatoes, cassava, and onions.

No one knows exactly how many USOs were eaten by different species of australopiths, but it is likely that tubers, bulbs, and roots constituted a substantial percentage of their calories and became even more important than fruits for some species. In fact, there is good reason to speculate that a diet rich in USOs—let's call it the Lucy Diet—was so effective that it partly made possible the remarkable radiation of these hominins. In order to appreciate the advantages of the Lucy Diet, it is useful to remember that about 75 percent of the plant foods that chimps eat is fruit, and the rest comes from leaves, piths, seeds, and herbs. If chimp fruits came with nutritional labels, you'd find that they are extremely high in fiber, but they are also moderately rich in starch and protein and low in fat.[12] As you might expect, chimp fallback foods are even higher in fiber and lower in starch, hence calories.[13] USOs, however, are more starchy and energy rich than many wild fruits, and they have about half the fiber content.[14] Chimpanzees infrequently dig for USOs, which are rare in forests, but when the australopiths started to dig for their dinner they would have been able to substitute USOs for the sorts of fallback foods that chimps eat when they can't get fruits.

To summarize, australopiths as a whole were gatherers who ate a varied diet that included fruit, but some of them also benefited strongly from digging frequently for tubers, bulbs, and roots. They almost certainly foraged for other fallback plant foods too, including leaves, stems, and seeds, and we can guess that, like chimps and baboons, they regularly enjoyed insects such as termites and grubs, and they must have eaten meat whenever it was possible, probably by scavenging, since being slow and unsteady bipeds likely made them ineffective hunters. However, what determined their menu choices? What evidence do we have? And, most important, how did the challenges of getting dinner—a major component of what Darwin termed the "struggle for existence"—influence the evolution of hominin bodies so they could get to these foods and eat them?

What Large Teeth You Have, Grandma!

Your body is replete with adaptations to help you acquire, chew, and then digest food. Of all these adaptations, none are as revealing as your teeth. You probably give your teeth little consideration except in terms of how they look or how much pain they cause and cost they incur, but before the era of cooking and food processing, losing your teeth could be a death sentence. Natural selection thus acts strongly on teeth because the shape and structure of each tooth largely determines an animal's ability to break down food into small particles, which are then digested to extract vital energy and nutrients. Since digesting smaller particles yields more energy, you can readily appreciate that the ability to chew as effectively as possible had substantial fitness benefits for animals like the australopiths, who, like apes, probably spent nearly half their day chewing.

Chewing USOs would have been a special challenge. The domesticated roots and bulbs we eat today have been bred to be low in fiber and tender, and cooking makes them even more chewable. In contrast, raw, wild USOs are extremely fibrous and unpleasantly tough to the modern palate. Unprocessed, they require lots of hard chewing—something you can appreciate by trying to munch a raw yam or a rutabaga. You need to chew it over and over, and with lots of force. In fact, some USOs are so fibrous that hunter-gatherers eat them in a special manner known as wadging: chewing them for a long time in order to extract any nutrients and juices and then spitting out the leftover pulp. Imagine wadging your food for hours upon hours because you are hungry and there is little else to eat.

If survival meant the ability to eat tough, hard foods effectively, natural selection would have favored australopiths better able to bite forcefully and to withstand the endless repetitions of powerful chews.

We can therefore infer a great deal about what foods, especially fallback foods, the australopiths and other hominins were selected to eat from the shape and size of their teeth. Most importantly, if there is any one defining characteristic of the australopiths it is big, flat cheek teeth with thick enamel. Gracile australopiths such as *Au. africanus* had molars that were 50 percent bigger than a chimp's, and the rocklike enamel crown of the tooth (the hardest tissue in the body) is twice as thick. Robust australopiths such as *Au. boisei* are even more extreme, with molars that were more than two times the size and up to three times the thickness. To put these differences into perspective, the area of your first molar is roughly the size of a pinky nail, about 120 square millimeters (0.19 square inches), but the same tooth in an *Au. boisei* is the size of a thumbnail, approximately 200 square millimeters (0.31 square inches). In addition to being expansive and thick, australopith teeth were very flat, much less cuspy than chimpanzee teeth, and they had long and wide roots, which helped anchor them in the jaws.[15]

Researchers have devoted much time to studying how and why the australopiths grew such big, thick, and flat cheek teeth, and the unsurprising answer is that these characteristics were adaptations to chew food that was tough and sometimes also hard.[16] Just as thicker, bigger soles make hiking boots more resilient on trails than thin-soled sneakers, thicker and larger teeth are better suited to breaking down harder, tougher foods. Having thick enamel helps teeth resist wear from high pressures and from grit that inevitably clings to foods. In addition, big, flat tooth surfaces are useful because they spread bite forces over a large area and allow you to grind foods with a partly sideways movement, ripping tough fibers apart. Basically, the australopiths, especially the robust species, had giant teeth shaped like millstones, well adapted for endlessly grinding and pulverizing tough food under high pressure. If you had to chew uncooked, unprocessed tubers for half of each day for your entire life, you'd appreciate having these humongous teeth, too. And to some extent, you still do, thanks to your australopith

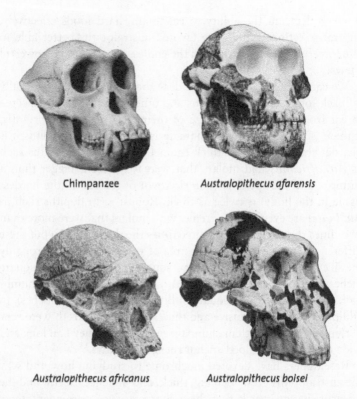

Chimpanzee

Australopithecus afarensis

Australopithecus africanus

Australopithecus boisei

FIGURE 6. Comparison of a chimpanzee skull with three species of australopiths. *Australopithecus afarensis* and *Australopithecus africanus* are both considered more gracile, while *Australopithecus boisei* is more robust, with bigger teeth, larger chewing muscles, and a larger face.

legacy. Although human cheek teeth are not as big and thick as those of australopiths, they are actually bigger and thicker than those of chimps.

Most things in life involve trade-offs, including tooth size. There is only so much room in the jaw for teeth, even if you have a long snout like an australopith. In terms of the front teeth, the earliest australopiths, such as *Au. afarensis,* have very apelike incisors that are broad and projecting, well adapted for sinking your teeth into fruits. But as australopith cheek teeth evolved to become bigger and thicker, their incisors became smaller and more vertical, and their canines also shrank to about the same size as their incisors. To some extent, smaller front teeth reflect the declining importance of fruit in these hominins' diet, but they also reflect the need to make room for bigger cheek teeth. Today, we still have small front teeth with incisor-like canines.

If your molars are big and thick in order to chomp for many hours a day on tough, hard, fibrous food, you also need big, strong chewing muscles. Not surprisingly, the australopith skulls such as those in figure 6 bear many traces of having had massive chewing muscles that could generate lots of bite force. The temporalis, the fan-shaped muscle along the side of the head, was so large in many australopiths that bony crests grew off the top and back of the skull to give the muscle more room to insert. In addition, this muscle's belly, which runs between the temples and the cheekbone to insert on the jaw, was so thick that the cheekbones (the zygomatic arches) of the australopiths were displaced far to the side, making their faces as wide as they were tall. The large cheekbones of the australopiths also provided plenty of room to vastly expand another major chewing muscle, the masseter, which runs from the cheekbone to the base of the jaw. In addition to being large, australopith chewing muscles were also configured to generate forces efficiently.[17]

Have you ever chewed something so hard and for so long that your jaw muscles ached? It turns out that when animals, including humans, generate such high bite forces they cause bones in the jaw and face to deform slightly, causing microscopic damage. Minor levels of deformation and damage are normal and cause bones to repair themselves and grow thicker.[18] Repetitive high deformations, however, can damage the bone seriously, potentially causing a fracture. Therefore, species that generate high chewing forces tend to have upper and lower jaws that are thicker, taller, and wider, thereby lowering the stresses caused by every bite, and the australopiths are no exception. As you can see in figure 6, the australopiths had massive jaws, and their large faces were heavily reinforced with thick pillars and sheets of bone that allowed them to chew tough, hard foods all day long without breaking their faces.[19] This facial buttressing is impressive in the gracile australopiths, but the robust australopiths have faces and jaws so heavily built they resemble armored tanks.

In short, australopiths, like chimps and gorillas, probably loved fruit, but they must have eaten whatever foods they could get their hands on. There was no single australopith diet, and the half dozen or so species that we know about undoubtedly ate varied diets that reflected the diverse ecological conditions in which they lived. But as climate change caused fruits to become rarer, tough fallback foods, especially USOs, must have become increasingly important resources for these ancient relatives—a heritage we still retain to some extent.[20] But how did they get these foods in the first place?

Tottering for Tubers

As you forage in a market, changing your diet mostly involves reaching for a different box of this or that, perhaps even venturing down an unfamiliar aisle. Hunter-gatherers, in contrast, spend hours every day traveling long distances in search of food. In this respect, chimpanzees and other forest-dwelling apes are more like modern shoppers than hunter-gathers because they rarely travel far to fill their bellies, regardless of whether they eat their preferred diet of fruit or "fall back" on less desirable leaves, stems, and herbs. A typical female chimpanzee walks about 2 kilometers (1.2 miles) a day, mostly going from one fruiting tree to another; male chimps walk an additional kilometer or so (closer to 2 miles) each day.[21] Otherwise, both sexes spend most of the day feeding, digesting, grooming, and otherwise interacting. When fruit is scarce, chimps and other apes resort to fallback foods that are ubiquitous, but doing so requires little change in how far they travel. In essence, apes are surrounded by foods they mostly choose to ignore.

Switching from a diet primarily of fruit to one chiefly of tubers and other fallback foods must have had an enormous impact on australopith travel needs. There were many species of australopiths, but all of them lived in partly open environments that ranged from woodlands adjoining rivers or lakes to grasslands. In addition to being less filled with fruit-bearing trees, these habitats were also more seasonal than the rain forests in which apes usually live. As a result, the australopiths must have foraged for foods that were more dispersed, and they almost certainly had to walk longer distances every day to find enough to eat, sometimes in open landscapes that would have exposed them to dangerous predators and withering heat. But at the same time, australopiths probably still had to climb trees, not just for food, but also to find safe places to sleep.

The demands of traveling far to get enough food and water are evident in many important adaptations for walking that evolved in several species of australopiths and which are still evident in humans today. As we saw before, early hominins like Ardi and Toumaï were bipeds of some sort, but Ardi (and thus perhaps Toumaï) did not walk entirely like us but probably took shorter strides using mostly the side of her foot to bear her weight. Ardi also retained lots of features for tree climbing, such as grasping feet with divergent big toes that likely compromised her ability to walk as efficiently as we do. However, a number of adaptations for more habitual and efficient bipedalism first appear starting about 4 million years ago in some australopiths, indicating that there was strong selection to make at least some of these species better long-distance walkers. These adaptations are such important features of the human body today that they are worth considering to help make sense of how and why we walk as we do.

Let's begin with efficiency. When apes walk, they are unable to stride like humans with relatively straight hips, knees, and ankles; instead they shuffle forward with these joints bent at an extreme angle. A gait that resembles the way Groucho Marx walked is amusing to watch, but it is also tiring and costly for reasons that help illuminate the fundamental mechanics of walking. Figure 7 illustrates how during walking, legs function like pendulums that alternate their center of rotation. When the leg is swinging forward, the center of rotation is the hip. But when the leg is on the ground and supporting the body above, it becomes an upside-down pendulum whose center of rotation is the ankle. This reversal allows us and other mammals to save energy with a clever trick. During the first half of every step, the leg's muscles contract to push the leg down, vaulting the body over the foot and ankle. This vaulting action raises the body's center of mass, storing up potential energy in the same way you build up potential energy in a weight by lifting it off the ground. Then, during the second half of each step, this stored energy is mostly returned in the form of kinetic energy as the body's center of mass falls (as if you were to drop the weight). Pendular walking is thus very efficient. However, walking becomes much more costly when you shuffle like a chimp with extremely bent hips, knees, and ankles because gravity is always pulling your body down, trying to flex those joints even more. Groucho gaits require you to contract your butt, thigh, and calf muscles constantly and forcefully to maintain your leg as a stiff, upside-down pendulum. In addition, flexing the leg's joints shortens your stride, so you travel less far per step. Experiments that measure the energy cost of walking show that a bent-hip and bent-knee gait is considerably less efficient than walking normally: a male chimp that weighs 45 kilograms (100 pounds) spends about 140 calories to walk 3 kilometers (nearly 2 miles), around three times as much as a 65 kilogram (145 pound) human requires to walk the same distance.[22]

Unfortunately, we'll never be able to watch australopiths walk or entice one to wear an oxygen mask to measure its cost of locomotion. Some researchers think these ancestors walked like upright chimps, with flexed hips, knees, and ankles.[23] Several lines of evidence, however, suggest that some species of australopiths strode efficiently, like you and I, with relatively straight (extended) joints. A number of these clues come from the foot, which has many features we retain today. Unlike apes and Ardi, whose big toes are long and diverge outward to help them grasp on to things and climb trees, species like *Au. afarensis* and *Au. africanus* had human-shaped big toes that were short, hefty, and in line with the other toes.[24] Like us, they also had a partial longitudinal arch in the foot, capable of stiffening the middle of the foot while they walked.[25] A stiffened arch and upwardly oriented joints at the base of the toes indicate that australopiths, like humans, were able to use their toes effectively to push the body forward and upward at the end of each step. And, crucially, some australopith species, such as *Au. afarensis,* had a big, flat heel bone, adapted for coping with high-impact forces caused by heel striking.[26] This kind of heel, characteristic of humans as well, tells us that when Lucy walked, she must have swung her leg forward in an extended, humanlike manner with a lengthy stride. However, at least one other australopith species,

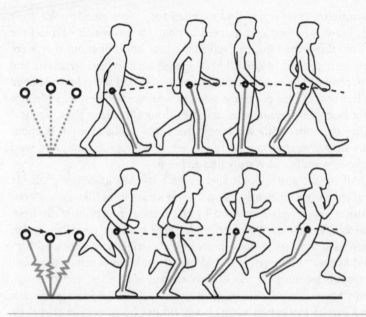

FIGURE 7. Walking and running. In walking, the leg functions during the stance like an upside-down pendulum, elevating the center of mass (circle) in the first half of the stance before it falls in the second half. In running, the leg acts more like a spring, stretching as the center of mass falls in the first half of the stance and then recoiling to help push the body up in the second half of the stance and then into a jump.

Au. sediba, had smaller, less stable heels and probably walked on a turned-in foot with a less marked heel strike and a shorter stride.[27]

Another set of adaptations for efficient walking that we still retain is evident in many of the lower limbs of australopith fossils.[28] Australopiths had femurs that were angled inward, placing their knees near the body's midline, so they didn't have to walk with a wide stance, swaying from side to side like a toddler or a drunk.[29] Their hip and knee joints were large and well buttressed, able to deal with the high forces caused when walking with just one leg on the ground. For the most part, their ankles had a nearly humanlike orientation with more stability but less flexibility than chimp ankles, presumably to help prevent dangerous ankle sprains.

Finally, it is clear that australopiths had several adaptations to stabilize their upper bodies when they were bipedal. We don't yet know whether long, curved lumbar spines, which position the trunk above the hips, evolved in the first hominins, but they were certainly present in australopith species such as *Au. africanus* and *Au. sediba.*[30] In addition, australopiths also had wide, basin-shaped pelves that curved out to the side. As we discussed before, wide hips that face sideways allow the muscles along the side of the hip to stabilize the upper body when only one leg is on the ground. Without this shape, we'd always be in danger of falling sideways, and we'd have to waddle awkwardly like a chimp.

All in all, australopith species such as *Au. afarensis* probably walked rather efficiently using a somewhat humanlike gait, a conclusion evocatively preserved by the famous footprint trails from Laetoli, Tanzania. Whoever made these trails (a good bet is *Au. afarensis*) appears to have been able to stride with extended hips and knees.[31] However, it would be a mistake to conclude that australopith locomotion was exactly the same as ours, and they still must have climbed trees to get fruit, to seek refuge from predators, and maybe to sleep at night. It should not be surprising that their skeletons retain some features inherited from apes that were useful for climbing trees. Like chimps and gorillas, they had relatively short legs and long arms with long, slightly curved toes and fingers. Many australopith species had powerful forearm muscles and upwardly oriented shoulders, well adapted for hanging or pulling themselves up. Adaptations for tree climbing are especially prominent in the upper body of *Au. sediba.*[32]

Selection for striding gaits in the australopiths left several legacies in the human body. Most important, their ability to walk effectively and efficiently played a key role in the arc of human evolution by transforming hominins into endurance walkers, well adapted for long-distance trekking through open habitats. Remember that selection to reduce the cost of walking is evidently of little consequence for chimps, probably because they walk only a mile or two in any given day, and they also need to climb and leap in trees. But if the australopiths had to travel long distances regularly in search of fruit or tubers, increased economy of locomotion would have been very advantageous. Imagine that a typical australopith mother weighed 30 kilograms (66 pounds) and had to travel 6 kilometers

(3.7 miles) a day, twice as far as a chimpanzee mother. If she walked as efficiently as a human female, she would save about 140 calories a day (which adds up to nearly 1,000 calories a week). If she were only 50 percent more economical than a chimp, she would still save 70 calories a day (nearly 500 calories a week). When food was scarce, such differences could have a large selective benefit.

As we have already discussed, being bipedal had other highly consequential costs and benefits for hominin bodies. The biggest disadvantage to being upright is the inability to run fast by galloping. The australopiths must have been slow. Whenever the australopiths ventured down from trees, they were easy pickings for such carnivores as lions, saber-toothed cats, cheetahs, and hyenas that hunt in open habitats. Perhaps they were able to sweat and thus could wait until midday to move about when these predators would have been unable to cool down as effectively. In terms of advantages, tramping around upright makes it easier to carry food, and a vertical posture exposes less surface area to the sun, which means that bipeds heat up less than quadrupeds from solar radiation.[33]

The final major advantage of being a biped, emphasized by Darwin, was that it freed the hands for other tasks, including digging. USOs often lie several feet belowground, and it can take twenty to thirty minutes of hard work to excavate them with a stick. I suspect that digging was not a problem for the australopiths. The shapes of their hands are intermediate between those of apes and humans, with longer thumbs and shorter fingers than apes,[34] and they must have been able to grasp a stick effectively. In addition, digging sticks require little skill to select or modify, and making them is certainly within the capabilities of chimps, which modify sticks to fish for termites and spear small mammals and select stones to break open nuts.[35] Perhaps selection for digging with sticks set the stage for later selection to make and use stone tools.

Your Inner Australopith

Why should anyone today care about the australopiths? Apart from being upright walkers, they seem so very different from you and me. How can we relate to these long-extinct ancestors whose brains were little bigger than a chimp's and who spent their days foraging for an unimaginably tough and unpleasant diet?

I think there are two good reasons to pay attention to the australopiths. First, these distant ancestors were a key intermediate stage in human evolution. Evolution generally occurs through a long series of gradual changes, each of which is contingent on previous events. Just as the australopiths would not have evolved had not early hominins such as *Sahelanthropus* and *Ardipithecus* become bipeds of a sort, the genus *Homo* would not have evolved if *Australopithecus* had not become less arboreal, more habitually bipedal, and less dependent on fruit, setting the stage for subsequent evolution occasioned by yet more climate change. Even more important, there is a lot of australopith in all of us. Humans are odd apes because we spend little to no time in trees (were you arboreal today?), we walk a lot, and we don't eat just fruit for breakfast, dinner, and lunch. These trends might have begun when we initially split from the apes, but they intensified remarkably over the millions of years during which various species of australopiths evolved. Many traces of these evolutionary experiments persist in your body. Compared to a chimp, your cheek teeth are thick and big. Your big toe is short, stubby, and woefully unable to grasp branches. You have a long, flexible lower back, an arch in your foot, a waist, a big knee, and many other features that help make you an excellent long-distance walker. We take these features for granted as normal, but they are actually very unusual, present in our bodies only because of strong selection for gathering and eating fallback foods millions of years ago.

Nevertheless, you are not an australopith. Compared to Lucy and her kin, your brain is three times bigger, and you have long legs, short arms, and no snout. Instead of eating lots of low-quality food, you rely on very high quality food like meat, as well as tools, cooking, language, and culture. These and many other important differences evolved during the Ice Age, which began around two and a half million years ago.

4

The First Hunter-Gatherers

*How Nearly Modern Bodies Evolved
in the Human Genus*

> A Hare one day ridiculed the short feet and slow pace
> of the Tortoise, who replied, laughing: "Though you
> be swift as the wind, I will beat you in a race."
>
> —AESOP, "The Tortoise and the Hare"

Are you worried about rapid global climate change today? If not, you should be, because rising temperatures, altered rainfall patterns, and the ecological shifts they cause imperil our food supply. Yet, as we have already seen, global climate change has long been a major impetus in human evolution because of its effects on the age-old problem of "what's for dinner?" It turns out that getting enough food in the face of global climate change also triggered the age of humans.

Getting dinner (or, for that matter, breakfast and lunch) probably does not dominate your list of daily concerns, yet most creatures are almost always hungry and preoccupied with the quest for calories and nutrients. To be sure, animals also need to find mates and avoid being eaten, but the struggle for existence is often a struggle for food, and until recently the vast majority of humans were no exception to this rule. Consider also that acquiring food is even more taxing when your habitat alters dramatically, causing the foods you normally eat to vanish or become less common. As we saw, the challenge of finding enough to eat sparked the first two major transformations in human evolution. As Africa became cooler and drier many millions of years ago, fruit became more scattered and scarce, favoring those ancestors who were better able to forage by standing and walking upright. Additional evolutionary responses were big, thick cheek teeth and large faces well suited to eating foods other than fruit, including tubers, roots, seeds, and nuts. Yet, as important as these transformations were, it is hard to think of Lucy and other australopiths as human. Although bipeds, they retained ape-sized brains, and they didn't speak, think, or eat like us.

Our bodies and the way we behave evolved to be much more recognizably "human" at the dawn of the Ice Age, a truly pivotal period of change in the earth's climate that was initiated by continued global cooling between 3 and 2 million years ago. Over this period, the earth's oceans cooled about 2 degrees Celsius (3.6 degrees Fahrenheit).[1] Two degrees may seem trivial, yet as an average of global ocean temperatures, it represents an enormous quantity of energy. Global cooling involved many back and forth shifts, but by 2.6 million years ago the earth had chilled sufficiently to cause ice caps at the poles to expand. Our ancestors had no idea that gigantic glaciers were forming thousands of miles away, but they certainly experienced cycles of habitat change that were intensified by tumultuous geological activity, especially in eastern Africa.[2] Because of a massive volcanic hotspot, the entire region was pushed upward like a soufflé, and then (like some soufflés) the central portion collapsed, forming the Great Rift Valley. The Rift Valley created an extensive rain shadow, drying out much of eastern Africa. The Rift Valley also harbored many lakes, which to this day continue to fill up and then drain out in cycles.[3] Although eastern Africa's climate was constantly changing, the overall trend was that forests shrank while woodlands, grasslands, and other more arid, seasonal habitats expanded. By 2 million years ago, the region looked much more like the set of The Lion King than Tarzan.[4]

Imagine being a hungry hominin about 2.5 million years ago, living in a shifting mosaic of grasslands and woodlands, and wondering what to eat. How would you cope as preferred foods, like fruits, became scarcer? One solution, which we saw in the big-faced and humongous-toothed robust australopiths, was to focus even more intensively on increasingly prevalent tough, hard foods like roots, tubers, bulbs, and seeds. These hominins must have spent many hours a day arduously chewing, chewing, and chewing. Fortunately for us, natural selection seems to have also favored a second, revolutionary strategy to cope with changing habitats: hunting and gathering. This innovative way of life involved continuing to gather tubers and other plants but incorporated several new, transformative behaviors that included eating more meat, using tools to extract and process foods, and cooperating intensively to share foods and other tasks.

The evolution of hunting and gathering underlies the evolution of the human genus, *Homo.* Moreover, the key adaptations that were selected to make this ingenious way of life possible among the first humans were not big brains, but modern-shaped bodies. More than anything else, the evolution of hunting and gathering spurred your body to be the way it is.

Who Were the First Humans?

The Ice Age precipitated the evolution of hunting and gathering along with modern bodies in several species of early *Homo,* but the most important is *H. erectus.* This consequential species has figured prominently in our understanding of human evolution since 1890, when Eugène Dubois, an intrepid Dutch army doctor inspired by Darwin and others, set off to Indonesia to find the true missing link between humans and apes. Blessed with good luck, Dubois found a fossil skullcap and a femur within months of arriving and promptly named it *Pithecanthropus erectus* ("upright apeman").[5] Then in 1929, comparable fossils were found in a cave near Beijing (then Peking), China, and named *Sinanthropus pekinensis.* In the ensuing decades more fossils of a similar nature started turning up in Africa, at Olduvai Gorge in Tanzania, and in places like

Morocco and Algeria in North Africa. As with the Peking Man fossils, many of these finds were initially given new species names, and it wasn't until after World War II that scholars came to the conclusion that the far-flung specimens actually belonged to a single species, *H. erectus*.[6] According to the best evidence currently available, *H. erectus* first evolved in Africa by 1.9 million years ago and then rapidly started to disperse from Africa into the rest of the Old World. *H. erectus* (or a closely related species) shows up in the Caucasus Mountains of Georgia by 1.8 million years ago and in both Indonesia and China by 1.6 million years ago. In parts of Asia, the species persisted until less than a few hundred thousand years ago.

As you might expect for a species that endured for almost 2 million years on three continents, *H. erectus* came in a variety of shapes, much as we still do. Table 2 summarizes some essential facts. They ranged from 40 to 70 kilograms (88 to 150 pounds) in weight, and from 122 centimeters to more than 185 centimeters (4 feet to almost 6 feet) in height.[7] Many of them were the size of humans today, but females were at the smaller end of the human range, as was an entire population discovered in Georgia (at a site named Dmanisi). If you met a group of *H. erectus* on the street, you'd probably recognize them as being extremely humanlike, especially from the neck down. As figure 8 depicts, unlike australopiths, their bodies had modern human proportions with relatively long legs and short arms. They had tall, narrow waists and completely modern feet, but their hips flared out more to the side than ours. Like us, they had low, wide shoulders and broad, barrel-shaped chests. But their heads were not entirely like ours. Although *H. erectus* didn't have snouts, their faces were tall and deep, and males especially had an enormous, barlike browridge above their eyes. *H. erectus* brains were intermediate in size between australopith and human brains, and their skulls were long and flat on top and angled out at the back instead of being round like ours. Their teeth were nearly identical to human teeth today, but just a little larger.

Of the many species in your family tree, *H. erectus* was one of the most important, yet its evolutionary origins are murky. There are at least two other early species in the genus *Homo*, also summarized in table 2, that might have been its ancestor. The first,

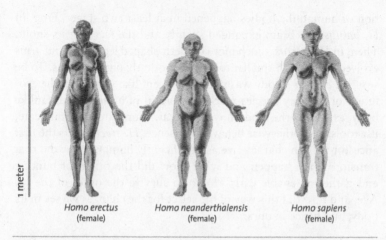

Homo erectus   Homo neanderthalensis   Homo sapiens
(female)   (female)   (female)

FIGURE 8. Reconstructions of females of three species of *Homo: H. erectus, H. neanderthalensis,* and *H. sapiens.* Note the general similarities in body proportions but the larger brain in the Neanderthal and the smaller face and more rounded head of the modern human. Reconstructions copyright © 2013 John Gurche.

H. habilis, which means "handy man," was discovered in 1960 by Louis and Mary Leakey and so named because it was presumed to be the maker of the first stone tools. *H. habilis* has uncertain dates, but it probably evolved by 2.3 million years ago and persisted until 1.4 million years ago. *H. habilis* apparently had an australopith's body: small, with long arms and short legs. It also had large and thickly enameled cheek teeth. However, its brain was a few hundred grams larger than any australopith's, and its skull was round and lacked a snout. Its hand was nearly modern and well adapted for making and using stone tools.

H. habilis had a less well known contemporary, *H. rudolfensis.* As far as we can tell, *H. rudolfensis* had a slightly bigger brain than *H. habilis,* but its teeth and face were larger, flatter, and more like those of australopiths.[8] It is plausible that *H. rudolfensis* was a large-brained *Australopithecus* and not actually a member of the genus *Homo.*[9]

Regardless of how many species of early *Homo* there were and how precisely they were related to one another, a general picture that emerges from the fossils so far discovered is that the evolution of humanlike bodies happened in at least two stages. First, in *H. habilis,* the brain expanded slightly and the face lost its snout. Then, in *H. erectus,* much more modern-shaped legs, feet, and arms evolved along with smaller teeth and modestly bigger brains. To be sure, *H. erectus*'s body was not 100 percent like yours, but the evolution of this key species marks the origin of a largely humanlike body, as well as the modern ways we eat, cooperate, communicate, use tools, and otherwise behave. In essence, *H. erectus* was the first ancestor we can characterize as significantly human. How did that transformation happen and why? How did the origin of hunting and gathering enable early *Homo* to survive the onset of the Ice Age, and how did this way of life select for the changes we see in its body, and hence in ours?

How Did H. erectus *Get Dinner?*

Barring the invention of time travel or the discovery of a relict species of early *Homo* on some as-yet uncharted island, we must piece together a picture of how the first members of the human genus eked out a living by studying their fossils and the artifacts they left behind in conjunction with what we know about how hunter-gatherers live today. Such reconstructions inevitably involve guesswork, but you may be surprised at how much we can reliably infer. This is because hunting and gathering is an integrated system with four essential components: gathering plant foods, hunting for meat, intensive cooperation, and food processing. How, when, and why did the first humans accomplish these behaviors?

Let's start with gathering. In the African habitats in which early *Homo* lived, foraging for plant foods undoubtedly contributed to the majority of the diet, probably 70 percent or more. If gathering seems easy, it isn't. In a rain forest, apes need to walk just 2 to 3 kilometers (about 1 to 2 miles) a day to *collect* enough food by simply picking the edible fruits and leaves they encounter. In contrast, hominins in more open habitats would have needed to trek much farther every day, at least 6 kilometers (nearly 4 miles) if modern hunter-gatherers are any guide, to find and then *extract* foods to make them digestible.[10] Extracted foods require getting access to the nutrient-rich parts of the plant that are protected, sometimes by being hidden underground (like tubers), encased in hard shells (like many nuts), or defended by toxins (like many berries and roots). In addition, since open habitats have low densities of edible plants and they are more seasonal than fruit-filled rain forests, the first hunter-gatherers would have needed to rely on a large range of extracted foods. Hunter-gatherers in Africa typically forage for many dozens of different plants, many of which are seasonal, hard to find, and challenging to extract. Underground storage organs, for example, constitute a large percentage of many African hunter-gatherer diets, but a single tuber takes ten to twenty minutes of hard labor to excavate, often requires removal of large, stubborn rocks that lie in the way, and then needs more effort to pound or cook it to render it digestible. Another highly valued food that hunter-gatherers extract is honey, which is sweet, tasty, and rich in calories but difficult and sometimes dangerous to acquire.

The advantages of eating plants are that one can reliably predict where to find them, they are often relatively abundant, and they don't run away. A big disadvantage of plant foods, especially non-domesticated plants, is that they are high in undigestible fiber and have a comparatively low nutrient density. Back-of-the-envelope calculations allow us to infer that early *Homo,* especially mothers, would have had a problem finding enough gathered food to survive and reproduce. A *H. erectus* female who weighed 50 kilograms (110 pounds) would have needed about 1,800 calories a day just for her body's needs, plus another 500 calories when she was nursing or pregnant, which was probably most often the case. In all likelihood, she also needed at least 1,000 to 2,000 additional calories every day for her older offspring who had been weaned but were not yet old enough to forage independently. If you add it all up, she probably needed about 3,000 to 4,500 calories on a typical day. Yet studies of contemporary hunter-gatherers in Africa show that mothers are able to gather between 1,700 and 4,000 calories of plant food per day, with nursing mothers encumbered by toddlers being at the lower end of that range.[11] Since *H. erectus* females are unlikely to have been better foragers than modern females, a typical *H. erectus* mother must frequently have been unable to gather enough calories to pay for her energy needs plus those of her dependent offspring. Solving this deficit required additional energy from other sources.

One of these sources was meat. Archaeological sites dated to at least 2.6 million years ago, possibly older, include animal bones with cut marks that were created when simple stone tools were used to cut the flesh away.[12] Some of these bones were also fractured in a distinctive way to extract the marrow inside. We therefore have irrefutable evidence that hominins started to consume meat by at least 2.6 million years ago. How much meat they ate is conjecture, but meat constitutes approximately one-third of the diet among hunter-gatherers in the tropics (more fish and meat are consumed in temperate habitats).[13] In addition, hunter-gatherers must have craved meat back then as much as chimps and humans still do today, and for good reason. Eating an antelope steak yields five times more energy than an equal mass of carrots, as well as essential proteins and fats. Other animal organs such as the liver, heart, marrow, and brain also provide vital nutrients, especially fat, but also salt, zinc, iron, and more. Meat is a rich food source.

Meat has been an important component of the human diet ever since early *Homo,* but being a part-time carnivore is time-consuming, chancy, dangerous, and difficult for hunter-gatherers today, and it must have been even more challenging and risky at the dawn of the Paleolithic, long before projectile weapons were invented. Although males hunted and scavenged, it is unlikely that early *Homo* mothers who were pregnant or nursing were able to hunt or scavenge on a regular basis, especially while taking care of toddlers. We can therefore infer that the origins of meat eating coincided with a division of labor in which females mostly gathered while males not just gathered but also hunted and scavenged. An essential hallmark of this ancient division of labor—still fundamental to the way hunter-gatherers survive today—is food sharing. Male chimps rarely if ever share food, and they never share food with their offspring. Hunter-gatherers, however, marry each other, and husbands invest heavily in their wives and offspring by provisioning them with food. A male hunter today can acquire between 3,000 and 6,000 calories a day—more than enough food to supply his own needs and to provision his family. Although hunters share the meat from large kills with the entire camp, they still provide the lion's share of the food they hunt to their family.[14] In addition, fathers hunt more frequently when they have wives with young children who need to be nursed and minded intensively. Fathers, in turn, frequently depend on the plants their mates gather, especially when they come home from a long hunt, hungry and empty-handed. The first hunter-gatherers would have benefited so strongly from sharing food that it is hard to imagine how they could have survived without both females and males provisioning each other and cooperating in other ways.

Food sharing, moreover, does not occur just between mates and between parents and offspring, but also between members of a group, highlighting the importance of intense social cooperation among hunter-gatherers. One basic form of cooperation is the extended family. Studies of hunter-gatherers show that grandmothers—capable, experienced older foragers, often without young children—provide critical supplements of food to mothers, as do sisters, cousins, and aunts. In fact, it has been argued that grandmothers are so important that human females were selected to live long past the age they can be mothers so they can help provision their daughters and grandchildren.[15] Grandfathers, uncles, and other males sometimes help as well. Sharing and other forms of cooperation also extend crucially beyond families. Hunter-gatherer mothers rely on one another to help watch children,[16] and males share meat extensively not just with their families but also with other men. When a hunter kills something large, like a several-hundred-pound antelope, he distributes meat to everyone in camp. This sort of sharing isn't just an effort to be nice and to avoid waste; it's a vital strategy to reduce the risk of hunger, because the chances of a hunter killing a large animal on any given day are small. By sharing meat on the days he hunts successfully, a hunter increases his chances of getting meat from fellow hunters on the days he comes home empty-handed. Men also sometimes hunt in groups to increase their probability of hunting success and to help one another carry home the bounty. Not surprisingly, hunter-gatherers are highly egalitarian and they place great stock in reciprocity, helping assure everyone a more regular supply of resources. Today we think of greed and selfishness as sins, but in the highly cooperative world of hunter-gatherers, not sharing and being uncooperative can mean the difference between life and death. Group cooperation has probably been fundamental to the hunter-gatherer way of life for more than 2 million years.

The final, essential component of hunting and gathering is food processing. Many of the plant foods that hunter-gatherers eat are difficult to extract, hard to chew, and unpleasant to digest, often because they are considerably more fibrous than the highly domesticated plants most of us eat today. A typical wild tuber or root is far harder to chew and digest than a raw turnip available from your supermarket. If early *Homo* had needed to eat large quantities of unprocessed wild plants, they would have needed to feed like chimps, spending half the day chewing and filling their stomachs with fiber-rich foods and the other half of the day waiting for their stomachs to empty so they could start the process again. Meat, although more nutritious, was also a challenge because early *Homo,* like apes and humans today, have low, flat teeth that are poorly adapted to chewing meat. If you've ever tried to chew raw game, you'll experience this problem quickly. Our flat teeth are unable to cut through the tough meat fibers, so you chew and chew and chew. It takes a chimp up to eleven hours to chew a few pounds of monkey flesh.[17] In short, if the first hunter-gatherers chewed only raw, unprocessed foods the way apes do, they would not have had enough time to be hunter-gatherers.

The solution to this problem was to process food, initially using very simple technology. The oldest stone tools are so primitive that you might initially fail to recognize some of them as tools at all. Known collectively as Oldowan industry tools (after Olduvai Gorge, in Tanzania), they were made by using one stone to knock off a few chips from another fine-grained stone. Most are just sharp stone flakes, but some are chopping tools with long, knifelike edges. Although these ancient artifacts are a far cry from the sophisticated tools we use today, they are beyond the capability of any chimp to make, and their simplicity should not detract from their significance. They are remarkably sharp and versatile. Every spring, students in my department make Oldowan tools and then butcher a goat in order to experience just how effective this technology is for skinning an animal, cutting the meat off its bones, and then removing the marrow.

Although the goat meat is hard to chew raw, the flesh becomes vastly easier to chew and digest if you first cut it into small pieces.[18] Food processing also works miracles for plant foods. The simplest forms of processing break down cell walls and other indigestible fibers, making even the toughest plants easier to chew. In addition, simply using stone tools to cut and pound raw foods like tubers or steaks substantially increases how many calories you obtain from each morsel.[19] This is because food that has been broken down before consumption is more efficiently digested. It should therefore hardly be surprising that studies of the oldest stone tools show that some were used for cutting meat and that most were used for cutting plants. People have been processing their food for at least as long as they have been hunting and gathering.

If we put together these many lines of evidence, we can conclude that the first species of the human genus solved the problem of "what's for dinner?" during a period of major climate change by adopting a radical, novel strategy. Instead of eating more low-quality food, these progenitors figured out how to procure, process, and eat more high-quality food by becoming hunter-gatherers. This way of life involves traveling long distances every day to forage for food and sometimes to scavenge or hunt. Hunting and gathering also requires intensive levels of cooperation and simple technology. Evocative traces of all these behaviors come from the oldest known archaeological sites, which date to 2.6 million years ago. If you were to come across one of these sites in eastern Africa, you might fail to recognize what you stumbled upon. The arid, semidesert landscape in which they are found is scattered with volcanic rocks, and there are plenty of fossils. But if you look carefully, you might find a sparse, small scattering (just a few square yards) of simple stone tools along with a few animal bones, some of which bear traces of butchery. Some of the stones were transported many miles from their source and then fabricated into tools at the spot. Many of the bones also have gnawing marks from hyenas, reminding us that our ancestors had to compete with nasty, dangerous carnivores to enjoy these precious meals. The first sites were probably ancient, ephemeral places of activity. Imagine a group of *H. habilis* or *H. erectus* individuals gathering under the shade of a tree, hastily sharing some meat, processing tubers, fruits, and other foods that were gathered elsewhere, and making simple tools. This combination of basic behaviors—eating meat, sharing, tool making, and food processing—may seem ordinary, but it is actually unique to hominins, and it transformed the human genus.

What were the effects of hunting and gathering on the evolution of the human body? What adaptations did this way of life select for to enable the first humans to be hunter-gatherers?

Trekking

Apes typically walk less than 3 kilometers (2 miles) a day, but humans are prodigious long-distance walkers. One extreme human, George Meegan, recently trudged all the way from the southern tip of South America to the northernmost part of Alaska, averaging 13 kilometers (8 miles) a day.[20] Although Meegan's trek was unusual, his mean daily distance was actually within the range of how far modern hunter-gatherers walk when they forage (females average 9 kilometers [5.6 miles per day] and males average 15 kilometers [9.3 miles]).[21] Since *H. erectus* adults were about the size of most modern human hunter-gatherers, needed the same number of calories, and lived in similar habitats, they, too, must have walked comparable distances on a daily basis in hot, open conditions to find enough food. As you might expect, this legacy of trekking is stamped in a series of adaptations throughout the human body that originated in early *Homo* and that helped to make the human genus even better at long-distance walking than the australopiths.

The most obvious of these adaptations, evident from figure 9, was long legs. A typical *H. erectus*'s legs are 10 to 20 percent longer than those of an australopith's after factoring in differences in body size.[22] When two people with markedly different leg lengths walk together, the longer-legged person travels farther with each step. Since the cost of moving the body a given distance is priced by the step, longer legs reduce the cost of walking; by some estimates, the longer legs of *H. erectus* would have nearly halved its travel costs compared to an australopith's.[23] The disadvantage of longer legs, however, is to make climbing trees more difficult (tree climbers benefit from short legs and long arms).

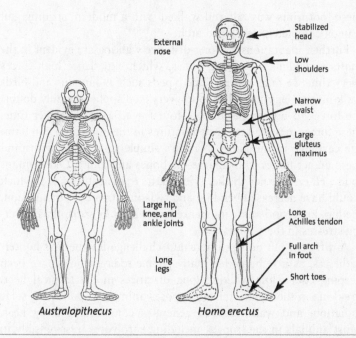

External nose

Stabilized head

Low shoulders

Narrow waist

Large gluteus maximus

Large hip, knee, and ankle joints

Long Achilles tendon

Full arch in foot

Short toes

Long legs

Australopithecus

Homo erectus

FIGURE 9. Some adaptations for walking and running in *Homo erectus* (compared to an *Australopithecus afarensis*). The features indicated on the left would have benefited both walking and running, but the features indicated on the right are primarily for running. The Achilles tendon does not preserve, so its length is a guess. Figure adapted from D. M. Bramble and D. E. Lieberman (2004). Endurance running and the evolution of *Homo*. *Nature* 432: 345–52.

Another important set of adaptations for walking in *H. erectus* can still be found your feet. We have already seen that some species of australopiths had a relatively modern foot with a robust big toe that was nearly in line with the other toes and a partial arch that was capable of stiffening the middle of the foot so the toes could push the body forward and up at the end of every step. But these creatures appear to have been slightly flat-footed when they walked. Although no one has yet found a complete *H. erectus* foot, 1.5-million-year-old footprints have been found in Kenya that were probably made by *H. erectus* and that are very similar to the footprints you and I leave when walking on a beach.[24] Whoever made these footprints was tall and walked with a modern, striding gait using a completely developed arch.

Further adaptations for long-distance walking are evident in the shafts and joints of our leg bones, which experience high forces every time we take a step. Since bipeds such as humans and birds walk on two legs instead of four, every step applies roughly double the forces to our legs than to a four-legged animal's. Over time, these forces may lead to stress fractures in the bones and can damage cartilage in the joints. Nature's simple solution to withstand these higher forces is to enlarge the bones and joints. Like humans today, H. erectus has thicker bone shafts than australopiths, which would have decreased bending and twisting stresses.[25] In addition, the hip, knee, and ankle joints are larger in H. erectus, thus lowering stresses in these joints.[26]

A different but no less important challenge for the first hunter-gatherers, as it remains for many people today, would have been keeping cool while walking long distances in the tropical heat. Trekking in the equatorial sun exposes animals to punishing solar radiation, and walking itself generates considerable body heat. Most animals in the tropics, including carnivores, rest sensibly in the shade during the middle of the day. Since bipedal hominins cannot sprint very fast, the ability to walk long distances during the day without overheating was probably a critical adaptation for early hunter-gatherers in Africa, allowing them to forage when carnivores were least likely to kill them. The English entertainer Noël Coward once quipped that only "mad dogs and Englishmen go out in the midday sun," but he should have written "mad dogs and hominins."

One simple way we keep cool is by being bipedal. Standing and walking upright greatly decreases how much of the body's surface gets maximally exposed to direct solar radiation, lessening how much the sun heats us up.[27] We mostly broil the top of our heads and shoulders, but quadrupeds also grill their entire backs and necks. Another adaptation is the taller, longer-limbed body shape of H. erectus compared to australopiths. Stretching out the body's shape helps us cool through sweating, when we secrete water on the skin's surface. When sweat evaporates, it cools the skin, and thus the blood below. For this reason, human populations that evolved in hot, arid habitats have been selected to have large surface areas relative to body mass by being taller, longer limbed, and more slender than populations adapted to colder habitats (think of a tall Tutsi compared to an Inuit). Just how slender-hipped *H. erectus* was remains a subject of debate, but their overall shape must have helped them dump heat in the midday sun.[28]

A last but especially alluring adaptation we inherited from early *Homo* to keep cool when trekking is a projecting external nose. Australopith faces reveal that they had flat noses very much like an ape or any other mammal, but the outwardly angled margins of the nasal cavity in *H. habilis* and *H. erectus* indicate the presence of a humanlike proboscis that stuck out from the face.[29] Aside from being attractive (to us), our unique outer nose plays an important role in thermoregulation by generating turbulence in the air we inhale through the inner nose. When an ape or dog breathes in through its nose, air flows in a straight line through the nostrils and into the inner nose. But when humans inhale nasally, the air goes up through the nostrils, takes a 90-degree turn, and then goes through another pair of valves to reach the inner nose. These unusual features cause the air to swirl in chaotic vortices. Although this turbulence requires the lungs to work a little harder, it increases contact between air and the mucus membranes that line the inner nose. Mucus holds lots of water but not very strongly. So when you inhale hot, dry air through an external nose, the resulting turbulent flow enhances the inner nose's ability to humidify the air. Such humidification is important because inspired air needs to be saturated with water to prevent the lungs from drying out. Just as importantly, the turbulence helps the nose recapture that moisture when we exhale.[30] The evolution of large external noses in early *Homo* is strong evidence for selection to walk long distances in hot, dry conditions without dehydrating.

Evolved to Run

Walking long distances is fundamental to being a hunter-gatherer, but people sometimes have to run. One powerful motivation is to sprint to a tree or some other refuge when being chased by a predator. Although you only have to run faster than the next fellow when a lion chases you, bipedal humans are comparatively slow. The world's fastest humans can run at 37 kilometers (23 miles) per hour for about ten to twenty seconds, whereas an average lion can run at least twice as fast for approximately four minutes. Like us, early *Homo* must have been pathetic sprinters whose terrified dashes were too often ineffective. However, there is plentiful evidence that by the time of *H. erectus* our ancestors had evolved exceptional abilities to run long distances at moderate speeds in hot conditions. The adaptations underlying these abilities helped transform the human body in crucial ways and explain why humans, even amateur athletes, are among the best long-distance runners in the mammalian world.

Today, humans run long distances to stay fit, commute, or just have fun, but the struggle to get meat underlies the origins of endurance running. To appreciate this inference, try to imagine what it was like for the first humans to hunt or scavenge 2 million years ago. Most carnivores kill using a combination of speed and strength. Large predators, such as lions and leopards, either chase or pounce on their prey and then dispatch it with lethal force. These dangerous carnivores can run as fast as 70 kilometers (43 miles) per hour, and they have terrifying natural weapons: daggerlike fangs, razor-sharp claws, and heavy paws to help them maim and kill. Hunters and scavengers, like hyenas, vultures, and jackals, also need to run and fight because carcasses are hotly contested, evanescent resources that quickly become a snarling focal point of fierce fighting as other dangerous scavengers vie for the chance to pick them to the bone.[31] Today, we hunt and defend ourselves using technologies such as projectile weapons, but the bow and arrow was invented less than 100,000 years ago, and the simplest stone spear points were invented about 500,000 years ago.[32] The most lethal weapons available to the first hunter-gatherers were sharpened wood sticks, clubs, and rocks. It must have been extremely perilous and difficult for slow, puny, weaponless hominins to enter into the rough, tough, and hazardous business of eating other animals for dinner.

An important solution to this problem was endurance running. Maybe the initial selection for running was to help early *Homo* to scavenge. Hunter-gatherers today sometimes power scavenge by watching for vultures circling in the sky, a sure sign of a kill below. They then run to the carcass and bravely chase off the lions or other carnivores to feast on whatever is left.[33] Another strategy is to listen intently at night to the sounds of lions hunting, and then dash first thing in the morning to the area of the kill before other scavengers arrive. Either type of scavenging requires hunter-gatherers to run long distances. Further, once hominins obtained meat, they probably benefited from running away with whatever they could carry to eat in safety far from other scavengers.

Hunter-gatherers have been scavenging for millions of years, but there is archaeological evidence that by 1.9 million years ago early humans were hunting large animals like wildebeest and kudu.[34] If running was important for scavenging, imagine how important running was for the first hunters, who were slow and poorly armed. If you were to try to kill a big animal like a zebra or kudu with nothing more lethal than a club or an untipped wooden spear, you'd often be better off as a vegetarian. Untipped spears cannot dispatch most animals without being thrust at close quarters.[35] In addition, early *Homo* hunters were certainly not fast enough to sprint close to their prey, and even if they managed to sneak up close to their quarry, they then risked being kicked or gored. My colleagues David Carrier, Dennis Bramble, and I have argued that the solution to this problem is an ancient method of hunting based on endurance running known as persistence hunting.[36] Persistence hunting takes advantage of two basic characteristics of human running. First, humans can run long distances at speeds that require quadrupeds to switch from a trot to a gallop. Second, running humans cool by sweating, but four-legged animals cool by panting, which they cannot do while galloping.[37] Therefore, even though zebras and wildebeest can gallop much faster than any sprinting human, we can hunt and kill these swifter creatures by forcing them to gallop in the heat for a long period of time, eventually causing them to overheat and collapse. This is just what persistence hunters do. Typically, a hunter or a group of hunters will single out a large mammal (often the biggest possible) to chase in the middle of the day when it's hot.[38] At the beginning of the chase, the animal gallops away to find a shady place to hide and cool by panting. But the hunters quickly follow by tracking, often at a walk, and then chase their prey again at a run, making the frightened creature gallop before it has had time to cool completely. Eventually, after many cycles of intermittent tracking and chasing—a combination of walking and running—the animal's body temperature rises to lethal levels, causing it to collapse from heatstroke. At this point, a hunter can dispatch the animal safely, easily, and without sophisticated weapons. All the hunter needs is the ability to both run and walk long distances (sometimes 30 kilometers, or about 19 miles), the intelligence to track, partly open habitats, and access to drinking water before and after the hunt.

Persistence hunting has become rare since the invention of the bow and arrow as well as other technologies like nets, domestication of dogs, and guns, but there are nonetheless recent accounts of persistence hunting in many parts of the world, including by Bushmen in southern Africa, native Americans in North and South America, and Aborigines in Australia.[39] The enduring traces of this legacy lie in the human body, which is replete with adaptations that make us exceptional at long-distance running, many of which first appear in *H. erectus*.

One of the most important adaptations for human running is our unique ability to cool by sweating instead of panting, thanks to millions of sweat glands combined with a lack of fur. Most mammals have sweat glands on just their palms, but apes and Old World monkeys have some sweat glands elsewhere on their bodies, and at some point in human evolution we exuberantly augmented the number of the glands to between 5 and 10 million.[40] When we heat up, sweat glands secrete mostly water onto the body's surface. When the sweat evaporates, it cools the skin, the blood beneath, and then the entire body.[41] Humans can sweat more than a liter per hour, enough to cool an athlete running hard in hot conditions. Even though the temperature at the 2004 Olympic women's marathon in Athens reached 35 degrees Celsius (95 degrees Fahrenheit), high sweat rates enabled the winner to run at an average speed of 17.3 kilometers (10.7 miles) per hour for more than two hours without overheating! No other mammal can do that because they lack sweat glands, and because most mammals are covered with fur. Fur is useful to reflect solar radiation, as a hat does, to protect the skin, and to attract mates, yet fur keeps air from circulating close to the skin, preventing sweat from evaporating. Humans actually have the same density of hairs as a chimpanzee, but most human hair is very fine, like peach fuzz.[42] We do not yet know when humans evolved lots of sweat glands and lost their fur, but I suspect these adaptations first evolved either in the genus *Homo* or they initially evolved in *Australopithecus* and then became elaborated in *Homo*.

Although fur and sweat glands don't fossilize, humans have dozens of additional adaptations in our muscles and bones for endurance running whose traces first appear in fossils of *H. erectus*. Most of these features allow us to use our legs like giant springs to jump efficiently from one leg to another in a manner totally different from walking, which uses the legs like pendulums. As figure 7 shows, when your foot hits the ground during a run, your hips, knees, and ankles flex during the first half of stance, causing your center of mass to drop, thus stretching many of the muscles and tendons in your legs.[43] When these tissues stretch, they store up elastic energy, which they release while recoiling during the second half of stance, helping you jump into the air. In fact, a running human's legs store and release energy so efficiently that running is only about 30 to 50 percent more costly than walking in the endurance-speed range. What's more, these springs are so effective that they make the cost of human endurance running (but not sprinting) independent of speed: it costs the same number of calories to run five miles at a pace of either 7 or 10 minutes per mile, a phenomenon many people find counterintuitive.[44]

Since running uses the legs like springs, some of our most important adaptations for running are literally springs. One key spring is the dome-shaped arch of the foot, which develops from the way ligaments and muscles bind together the foot's bones as children start to walk and run. As discussed before, australopith feet had a partial arch to help them stiffen the foot for walking, but their arches were probably neither as convex nor as stable as ours, which means they could not function as effectively as springs. Although we have no whole feet of early *Homo*, footprints and partial feet indicate that *H. erectus* had a completely humanlike arch. A full and springy arch is not necessary for walking (ask anyone with flat feet), but its springlike action helps lower the cost of running by about 17 percent.[45] The human leg's other major, novel spring is the Achilles tendon. This tendon is less than a centimeter long (about a third of an inch) in chimps and gorillas, but is usually more than 10 centimeters (4 inches) long and very thick in humans, storing and releasing almost 35 percent of the mechanical energy generated by the body during running but not walking. Unfortunately, tendons do not fossilize, but the small size of the Achilles tendon's attachment site in australopith heel bones suggests that this tendon was as diminutive in australopiths as it is in African apes, and that it was first expanded in the genus *Homo*.

Many telltale adaptations that the human genus evolved for running function to stabilize the body. Running is essentially jumping from one leg to the other, making it a much less stable gait than walking; even a small nudge or landing on uneven ground or a banana peel can easily topple and injure a runner. Although injuries like sprained ankles are problems today, they were potential death sentences on the savanna 2 million years ago. Thus ever since *H. erectus* we have benefited from a series of novel features, from head to toe, that help us keep from falling while running. None is more prominent than the gluteus maximus, the largest muscle in the human body. This enormous muscle is barely active during walking but contracts very forcefully during running to prevent the trunk from toppling forward with every step.[46] (You can test this yourself by walking and running while grasping your buttocks: feel how much more intensely the muscle clenches with every step when running.) Apes have a small gluteus maximus, and we can tell from fossil hip bones that the muscle was relatively modest in australopiths and first became enlarged in *H. erectus*. Big butt muscles also help one climb and sprint, but since australopiths must have engaged in these activities as much if not more than *H. erectus,* the muscle's expansion probably was primarily for long-distance running.

Another vital set of adaptations that first appear in early *Homo* function to help stabilize the head when we run. Unlike walking, running is a jolting gait that causes your head to jerk around rapidly enough to blur your vision if unchecked. To appreciate this problem, watch a runner with a ponytail: the forces acting on the head oscillate the ponytail in a figure-eight motion with each step even as the head remains fairly still—evidence of unseen stabilizing mechanisms at work. Since humans have short necks that attach to the center of the skull base, we cannot flex and extend our necks to stabilize the head as quadrupeds do. Instead, we evolved a novel set of mechanisms to keep our gaze steady. One of these adaptations is enlarged sensory organs of balance, the semicircular canals of the inner ear. These canals function like gyroscopes, sensing how fast the head pitches, rolls, and yaws and then triggering reflexes that cause the eye and neck muscles to counter those movements (even when your eyes are closed). Since bigger semicircular canals are more sensitive, animals like dogs and rabbits whose heads encounter lots of jiggling tend to have larger semicircular canals than more sedentary animals. Fortunately, the skull preserves these canals' dimensions, so we know that they evolved to be much larger relative to body size in *H. erectus* and modern humans than in apes and australopiths.[47] One more special adaptation for damping your head's jiggling motions is the nuchal (neck) ligament. This strange bit of anatomy, first detectable in early *Homo* but absent in apes and australopiths, is like a rubber band that connects the back of your head to your arms along the midline of your neck. Every time your foot hits the ground, the shoulder and arm from that side of the body fall just as your head pitches forward. By connecting the head to the arm, the nuchal ligament allows your falling arm to gently pull your head back, keeping it stable.[48]

As you might expect, there are additional features in the human body that help us run effectively and that appear to have first evolved in the genus *Homo*.[49] These features, summarized in figure 9, include relatively short toes (which stabilize the foot)[50]; narrow waists and low, wide shoulders (both of which help a runner's torso to twist independently from the hips and head)[51]; and a predominance of slow-twitch muscle fibers in the legs (which give us endurance but compromise speed)[52]. Many of the traits benefit both walking and running, but some, such as a large gluteus maximus, the nuchal ligament, big semicircular canals, and short toes don't affect how well we walk and are primarily useful when we run, which means they are adaptations to running. These traits suggest that there was strong selection in the genus *Homo* not just for walking but also for running, presumably for scavenging and hunting. Consider also that a few of these adaptations, especially long legs and short toes, compromise our ability to climb trees. Selection for running may have caused the human genus to become the first primates that are clumsy in trees.

In short, the benefits of acquiring meat through scavenging and hunting account for many transformations of the human body first evident in early *Homo* that enabled early hunter-gatherers not only to walk but also to run long distances. Whether an *H. erectus* could outrun a human today is impossible to know, but there is no doubt these ancestors left a legacy of adaptations throughout our bodies that explain how and why humans are one of the few mammals that can and do run long distances with ease, and why we are the only mammal that can run marathons in the heat.

Tooling Around

Could you live without tools? It used to be thought that only humans made tools, but actually a few other species like chimps occasionally use simple tools like rocks to smash nuts, or modify twigs to fish for termites.[53] However, ever since hunting and gathering evolved, human survival has depended heavily on tools to dig plants, hunt and butcher animals, process foods, and more. Humans have been making stone tools for at least 2.6 million years (maybe even longer), and a wide range of sophisticated tools are now ubiquitous in every human population in every corner of the earth. It should hardly be surprising that selection for making and using tools accounts for a number of distinctive features in the human body that first evolved in the genus *Homo*.

If there is any one part of the human body that most directly reflects our dependency on tools, it is the hand. Chimps and other apes generally hold objects the way you might grip a hammer's handle, using the fingers to squeeze it into the palm (a power grip). Sometimes chimps will hold a small object between the side of the thumb and the side of the first finger, but they cannot grip pencils or other tools precisely between the fleshy pad of the thumb and the tips of opposing fingers.[54] Humans can grip this way because we have relatively long thumbs and short fingers as well as extremely strong thumb muscles and robust finger bones with large joints.[55]

If you've ever tried making stone tools and using them to butcher an animal, you'll quickly appreciate just how important the combination of precision and strength must have been to early hunter-gatherers. You need strength to whack stones together repeatedly to make tools, and holding stone tool flakes in a precision grip while you skin and deflesh a carcass requires extraordinary finger strength as the tool becomes dull with use and slippery from fat and blood.[56] Gracile australopiths like Lucy had hands that were intermediate between apes and humans, and they were certainly able to hold and use digging sticks, but hands capable of powerful precision grips are unambiguously evident by about 2 million years ago.[57] In fact, it was the fossil of a nearly modern hand from Olduvai Gorge that helped inspire Louis Leakey and colleagues to name the oldest species in the human genus *Homo habilis* ("handy man").

Another tool-related skill that apparently evolved in the genus *Homo* and that helped change our bodies is throwing. Even if the first hunters lacked tipped spears to kill animals from a distance, they still had to throw or thrust simple javelin-like weapons. Only humans can do this. Chimps and other primates sometimes toss rocks, branches, and nasty stuff like feces with reasonable aim, but they cannot throw anything with a combination of speed and accuracy. Instead, they hurl clumsily with a straight elbow, using just their upper bodies. We throw in a totally different manner, usually beginning with a step in the direction of the throw, with the torso facing sideways, the elbow flexed, and the arm cocked behind the rest of the body. We then generate massive amounts of energy in a whiplike fashion by rotating the waist and then torso, which unleashes forward motions in the shoulder, elbow, and finally wrist. Although the legs and waist are important for throwing hard, the majority of a throw's energy comes from the shoulder, which we load like a catapult by cocking the arm behind the head.[58] By releasing at just the right moment, humans can throw projectiles like spears, rocks, and baseballs at up to 100 miles an hour with pinpoint accuracy. Performing this sequence of motions correctly requires much practice as well as the appropriate anatomy, some of which first evolved in australopiths, but that don't all appear in combination until *H. erectus*. These include a highly mobile waist, shoulders that are low and wide, a shoulder joint that is oriented to the side rather than more vertically, and a highly extensible wrist.[59] *H. erectus* hunters were probably the first good throwers.

Humans need tools not only to hunt and butcher but also to process food. Try eating a raw meal without using tools to cut, grind, or tenderize anything. You'll be able to eat foods like lettuce, carrots, and apples, but you'd find tough foods such as meat or tubers to be hard to swallow. Cooking was probably not invented until less than a million years ago, but stones and bones from the oldest archaeological sites show that early *Homo* started to cut and bash many foods prior to chewing.[60] Even such basic food processing yields benefits. One is to reduce the time and effort needed to chew and digest. Unlike chimps, which spend more than half the day eating and digesting, tool-using hunter-gatherers have more free time to forage, hunt, and do other useful things. In addition, simply tenderizing a tuber or a steak before you chew it improves its digestibility and significantly augments how many calories it yields.[61] Finally, food processing allows teeth and chewing muscles to be smaller. As we saw earlier, the australopiths evolved extremely thick cheek teeth and massive chewing muscles to break down large quantities of tough, hard food. However, the cheek teeth of *H. erectus* shrank about 25 percent to nearly the size of a modern human's molars,[62] and their chewing muscles also dwindled to almost modern size. These reductions, in turn, allowed selection to shorten the lower face in the genus *Homo*. We are the only snoutless primates, in part thanks to tools.

Guts and Brains

Most often you think with your brain, but sometimes the digestive system seems to take over and makes decisions on behalf of the rest of the body. Gut instincts are actually more than just urges or intuitions, and they highlight vital links between the brain and the gut that changed critically in the genus *Homo* following the origins of hunting and gathering.

To appreciate how selection for hunting and gathering favored changes to our brains and guts and the relationship between these two parts of the body, it helps to consider that these organs are both expensive tissues that cost lots of energy to grow and maintain. In fact, brains and guts each consume about the same amount of energy per unit mass, each expend about 15 percent of the body's basal metabolic cost, and each requires similar amounts of blood supply to deliver oxygen and fuel and to remove wastes.[63] Your guts also have about 100 million nerves, more than the number of nerves in your spinal cord or your entire peripheral nervous system. This second brain evolved hundreds of millions of years ago to monitor and regulate the gut's complex activities, which include breaking down food, absorbing nutrients, and passing food and waste from the mouth to the anus.

One odd characteristic of humans is that our brains and gastrointestinal tracts (when empty) are similarly large, weighing slightly more than a kilogram each. In most mammals of the same body mass, the brain is about a fifth of the size of a human's, while the guts are twice as large.[64] In other words, humans have relatively small guts and big brains. In a landmark study, Leslie Aiello and Peter Wheeler proposed that our unique ratio of brain to gut size is the result of a profound energetic shift that began with the first hunter-gatherers, in which early *Homo* essentially traded off large guts for large brains by switching to higher-quality diets.[65] According to this logic, by incorporating meat in the diet and relying more on food processing, early *Homo* was able to spend much less energy digesting its food and could thus devote more energy toward growing and paying for a larger brain. In terms of actual numbers, australopith brains were about 400 to 550 grams; *H. habilis* brains were slightly larger, about 500 to 700 grams; and early *H. erectus* brain sizes ranged from 600 to 1,000 grams. When adjusted for body size, which also got bigger, a typical *H. erectus*'s brain was 33 percent larger than an australopith's.[66] Although intestines do not preserve in the fossil record, some theorize that *H. erectus* had smaller guts than australopiths. If so, the energetic benefits of hunting and gathering appear to have made possible the evolution of bigger brains in part by allowing the first humans to make do with smaller guts.

Larger brains must have been an advantage among the first hunter-gatherers despite their greater energetic cost. Effective hunting and gathering requires intense cooperation through sharing food and information and other resources. Further, cooperation among hunter-gatherers occurs not just among kin but also among unrelated members of the same group.[67] Everyone helps everybody. Mothers help one another forage, process food, and take care of one another's children. Fathers help one another hunt, share the spoils of their successes, and work together to build shelters, defend resources, and more. These and other forms of cooperation, however, require complex cognitive skills beyond those of apes. To cooperate effectively one needs a good theory of mind (to intuit what another person is thinking), the ability to communicate through language, the faculty to reason, and an ability to suppress one's urges. Hunting and gathering also requires good memory to remember where and when to find different foods, as well as a naturalist's mind to predict where foods will be. Tracking in particular requires many sophisticated cognitive skills, including both deductive and inductive thinking.[68] To be sure, the first hunter-gatherers 2 million years ago were not as cognitively advanced as people today, but they must have benefited from having bigger, better brains than australopiths. Then, once hunting and gathering became successful enough to make more energy available, this way of life permitted selection for the evolution of yet larger brains. It is not coincidental that major increases in brain size occurred after the origins of hunting and gathering.

Have you ever worried that you might get stranded on a desert island and have to become a hunter-gatherer to survive? Every once in a while this actually happens, most famously to Alexander Selkirk, the inspiration for Robinson Crusoe, who learned to chase down wild goats in his bare feet while stranded on a tiny island 400 miles west of Chile.[69] Another exemplar is Marguerite de La Rocque, a French noblewoman who was marooned for several years on an island off the coast of Quebec in 1541 along with her lover, a maidservant, and soon thereafter Marguerite's new child. Alas, of this unhappy foursome, only Marguerite survived; she lived in a makeshift hut, gathered edible plants, and hunted wild animals with simple weapons until she was eventually rescued.[70] These and other stories of survival illustrate several unique human characteristics that most of us take for granted: the capacities to hunt for meat and gather plants, the ability to make and use tools, and endurance. All of these distinctive qualities trace back to the origins of the human genus, especially to *H. erectus*.

But Alexander and Marguerite were not *H. erectus*. They not only had much larger brains, but they also reproduced and grew up very differently from their ancient progenitors, and they thought, communicated, and behaved in other profoundly different ways. These differences highlight how the success of hunting and gathering, once it evolved, then set in motion further important changes to the human body as the vicissitudes of the Ice Age continued to alter, rapidly and repeatedly, the habitats in which the human genus still struggled to survive.

5

Energy in the Ice Age

How We Evolved Big Brains Along with
Large, Fat, Gradually Growing Bodies

We simply must balance our demand for energy with
our rapidly shrinking resources. By acting now we
can control our future instead of letting the future
control us.

—JIMMY CARTER (1977)

Imagine that a *H. erectus* family from 2 million years ago was
somehow cloned or transported to the twenty-first century and
allowed to hunt and gather in the Serengeti. If you could glimpse
them while you were on a safari you'd think their bodies sort of
resembled your family's from the neck down, but you'd also perceive that these primordial humans were significantly different in
several key respects. Most obviously, their brains would be much
smaller, and their large, chinless faces would be topped by massive
browridges perched in front of long, sloping foreheads. If you could
observe them for many years, you'd discover that their children
matured much faster than modern humans, becoming fully adult

by age twelve or thirteen, and it is possible that they had babies
at a slower rate than hunter-gatherers today. I also suspect they
would be scrawny, with much less body fat than even the skinni-
est of today's supermodels. These differences highlight how, after
the genus *Homo* first evolved, our ancestors continued to evolve in
important ways, eventually becoming big-brained, slow-maturing,
fast-breeding people with more body fat than any other species of
primate. These shifts probably happened gradually, but they reflect
a profound revolution in how our bodies use energy that set the
stage for the evolution of our species, *Homo sapiens*.

You may not realize that your body uses energy in a special way,
but it really does. To appreciate the exceptional way we acquire,
store, and spend energy, consider that life is fundamentally a way
of using energy to make more life. All organisms—from bacteria to
whales—pass their days mostly getting energy from food and then
spending that energy to grow, survive, and reproduce. Since natural
selection favors individuals with adaptations that help them have
more surviving offspring than others in their population, evolution inevitably drives organisms to acquire and use energy in ways
that increase their number of surviving children and grandchildren.
Most organisms, such as mice, spiders, and salmon, do this by
spending as little energy as possible growing and as much energy as
possible reproducing. These species mature rapidly, and they produce dozens, hundreds, or even thousands of eggs or babies in their
short lives. Although most of the progeny perish, a very few lucky
ones survive. Such a strategy of minimal investment—live fast, die
young, and breed profligately—makes sense when resources are
unpredictable and mortality is high. If life is chancy, go for quick,
cheap returns.

In most respects, humans are one of a relatively small number
of species that evolved a very different strategy of investing more
energy to reproduce more slowly. Like apes and elephants, we
mature at a leisurely pace, grow large bodies, and have few babies
but devote much time and energy to raising them well. This unusual
strategy succeeds because while apes and elephants produce fewer
babies than mice, a larger percentage of their offspring survive to
then reproduce. A house mouse can become a mother when she is

just five weeks old, has four to ten pups per litter, and can have a new litter every two months over the course of her approximately twelve-month life. However, the vast majority of her pups die young. In contrast, a chimp or elephant mother does not reproduce until she is at least twelve years old, and she gives birth to only one infant every five or six years over the next thirty or so years. About half of these offspring make it to becoming parents. This strategy of high investment—live slow, die old, and reproduce conservatively—can evolve only when resources are predictable and infant mortality is low.[1]

Humans obviously use energy and reproduce much more like chimps than mice, but over the course of the Ice Age the human genus altered this strategy in a remarkable, astonishing, and consequential way. On the one hand, our ancestors intensified the ape strategy by evolving to spend even more time and energy to grow their bodies. Whereas chimps mature in twelve or thirteen years, humans take about eighteen years to mature, and we expend considerably more energy to grow bigger, costlier bodies with vastly expanded brains that consume a greater percentage of our daily energy budget. In other words, humans invest an absolutely greater amount of energy than apes simply to grow and maintain their bodies. Yet, at the same time, we evolved to speed up the rate of reproduction. Hunter-gatherers typically have babies every three years, nearly twice the rate of apes. Further, since human babies take so much longer to mature, hunter-gatherer mothers must nurse and care for young infants at the same time they continue to feed and care for older, still immature children who are not yet ready to forage on their own. No ape mother ever has to cope with this sort of child-care challenge. In essence, we evolved to combine successfully the ape and mouse strategies in a completely novel way. To do this, however, required an energetic revolution that still has profound ramifications for human health.

How the human genus evolved the unique strategy of using more energy to grow bigger, brainier bodies for longer life spans while reproducing even faster is the next key transformation in the story of the human body. This part of the human body's story begins approximately at the dawn of the Ice Age, just after the invention of hunting and gathering and the origins of *H. erectus*.

Getting By and Getting Around in the Ice Age

When we last left our hero, *H. erectus*, it had just evolved. The oldest *H. erectus* fossils so far unearthed come from Kenya and date to 1.9 million years ago, but the species (or very closely related variants)[2] shows up shortly afterward in other parts of the Old World. The oldest fossils outside of Africa currently come from a 1.8-million-year-old site, Dmanisi, nestled in a hilly region of Georgia, between the Caspian and Black Seas. If the half dozen individuals so far excavated there are truly *H. erectus*, they are among the smallest fossils of this species ever found. They also include a toothless old man who probably needed help chewing his food.[3] Other discoveries indicate that *H. erectus* spread eastward into South Asia, probably below the Himalayas, showing up in Java by 1.6 million years ago, and in China about the same time.[4] *H. erectus* also dispersed to the west along the Mediterranean coast into southern Europe by at least 1.2 million years ago.[5] *H. erectus* is thus the first intercontinental hominin (although some have speculated that *H. habilis* also got out of Africa, an idea we'll discuss at the chapter's end).

How and why did *H. erectus* go global quickly? Cecil B. DeMille might have dramatized this event as a migration, maybe with a long line of bedraggled, big-browed hominins looking homesick as they trudge northward out of Africa with a swelling orchestral accompaniment. One can even imagine an early *H. erectus* Moses parting the Red Sea to lead his clan into the Middle East. In actual fact, this was not a migration, but a gradual dispersal. Dispersals occur as populations expand without increasing their density, which is just what one expects for the first modestly successful hunter-gatherers. Remember that hunter-gatherers live in small groups at low population densities within huge territories. If they were like modern hunter-gatherers, we can estimate they lived in groups of approximately twenty-five people (about seven or eight families) that inhabited territories of 250 to 500 square kilometers (96.5 to 193 square miles). At such densities, only between six and twelve people would be living on the island of Manhattan! Further, a *H. erectus* female who survived childhood was probably able to have a total of four to six offspring, of which only half survived to adulthood. If we use these numbers to estimate an average rate of population growth of approximately 0.4 percent per year, a *H. erectus* population would double in 175 years, and after a mere 1,000 years it would increase by more than fifty times. Since these hunter-gatherers did not live in towns or cities, the only way a population could grow while staying at an appropriately low density would be for overly populous groups to split and disperse into new territories. If an initial band of *H. erectus* foragers living near Nairobi, Kenya, split off a new band to the north every 500 years, and if each new band's territory was 500 square kilometers (about 190 square miles) and roughly circular, then it would take less than 50,000 years for the species to disperse in this manner up the Nile Valley to Egypt, then up the Jordan Valley and all the way to the Caucasus Mountains.[6] Even if groups split every 1,000 years, it would still take less than 100,000 years for *H. erectus* to disperse from East Africa to Georgia.

We should not be surprised that *H. erectus* rapidly dispersed far and wide. What is more noteworthy is that these hunter-gatherers started to colonize temperate habitats during the Ice Age. Many people think of the Ice Age as a period when vast glaciers covered much of the planet, but it was actually characterized by repeated cycles of extreme cooling when glaciers expanded, followed by rapid warming in which they contracted (these cycles account for the zigs and zags of figure 4). At first, these cycles were of moderate intensity and lasted about 40,000 years. Then, starting about 1 million years ago, the cycles became more intense and longer, enduring about 100,000 years. Each cycle had major effects on the habitats in which early humans were trying to survive. During maximum cold snaps (which started getting extreme about 500,000 years ago), the average temperature of the oceans dropped several degrees, and ice sheets blanketed one-third of the earth's surface, incorporating more than 50 million cubic meters (13 billion gallons) of water. The glaciers lowered the sea level by many feet, exposing the continental shelves. When the glaciers were at their maximum, one could walk from Vietnam to Java and Sumatra, or stroll across the English Channel from France to England. Each cycle of Ice Age climate change also altered distributions of plants and animals. During cold periods, most of central and northern Europe became an inhospitable arctic tundra with little to eat besides moss and reindeer, and southern Europe became a pine forest replete with bears and wild boars. Such conditions would have been hellish for early hunter-gatherers, especially before the invention of fire, and the evidence suggests that early humans were never present north of the Alps and Pyrenees during these cold snaps. Between glacial periods, however, the ice sheets retreated to the poles, rich Mediterranean forests returned to southern Europe, and hippos frolicked in the Thames.[7] Humans occupied much of the temperate Old World during these milder, more benevolent periods.

Populations living in Africa were not directly affected by glaciers, but they too experienced cycles of climate change. As moisture and temperature levels fluctuated, the Sahara as well as open habitats like the savannas alternately expanded and contracted relative to forests and woodlands.[8] These cycles acted like a giant ecological pump. During wetter periods when the Sahara shrank, hunter-gatherers likely prospered and dispersed from sub-Saharan Africa up the Nile Valley, through the Middle East, and then into Europe and Asia. But during drier periods when the Sahara expanded, hunter-gatherers in Africa were cut off from the rest of the world. Further, during colder, drier glacial periods in Europe and Asia, *H. erectus* would have faced severe hardships and they probably went extinct or were pushed southward, back toward the Mediterranean or southern Asia.

In short, *H. erectus* had the misfortune of evolving in Africa at the beginning of an intensely dynamic and challenging phase in the earth's history. Yet, instead of just enduring in Africa, the species quickly went global and continued to evolve over the vast expanse of Africa and Eurasia. Let's now look more closely at who these humans were and how they not only coped but also thrived during the Ice Age's dramatic fluctuations.

Archaic Humans of the Ice Age

When families or college roommates break up, they often lose contact with one another, but when species disperse this isolation is even more intense and consequential. As far-flung populations become reproductively separated, natural selection and other random evolutionary processes cause them to differ over time. Visitors to the Galápagos Islands can easily observe this phenomenon among the marine iguanas, which vary enough in size and color that experts can sometimes tell which island they come from with just a glance. The same process probably happened to *H. erectus*. As hunter-gatherer populations spread out over several continents and confronted the vicissitudes of the Ice Age, they started to vary and change, especially in size. For the most part, they got bigger, but in some cases they became smaller. On average, *H. erectus* individuals weighed between 40 and 70 kilograms (88 to 154 pounds) and they were between 130 and 185 centimeters tall (between 4.3 and 6.1 feet), but the population mentioned above from Dmanisi were at the low end of this range, with bodies and brains that were about 25 percent smaller than their African cousins.[9] A more common trend within the species, however, was that brains became both absolutely and relatively bigger over time. As figure 10 plots, brain size nearly doubled over the course of the species' duration, reaching nearly modern levels after a million years.[10] Yet despite these and other variations, *H. erectus* fossils from different times and places consistently share a common suite of features, as shown in figure 11. Their skulls are always long and flat, with low foreheads, a big browridge, and another horizontal ridge of bone at the back end of the skull. All of them had a large and vertical face, with big orbits and a capacious nose. Many of them had a slight ridge of bone (a keel) along the midline at the top of the skull. As we have discussed, the overall shape of the *H. erectus* body was much like that of modern humans, but with wider, more flared hips and thicker bones throughout the body.

By 600,000 years ago, some of the descendants of *H. erectus* had evolved sufficiently from their ancestors to merit being classified in different species. The best known is *Homo heidelbergensis,* also shown in figure 11, which extended from southern Africa to England and Germany. The most spectacular trove of *H. heidelbergensis* fossils comes from a single site in northern Spain, the Sima de los Huesos (the Pit of Bones). Here, between 530,000 and 600,000 years ago, at least thirty people were dragged many yards through a winding natural tunnel deep within a cliff and dumped

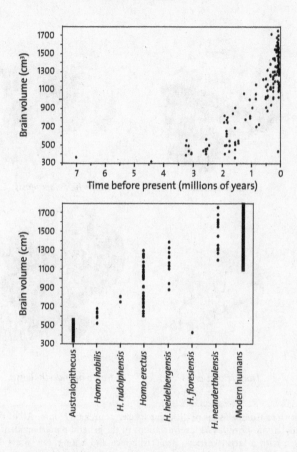

FIGURE 10. Brain size. The top graph plots how brain volume increased during human evolution. The bottom graph plots the range of brain volumes in different species of hominins.

into a pit (presumably after they died). Their skeletons provide a unique snapshot of a population of this species. Like *H. erectus,* they had long, low skulls with massive browridges, but their brains were bigger, ranging from 1,100 to 1,400 cubic centimeters, and their faces were larger, with especially capacious noses.[11] They were also big people, weighing between 65 to 80 kilograms (143 to 176 pounds).[12] At the same time, *H. erectus* either persisted in Asia or

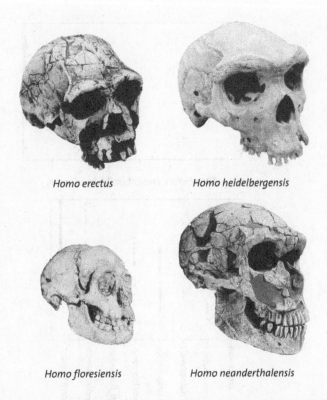

Homo erectus   *Homo heidelbergensis*

Homo floresiensis   *Homo neanderthalensis*

FIGURE 11. Comparisons of different species of archaic *Homo*. All of them, even the tiny *Homo floresiensis*, are variations of the general pattern evident in *Homo erectus*, with a large, vertical, projecting face and a long, low skull. Brain and face size, however, vary between the species, as do some other features. Image of *H. floresiensis* courtesy of Peter Brown.

perhaps evolved into another closely related species that also had large brains and faces. One intriguing remnant of this group is a well-preserved finger bone from a cave in the Altai Mountains of Siberia, about 2,000 miles north of Bangladesh. DNA from this scrap of bone indicates that she was the descendant of a lineage, currently known as the Denisovans, which presumably descended from *H. erectus* and shares a last common ancestor with humans and Neanderthals between 1 million and 500,000 years ago.[13]

Who the Denisovans were remains a mystery, but when modern humans migrated into Asia, they interbred with some of us in very small numbers.[14]

It is often difficult to classify fossils into species correctly, and there is no consensus on exactly how many species descended from *H. erectus*, and who begat whom. The important point is that they are essentially big-brained variants of *H. erectus*, and when thinking about the evolution of the human body, it is both convenient and sensible to group them together using the term "archaic *Homo*" (colloquially, archaic humans). Archaic *Homo*, as you might expect, were skilled hunter-gatherers. The stone tools they made were slightly more sophisticated and diverse than the tools made by *H. erectus*,[15] but their biggest innovation in weaponry was the spear point. Untipped spears were probably made since the beginning of the Stone Age, but they are almost never found because wood rarely preserves.[16] However, around 500,000 years ago archaic *Homo* invented a new and ingenious method of fabricating very thin stone tools with predetermined shapes, including triangular points.[17] This method takes great skill and much practice to master, but it revolutionized projectile technology because stone points made in this fashion are light and sharp enough to haft onto spears using pitch or sinew. Imagine what a difference these stone points made to hunters! Spears suddenly became much sharper: instead of bouncing off their prey, they could penetrate tough animal hides and even ribs, and once lodged inside, their jagged edges inflicted horrible, lacerating wounds. Armed with thin stone points, hunters could now kill prey from greater distances, decreasing a hunter's chances of getting injured while increasing the chances of success. Other tools made from this prepared-core technique were also better for skinning hides and performing other tasks.

An even more important invention was the control of fire. No one is quite sure when humans first managed to regularly create and use fire. Currently, the earliest evidence for the controlled use of fire by humans comes from a million-year-old site in South Africa and from a 790,000-year-old site in Israel.[18] Traces of fire, however, remain rare until 400,000 years ago, when fireplaces and burnt bones start showing up regularly in sites, suggesting that archaic *Homo*, unlike *H. erectus*, habitually cooked its food.[19] Cooking, when it did catch on, was a transformative advance. For one, cooked food yields much more energy than uncooked food and is less likely to make you sick. Fire also allowed archaic humans to keep warm in cold habitats, to fend off dangerous predators, like cave bears, and to stay up late at night.

Despite sometimes having fire, the extremes of the Ice Age must have been tough on archaic humans, especially populations in northern Europe and Asia. For example, during periods when glaciers covered northern Europe, *H. heidelbergensis* disappeared from all but the margins of the Mediterranean, probably because more northern populations went extinct or moved southward. But when the climate ameliorated, they dispersed northward again. If these dispersals were substantial, then populations of *H. heidelbergensis* in Europe and Africa were not totally isolated from one another genetically. Molecular and fossil data, however, indicate that they diverged into several partially separate lineages by 400,000 to 300,000 years ago.[20] The African lineage evolved into modern humans (whose origin we'll discuss in chapter 6). Another lineage evolved into the Denisovans in Asia, and in Europe and western Asia the most famous species of archaic *Homo* evolved, the Neanderthals.

Our Neanderthal Cousins

No ancient species arouses more passions than the Neanderthals. A few Neanderthal fossils had been discovered before 1859, when *On the Origin of Species* was published, but the species was not formally recognized until 1863. Since then, so much has been written and debated about these archetypal cavemen that they have become something of a mirror: our views of them sometimes reveal more about our conception of ourselves. At first, the Neanderthals were erroneously considered to be a missing link: nasty, brutish, primitive ancestors. After World War II, there was a healthy but extreme reaction to these views, partly motivated by widespread revulsion to Nazi pseudoscientific racism and partly because Neanderthals were correctly recognized to be close cousins that managed to survive in Europe during harsh glacial conditions, with brains that were as large as or larger than those of modern humans. Starting in the 1950s, many paleontologists classified Neanderthals as a human subspecies (a geographically isolated race) rather than as a separate species. Recent data, however, show that Neanderthals and modern humans were indeed separate species that diverged genetically at least 800,000 to 400,000 years ago.[21] Although there was a modicum of interbreeding between the two species, they are really close cousins, not ancestors.[22]

The most important facts about Neanderthals are that they were a species of archaic *Homo* that lived in Europe and western Asia between about 200,000 and 30,000 years ago. They were skilled and intelligent hunters, well adapted by natural selection and well supported by their wits to survive the cold, semi-arctic conditions of the Ice Age. As figure 11 illustrates, Neanderthal skulls have the same general configuration we see in *H. heidelbergensis:* a long, low cranium with an enormous face, a large nose, marked browridges, and no chin. However, they were bigger-brained, with average brain volumes of nearly 1,500 cubic centimeters. Their skulls also have a suite of distinctive features that enable anyone with a little practice to spot a Neanderthal easily. Classic Neanderthal features include a massive face that is especially inflated on either side of the nose, an egg-sized bulge at the back of the skull, a shallow groove on the back of the skull, and a space on the lower jaw behind the lower wisdom teeth. The rest of their bodies were much like those of other archaic *Homo,* but they were especially muscular and stocky, with short forearms and shanks. This sort of body shape, typical of arctic peoples like Inuit and Laplanders, helped them conserve body heat.

The Neanderthals were successful and talented hunter-gatherers who would probably still exist had it not been for *H. sapiens.* The Neanderthals made complex and sophisticated stone tools that they fashioned into a wide variety of tool types such as scrapers and points. They cooked their food and hunted big animals like wild aurochs, deer, and horses.[23] But the Neanderthals, in spite of their accomplishments, were not entirely modern in their behavior. They made few tools out of bone, including needles, even though they must have made clothes from pelts. They buried their dead simply, and they left almost no traces of symbolic behavior such as art.

They rarely ate fish or shellfish, even though these were abundant in some of the habitats in which Neanderthals lived. They seldom transported raw materials more than 25 kilometers (15.5 miles). As we will see, when modern humans did arrive in Europe starting about 40,000 years ago, they mostly replaced the Neanderthals.

Big Brains

Of all the changes evident in *H. erectus* and its archaic human descendants, the most obvious and impressive is the brain's enlargement. Figure 10 illustrates how brain size nearly doubled in the human genus over the Ice Age, and species such as the Neanderthals had brains that were actually slightly larger than the average brain size of people today. Enormous brains presumably evolved because they help us think, remember, and do other complex cognitive tasks, but if being smart is such a good thing, why didn't big brains evolve earlier, and why don't more animals have brains as large as ours? The answer, as I suggested earlier, has to do with energy. Big brains are prohibitively energy consuming for most species, but the dividends of hunting and gathering enabled *H. erectus* and archaic *Homo* to grow larger, costlier brains than was previously possible.

To evaluate how brains evolved to be larger, we first need to consider the tricky issue of how to assess brain size in the first place. Assuming you're an average human, your brain's volume is approximately 1,350 cubic centimeters. For comparison, a macaque's brain is 85 cubic centimeters, a chimp's is 390 cubic centimeters, and an adult gorilla's is 465 cubic centimeters. Human brains are thus voluminous compared to monkeys and at least three times bigger than those of the other great apes. But how much bigger is the human brain after accounting for differences in body size? The answer to this question is shown in figure 12, which plots brain size relative to body weight for several primate species. As you can see, this relationship is nonlinear: as bodies get bigger, brains get *absolutely* bigger but *relatively* smaller.[24] This relationship between brain and body size turns out to be highly correlated and consistent. Therefore, if you know a species' average body mass, you can compute its relative

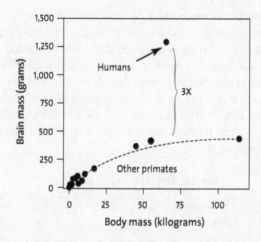

FIGURE 12. Brain size relative to body size in primates. Species with bigger bodies have larger brains, but the relationship is not linear. Compared to apes, humans have brains that are about three times larger than predicted by body size; compared to mammals in general, our brains are about five times larger.

brain size by dividing its actual brain size by the size you would predict from its body mass. This ratio, known as the encephalization quotient (EQ), is 2.1 for chimps and 5.1 for humans. Those numbers mean that chimps have brains about twice the size of a typical mammal of the same weight, and humans have brains about five times bigger than similar-sized mammals; compared to other primates, humans have brains about three times bigger than expected.

Let's now reconsider how brain size evolved using estimates of body mass from skeletons and measurements of brain volume from skulls.[25] These estimates, summarized in table 2, indicate that the earliest hominins were about as brainy as apes, but that absolute and relative brain size in early *H. erectus* was moderately bigger. A male *H. erectus* from 1.5 million years ago with a brain of 890 cubic centimeters who weighed 60 kilograms (132 pounds) had an EQ of 3.4, roughly 60 percent bigger than a chimp's. In other words, the initial evolution of the genus *Homo* involved a modest increase in brain size, but then brains accelerated relative to bodies. By 1 million years ago, our ancestors' brain volumes exceeded

1,000 cubic centimeters, and by 500,000 years ago they were within the size range of modern humans, as shown in figure 10. In fact, brains tended to be even larger at the end of the Ice Age than today because bodies were also bigger. As the world warmed up over the last 12,000 years, bodies shrank slightly, but so did brains, keeping relative brain sizes about the same in both recent and early modern humans.[26] After accounting for slight differences in body weight, an average modern human is just a tiny bit brainier than an average Neanderthal.

TABLE 2. **Species in the genus** *Homo*

Species	Date (millions of years ago)	Locations found	Brain size (cm³)	Body mass (kg)
Homo habilis	2.4–1.4	Tanzania, Kenya	510–690	30–40
Homo rudolfensis	1.9–1.7	Kenya, Ethiopia	750–800	?
Homo erectus	1.9–0.2	Africa, Europe, Asia	600–1,200	40–65
Homo heidelbergensis	0.7–0.2	Africa, Europe	900–1,400	50–70
Homo neanderthalensis	0.2–0.03	Europe, Asia	1,170–1,740	60–85
Homo floresiensis	0.09–0.02	Indonesia	417	25–30
Homo sapiens	0.2 – present	Everywhere	1,100–1,900	40–80

How did brains get bigger in the human genus? There are two major ways to grow a larger brain: grow it for longer or grow it faster. Compared to apes, we do both.[27] At birth, a chimpanzee's brain is 130 cubic centimeters, and then it triples in volume over the next three years.[28] Human newborn brains are 330 cubic centimeters and then quadruple over the next six to seven years. So we grow our brains twice as fast as chimps before birth, and then we grow them both longer and faster after we are born. Much of the extra size comes from having about twice as many brain cells, called neurons.[29] The cell bodies for these extra neurons mostly lie in the outer layer of the brain, a region called the neocortex, where almost all complex cognitive functions, such as memory, thinking, language, and awareness, occur. Even though the human neocortex is just a few millimeters wide, unfolded it would cover 0.25 square meters (2.5 square feet). More neurons create millions more connections than in a chimp's brain.[30] Since the brain functions through its network of connections, the human neocortex, by virtue of being bigger and more connected, has far more potential for doing complex tasks like remembering, reasoning, and thinking. If bigger brains make you smarter, then Neanderthals and other big-brained archaic humans were pretty intelligent.

Bigger brains, however, come with sizeable costs. Even though your brain constitutes only 2 percent of your body's weight, it consumes about 20 to 25 percent of your body's resting energy budget, regardless of whether you are sleeping, watching TV, or puzzling over this sentence. In absolute numbers, your brain costs 280 to 420 calories per day, whereas a chimpanzee's brain costs about 100 to 120 calories per day. In our modern world of energy-rich food, you can supply this amount of energy with about one donut per day, but a donut-deprived hunter-gatherer needs to forage an extra six to ten carrots to get the same number of added calories. These costs, moreover, are amplified if you are feeding children. A pregnant human mother caring for a three-year-old plus a seven-year-old needs about 4,500 calories a day to feed herself plus her fetus and children.[31] If her kids had chimp-sized brains, then she'd need approximately 450 fewer calories each day—no small quantity in the Paleolithic.

There are other, major challenges to having a large brain. Nearly a quart of blood, 12 to 15 percent of the body's total supply, flows through the brain at any given moment to provide fuel, remove waste, and keep it at just the right temperature. Consequently, the human brain requires special plumbing to deliver oxygenated blood and then return it to the heart, liver, and lungs. Brains are also fragile organs that need plenty of protection to keep them from being damaged when you fall or get hit on the head. Imagine shaking two brain-shaped mounds of Jell-O, one twice as big as the other. Because forces that break the Jell-O apart increase exponentially with size, the bigger Jell-O brain is far more likely to shear apart near its surface. Larger brains therefore need much more protection from concussions.[32] Big brains also complicate birth. A human newborn's head is about 125 millimeters long by 100 millimeters wide (4.9 inches by 3.9 inches), but the minimum dimensions of a mother's birth canal average 113 millimeters long and 122 millimeters wide (4.5 inches by 4.8 inches).[33] To pass through, the human neonate must enter its mother's pelvis facing sideways and then make a 90-degree turn within the canal so that it inconveniently faces downward rather than upward when it emerges.[34] Under the best of circumstances, the trip is a tight squeeze, and human mothers almost always require assistance to give birth.

If you add up all the costs, it's no wonder that most animals don't have very large brains. Big brains may make you smarter, but they cost a lot and cause many problems. The fact that brains have gotten bigger since *H. erectus* first evolved not only means that archaic humans were getting enough energy but also that the benefits of increased intelligence outweighed the costs. Sadly, we have few direct traces of the intellectual feats these archaic humans accomplished beyond mastering fire and making more complex tools, such as projectile points. The biggest benefits of bigger brains were probably for behaviors we cannot detect in the archaeological record. One set of added skills must have been an enhanced ability to cooperate. Humans are unusually good at working together: we share food and other crucial resources, we help raise one another's children, we pass on useful information, and we even sometimes risk our lives to aid friends or even strangers in need. Cooperative behaviors, however, require complex skills such as the ability to communicate effectively, to control selfish and aggressive impulses, to understand the desires and intentions of others, and to keep track of complex social interactions in a group.[35] Apes sometimes cooperate, such as when hunting, but they cannot do so very effectively in many contexts. For example, chimp females share food only with their infants, and males almost never share food.[36] Thus one of the apparent benefits of bigger brains is to help humans interact cooperatively with one another, often in large groups. In a famous analysis, Robin Dunbar showed that the size of the neocortex among primate species correlates reasonably well with group size.[37] If this relationship holds true for humans, then our brains evolved to cope with social networks of about 100 to 230 people, which is not a bad estimate of how many people a typical Paleolithic hunter-gatherer might have encountered in a lifetime.

Another principal benefit of bigger brains must have been an enhanced ability to be natural scientists. Today, few people know much about the animals and plants that live around them, but such knowledge used to be vital. Hunter-gatherers eat as many as a hundred different plant species, and their livelihoods depend on knowing in which season particular plants are available, where to find them in a large and complex landscape, and how to process them for consumption. Hunting poses even greater cognitive challenges, especially for weak, slow hominins. Animals hide from predators, and since archaic humans couldn't overpower their prey, early hunters had to rely on a combination of athleticism, wits, and naturalist know-how. A hunter has to predict how prey species behave in different conditions in order to find them, to get close enough to kill them, and then to track them when wounded. To some extent, hunters use inductive skills to find and follow animals, using clues such as footprints, spoor, and other sights and smells. But tracking an animal also requires deductive logic, forming hypotheses about what a pursued animal is likely to do and then interpreting clues to test predictions. The skills used to track an animal may underlie the origins of scientific thinking.[38]

Whatever the initial advantages of big brains, they must have been worth the cost or they wouldn't have evolved. But why do humans take so many extra years to grow them along with the rest of our bodies? When and why did we draw out the pace of brain and body growth?

Growing Gradually

Being a kid is fun, but from an evolutionary perspective, humans pay a high price for the excruciatingly drawn-out pace at which we mature. Your lengthy upbringing, which lasted approximately eighteen years, cost your parents lots of money and was a substantial fitness cost to them, especially your mother, by limiting how many other children she could have. Had you and your siblings matured twice as fast, your mother might have had twice as many offspring. By maturing gradually you also incurred some fitness costs yourself: you delayed when you could reproduce, shortened your reproductive life span, and increased the chances that you might fail to have children altogether. Moreover, from an energetic standpoint, a humanlike schedule of slow growth inflates the energy cost per offspring. It takes a whopping twelve million calories to grow a human into an eighteen-year-old adult, roughly twice as many calories as it takes to grow an adult chimpanzee. To a large extent, we can thank archaic *Homo* for the fact that we spend so much extra time and energy growing up.

To make sense of how and why big-brained archaic humans prolonged their development at such a cost, let's first compare the major stages that most large-bodied mammals go through before they become adults, shown in figure 13. First, during the *infant stage,* mammals are dependent on their mothers for milk and other kinds of support as their brains and bodies grow rapidly. After weaning (which is actually a gradual process), mammals go through a second *juvenile stage,* when they are no longer dependent on their mothers for survival, their bodies continue to grow gradually, and they continue to develop social and cognitive skills. The final stage before adulthood is *adolescence,* which begins as the testes or ovaries mature and initiate a growth spurt.[39] Adolescence is essentially that awkward, usually infertile period between the start of puberty and the end of skeletal growth, when reproductive maturity occurs. During human adolescence, secondary sexual features like breasts and pubic hair appear, the body finishes growing, and many social and intellectual skills develop fully.

Figure 13 also illustrates how human ontogeny is drawn out in several special ways. The most significant difference is that we have added a novel stage, *childhood.*[40] Childhood is a uniquely human period of dependency that occurs after weaning but before a child can feed entirely on its own and before its brain has finished growing. A chimpanzee infant completes its brain growth and erupts its first permanent teeth at about three years of age, yet continues to nurse (albeit with decreasing frequency) until it is four to five years old.[41] In contrast, human hunter-gatherers usually wean their infants by age three, at least three years *before* the brain ceases to grow and the permanent teeth start to erupt. Then follow about three years of childhood, usually until age six or seven, in which the child remains extremely immature and needs to be provisioned with lots of high-quality food. No child can survive without intensive levels of adult investment and patience. However, because hunter-

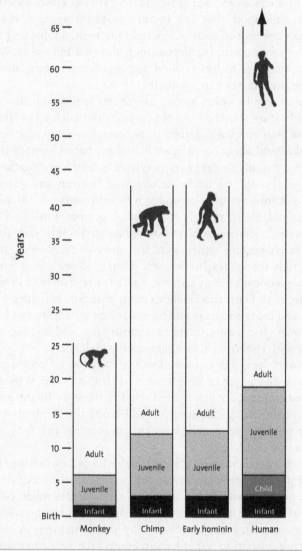

FIGURE 13. Different life histories. Humans have a more prolonged life history with an added stage of childhood and a longer period of being a juvenile prior to adulthood. Australopiths and early *Homo erectus* had a generally chimplike life history. Life history probably slowed down in species of archaic *Homo*, but exactly when and how much is still unclear.

gatherer mothers wean their offspring so early, in effect ushering them into childhood, they can become pregnant again relatively earlier than ape mothers. Over a normal life span, adding a childhood stage of postweaning dependency allows a hunter-gatherer mother with access to lots of food and assistance to have nearly twice as many babies as an ape mother.[42]

The other way in which human life history is special is that we have significantly stretched out the juvenile and adolescent stages that follow childhood. Altogether, these stages last about four years in a monkey and about seven years in an ape, but in humans they carry on for roughly twelve years. A typical human hunter-gatherer girl will go through menarche between age thirteen and sixteen, but she is not fully mature—reproductively or socially—for another five years, and she is unlikely to become a mother until she is at least eighteen.[43] Hunter-gatherer boys hit puberty a little later than girls but rarely become fathers until they are twenty. As every parent and high school teacher knows, human adolescents are not totally independent of their parents, but they can help take care of their younger siblings, contribute to many domestic activities, like cooking, and begin to forage and hunt—at first with help, and then on their own. Today's teenagers have mostly replaced hunting and gathering with either secondary school or farm labor.

When and why did our development become so prolonged? Why double the length of time it takes to grow a brain? Why add a period of childhood when a mother has to nurse an infant while simultaneously taking care of older, still immature children? Further, why prolong the juvenile stage, not to mention that long and painful period of adolescence?

Although bigger animals generally take longer to mature, the extended pace of development cannot be explained by body size increases in the genus *Homo*. After all, male gorillas weigh twice as much as humans but take only thirteen years to finish growing (about the same time it takes a 5-ton elephant to mature). A much more likely explanation is that human brains take longer to mature because they are so large and require such complex wiring. One factor is the size of the brain itself. Among primates, bigger brains take longer to reach full size: a tiny macaque brain takes one and a half years to grow, a chimp's brain is five times larger and takes three years to grow, and a human brain is four times larger than the chimp's and takes at least six years to reach full size. We can also estimate reasonably well how long it took extinct hominins to grow an adult-sized brain (amazingly, by using their teeth).[44] Australopiths like Lucy grew their brains about as fast as chimps, which makes sense since their brains were about the same size. Early *H. erectus* took about four years to grow a brain size of 800 to 900 cubic centimeters.[45] By the time bigger-brained archaic *Homo* species evolved, the pattern of early life history appears to have been roughly similar to ours. Neanderthals, whose brains were as large or sometimes larger than modern humans', attained adult-sized brains between five and six years, just a little faster than most but not all people today.[46]

A human brain is full-sized by six or seven years (which explains why kids and adults can share the same hats), but obviously a six-year-old child's brain and body require another dozen years or more to develop completely. When the juvenile and adolescent stages lengthened in our history is harder to determine, but we have some intriguing clues. One of the best lines of evidence is the Nariokotome Boy, a nearly complete skeleton of an immature *H. erectus* male who died 1.5 million years ago (probably from an infection) near a swamp, which covered him up and preserved most of his skeleton. His teeth indicate that he was about eight or nine years old when he died, but his skeletal age was typical of a thirteen-year-old human.[47] Because his second molars had just erupted, we know that he probably had a few years to go before becoming an adult. We can therefore infer that early *H. erectus* matured only slightly slower than chimps, which means that prolonged juvenile and adolescent periods developed more recently in human evolution. There are hints that Neanderthals might have been like *H. erectus* in this respect. One teenage Neanderthal from the site of Le Moustier was twelve years old when he died (we know this from his teeth), but his wisdom teeth had not yet erupted, indicating that he had another year or two of growth.[48] More data are needed, but it is possible that a very prolonged period of post-childhood development is unique to modern humans. Perhaps archaic humans didn't spend as much time as teenagers.

If we put together all the available evidence, it seems probable that as brains got bigger in the human genus, the critical period of early development (infancy and childhood) extended to enable larger brains to grow. Even if the pace of juvenile and adolescent development didn't lengthen fully until modern humans evolved, archaic human mothers surely confronted an energetic double whammy. First, because of childhood, most mothers had to nurse infants at the same time they were taking care of toddlers. Therefore, archaic human mothers needed lots of extra energy and help. A typical nursing mother would need about 2,300 calories a day for her own body's needs, plus several thousand more calories to feed her kids. There is no way she could succeed without access to high-quality food, including meat and cooking. In addition, she needed to live in a highly cooperative group with regular help from her children's father, grandparents, and others.

The second energy predicament faced by big-brained mothers and their offspring was how to pay for those large, relentlessly costly brains. Brain tissue cannot store its own energy supplies but must receive an unceasing, plentiful supply of sugar from the bloodstream. Short interruptions or deficits of blood sugar that last more than a minute or two cause irreparable, often lethal damage. Big-brained human mothers therefore need to stockpile lots of energy to pay for their voracious brains as well as those of their big-brained children during the inevitable times—sometimes lengthy—when they are taking in little or no energy, perhaps because of famine or illness. How did early human mothers survive such shortages, which were probably periods of very intense natural selection?

The answer is lots of fat. Like other animals, we mostly store surplus energy as fat, always keeping a reserve supply for times of need. However, humans are unusually fat compared to most mammals, and there is good reason to believe that ever since brains expanded and development slowed down in archaic *Homo*, we also became relatively fat.

Fat Bodies

One topsy-turvy characteristic of the modern world is how many people worry about fat. Although fat and weight have probably

obsessed humans for millions of years, until recently our ancestors mostly obsessed about not having enough fat in their diets and
insufficient weight on their bodies. Fat is the most efficient way
of storing energy, and at some point our ancestors evolved several
key adaptations for amassing larger quantities of fat than other
primates. Because of these ancestors, even the leanest among us
is relatively fat compared to other wild primates, and our babies
are especially fat in comparison to other primate infants. There is
good reason to hypothesize that without our ability and proclivity
to stockpile fat, archaic humans could never have evolved big brains
and slow-growing bodies.

We'll focus more on how your body uses and stores fat in later
chapters, but there are two key facts to know about this vital substance for now. The first is that the components of each molecule
of fat can come from digesting fat-rich foods, but our bodies also
synthesize them just as easily from carbohydrates (which is why
fat-free foods still make you fat).[49] Second, fat molecules are useful, highly concentrated stores of energy. A single gram of fat stores
nine calories, more than twice the energy per gram of carbohydrate
or protein. After you eat a meal, hormones cause you to convert
sugars, fatty acids, and glycerol into fat within special fat cells, of
which you have about thirty billion. Then, when your body needs
energy, other hormones break down fat into its components, which
your body can burn (more on this in chapter 10).

All animals need fat, but humans have a special need for lots of
fat right from the moment of birth, largely because of our energy-
hungry brains. An infant's brain is a quarter the size of an adult's,
but it still consumes about 100 calories per day, about 60 percent
of the tiny body's resting energy budget (an adult's brain consumes
between 280 and 420 calories per day, 20 to 30 percent of the
body's energy budget).[50] Since brains require sugar incessantly, having plenty of fat ensures our brains an unending, reliable supply of
energy. A monkey infant has about 3 percent body fat, but healthy
human infants are born with about 15 percent body fat.[51] In fact,
the last trimester of pregnancy is largely devoted to fattening up the
fetus. During these three months, the fetal brain triples in mass,
but fat stores increase one hundred–fold![52] Furthermore, a healthy
human's percentage of body fat rises to 25 percent during child-

hood, settling back down in adult hunter-gatherers to about 10 percent in males and 15 percent in females. Fat is more than an energy reservoir for the brain and for pregnancy and breast-feeding; it is also essential to fuel the endurance athleticism necessary to be a hunter-gatherer. When you walk and run, much of the energy you burn comes from fat (though as you speed up, you also burn more carbohydrates).[53] Fat cells also help to regulate and synthesize hormones such as estrogen, and skin fat functions as an excellent insulator, helping keep us warm.

Altogether, without lots of fat, human brains could not be so big, hunter-gatherer mothers would be less able to provide enough high-quality milk to nourish their big-brained offspring, and we would have less endurance. Unfortunately, fat doesn't preserve in the fossil record, so we cannot be sure when our ancestors started to fatten up compared to other primates. Perhaps the trend began with *H. erectus,* which helped fuel their slightly bigger brains as well as their long-distance treks and runs. High percentages of body fat, especially in babies, were probably even more important in archaic *Homo.* Were I a Neanderthal living during a glacial European winter, I'd also like plenty of body fat to help stay warm. Eventually, we may be able to test this hypothesis by figuring out which genes increase fat stores in humans and then determining when these genetic adaptations evolved.

The paradoxical legacy of fat's vital role in human evolution is that many of us are now too well adapted to craving and storing fat. In the documentary film *Super Size Me,* Morgan Spurlock gained about 11 kilograms (24 pounds) in just twenty-eight days by eating only McDonald's food (averaging 5,000 calories a day)! Such extreme feats are the legacy of thousands of generations of selection on humans for adaptations to store as much fat as possible on the rare occasions when they could indulge. A half pound of fat stored on Tuesday might have paid for a persistence hunt on Wednesday. And storing up a few pounds of fat when food was plentiful must have been essential during the inevitable lean seasons. Like money in the bank, fat reserves enable humans to stay active, maintain their bodies, and even reproduce during lean seasons.[54] Unfortunately, natural selection never prepared us to cope with endless seasons of plenty, let alone fast-food restaurants—a topic we'll consider in chapter 10.

Where Did the Energy Come From?

How did archaic *Homo* get the energy necessary to grow bigger bodies with even bigger brains, to extend their duration of growth, and perhaps to wean their children at younger ages and to accumulate more fat? There are only two ways to accomplish these feats. The first is simply to acquire more energy overall. The second is to allocate energy differently, spending more on brain growth and reproduction by spending less on other functions. The evidence suggests that they did both.

In order to make sense of these energetic strategies, consider your body's total energy budget as having several different accounts. The first is your basal metabolic rate (BMR), the energy you need to take care of your body's many tissues without having to move, digest food, or do anything else. For all mammals, BMR is mostly a function of body mass,[55] and humans do not appear to be exceptional in this regard. A typical chimpanzee weighing 40 kilograms (88 pounds) has a BMR of about 1,000 calories a day, and a typical hunter-gatherer of 60 kilograms (132 pounds) has a BMR of roughly 1,500 calories a day.[56] Yet, as chapter 4 discussed, humans have altered the percentage of energy we allocate to different portions of our BMR. It's a good guess that *H. erectus* and archaic *Homo* individuals were able to sustain a disproportionately larger brain in part by having a relatively smaller gut. Smaller intestines (as well as smaller teeth) could only have been possible if these species were eating a high-quality diet with lots of meat and lots of food processing.

Although your small gut helps you afford your large brain, it is also necessary to consider how much energy your body actually spends each day (your total energy expenditure, TEE) versus how much energy you acquire (your daily energy production, DEP). Humans are unusual in both respects, and archaic humans probably were too. Chimp TEEs likely average about 1,400 calories per day, but modern hunter-gatherer TEEs range between 2,000 and 3,000 calories per day, higher than body size alone predicts.[57] Hunter-gatherer TEEs are relatively high because they lead moderately active lives by walking and sometimes running long distances, carrying children and food, digging for plants, processing food, and performing other daily chores without assistance from any machines or beasts of burden. Since archaic humans probably had to travel and work as much as similar-sized modern human hunter-gatherers, their TEEs were probably not much different. What is more important, however, is that adult hunter-gatherer DEPs are generally higher than their TEEs. Although DEP is difficult to measure and highly variable on a daily, seasonal, individual, and even population-level basis, studies of many societies indicate that a typical adult hunter-gatherer acquires about 3,500 calories a day.[58] This is a crude estimate with much variation and many sources of error, but the bottom line is that adult hunter-gatherers usually collect a daily surplus of between 1,000 and 2,500 calories. This substantial surplus comes from several sources, including hunting for meat and foraging more widely for high-quality resources such as honey, tubers, nuts, and berries that yield more energy than they cost to procure.[59]

Two other key factors that likely helped archaic humans acquire modest energy surpluses were cooperation and technology. Hunter-gatherers cannot survive without some division of labor, lots of sharing between kin and nonkin, and other ways of working together. We cannot determine if the first hunter-gatherers cooperated as intensely as hunter-gatherers today, but selection would have quickly driven them to do so. The role of technology is easier to trace. We have already discussed how the first stone tools certainly helped early *Homo* cut and pound foods, and how archaic *Homo* later invented stone-tipped projectiles, which made it considerably easier and safer to kill animals. Cooking was an equally profound technological advance. Every time you eat something, you have to spend energy to chew and digest it (this is why your pulse and body temperature rise after a meal). Mechanically processing food by cutting, grinding, and pounding significantly lowers the cost of digesting both plant and animal foods. Cooking has even more substantial effects. Some foods, like potatoes, yield roughly twice as many calories or other nutrients if you eat them cooked versus raw.[60] Another benefit of cooking is that it kills germs that can make you sick, substantially reducing the cost to your immune system.

Regardless of precisely how archaic humans were able to acquire regular, dependable surpluses of high-quality food, these positive balances clearly set in motion a positive feedback loop. There are several different theories about how this feedback loop worked, but all are based on the same basic principle: once you take care of your body's basic needs, you can spend surplus energy in four different ways. You can use it to grow if you are young, you can store it as fat, you can be more active, or you can spend it on having and raising more offspring.[61] If life is chancy and rates of infant mortality are high, then the best evolutionary strategy is to be more like a mouse than an ape and plow as much surplus energy as possible into reproducing. However, if your children are thriving and surviving, then there is a strong benefit to evolve as archaic *Homo* evidently did: invest more energy in fewer, better-quality offspring by extending their development so they can grow larger brains. Since bigger brains permit more learning and more complex cognitive and social behaviors, including language and cooperation, these offspring have a better chance of surviving and reproducing because they develop into better hunter-gatherers. Then, when these smarter, more cooperative hunter-gatherers generate even bigger surpluses, selection will continue to favor even larger, slower-growing brains along with longer-growing, fatter bodies. In addition, mothers with adequate food supplies and strong social support would have benefited from weaning their infants at a younger age, because they could then have more children.

We cannot yet test many aspects of this scenario directly because we cannot prove when humans got fatter or when humans started to wean their offspring at a younger age than apes. However, we can measure when brains and bodies got bigger and when early stages of growth extended. These lines of evidence suggest a gradual evolutionary process, exactly what the feedback hypothesis predicts. As figure 10 shows, brain size did not suddenly shoot up in the human genus but increased steadily over more than a million years following the origins of *H. erectus*. A similarly gradual trajectory of change is probably true for the prolonging of human development. More data are needed to test these inferences, but it is a good bet that changing energy budgets fueled by energy surpluses were a key driving force behind evolution of the body in archaic human hunter-gatherers during the Ice Age.

The genus *Homo*'s trend toward acquiring and using more energy, however, was not universal. As you might expect, not all populations during the Ice Age enjoyed energy surpluses, and the fossil record is replete with evidence that the struggle for existence during certain periods was demanding and precarious, sometimes ending in disaster. When food became scarce our lineage's dependence on high fuel consumption turned from an asset to a liability, not unlike the way gas-guzzling cars become a costly encumbrance when fuel prices rise. Archaic human populations suffered and many of them probably went extinct in temperate Europe during periods when the glaciers expanded. Food can also become scarce in the tropics, especially on islands. In fact, the most illustrative example of how our dependence on energy can backfire is the case of *Homo floresiensis*, otherwise known as the Hobbit, a dwarfed species of archaic humans from Indonesia.

An Energetic Twist: The Tale of the Hobbits of Flores

Strange evolutionary events often happen on islands. Large animals on small, remote islands often confront energy crises because there are typically fewer plants and less food than on larger landmasses. In these settings, very large animals struggle because they need more food than the island can provide. In contrast, small animals frequently do better than their mainland relatives because they have enough food, they face less competition from other small species, and because islands often lack predators, releasing them from the need to hide. On many islands small species become larger (gigantism) and large species become smaller (dwarfism). Islands such as Madagascar, Mauritius, or Sardinia were thus hosts to giant rats and lizards (Komodo dragons) along with miniature hippos, elephants, and goats.

The same energetic constraints and processes also affect hunter-gatherers,[62] and the most extreme example apparently occurred in our genus on the remote island of Flores. Flores is part of the Indonesian archipelago, on the eastern side of a deep oceanic trench that separates Asia from a group of islands that also includes Bali, Borneo, and Timor. Even when sea levels were at their lowest during the Ice Age, many miles of deep sea separated Flores from the next closest island in Indonesia. Yet a few animals, including rats, varanid lizards, and elephants, somehow apparently managed to swim this distance and then underwent either gigantism or dwarfing. The island now has giant rats along with Komodo dragons, and until recently the island hosted a species of dwarfed elephant (*Stegodon*).

And then there is the Hobbit. In the 1990s, archaeologists working on Flores found primitive tools dated to at least 800,000 years ago,[63] indicating that hominins, maybe *H. erectus,* had rafted or swum to Flores even earlier. Then in 2003, a team of Australian and Indonesian researchers digging in the cave of Liang Bua made headlines around the world when they found a partial skeleton of a tiny fossil human dated to between 95,000 and 17,000 years ago. They named it *H. floresiensis* and proposed that it was the remnant of a dwarfed species of early *Homo*.[64] The media quickly nicknamed the species the Hobbit. Further excavations recovered the remains of at least six more tiny individuals.[65] These were small people, about a meter (3 feet) tall and weighing between 25 and 30 kilograms (55 to 66 pounds), with minuscule brains, about 400 cubic centimeters, the size of an adult chimp's. The fossils have an odd mix of features such as big browridges, no chin, short legs, and long feet without a full arch. Several studies suggest that the brain and skull of the Hobbit (shown in figure 11) most closely resembles *H. erectus* after correcting for the effects of size.[66] If so, then a reasonable scenario is that *H. erectus* got to the island at least 800,000 years ago and was driven by natural selection to become small-brained and small-statured in order to cope with a lack of food.

Needless to say, *H. floresiensis* has been controversial. Some scholars argued that the species' brain was simply too diminutive to fit on a body that size. When you compare animals of different body mass, larger species or individuals tend to have brains that are absolutely bigger but relatively smaller. Gorillas are three times as massive as chimps, but their brains are only 18 percent bigger. According to typical scaling laws, if the Hobbit were a half-sized human (a pygmy), you'd expect its brain to be about 1,100 cubic centimeters; if it were a dwarfed *H. erectus,* then you'd expect its brain to be about 500 to 600 cubic centimeters.[67] These predictions led several researchers to conclude that the Hobbit remains must come from some population of modern humans that suffered from a disease that causes dwarfism as well as a pathologically small brain. However, careful analyses of the species' brain shape, skull shape, and limbs indicate that *H. floresiensis* does not look like it had any known disease or suffered from abnormal growth.[68] In addition, studies of dwarfed hippos on other islands show that during the process of insular dwarfing, natural selection can shrink species' brains quite radically, more than enough to account for the tiny brains of *H. floresiensis.*[69] Apparently, when the going gets tough on small islands, big and expensive brains may be too costly a luxury to afford.

As Sherlock Holmes once remarked (albeit fictionally), "When you have eliminated the impossible, whatever remains, however improbable, must be the truth." If the Hobbit isn't a dwarfed, tiny-brained human, then it must be a real hominin species. There are really two possibilities. The first is that it's a descendant of *H. erectus.* Another more astonishing possibility, suggested by its primitive hands and feet, is that it's a relict of an even more primitive species like *H. habilis* that left Africa very early, somehow made it all the way to Indonesia, and then swam to Flores, leaving no other fossil traces outside of Africa. Either scenario requires considerable reductions in brain size. The smallest *H. erectus* brain ever found is 600 cubic centimeters and the smallest *H. habilis* brain is 510 cubic centimeters. So selection would have required at least a 25 percent reduction in brain size to account for the Hobbit's tiny brain.

For me, what's most important about the Hobbit is what this surprising species reveals about how important energy was in human evolution. In the context of an island with limited resources, reductions in brain and body size are hardly far-fetched but instead are exactly what one predicts for some sort of early or archaic *Homo* faced with insufficient energy supplies. Big bodies and brains are expensive, making them prime targets for natural selection to cut costs. By shrinking, *H. floresiensis* was probably able to survive on 1,200 calories per day, maybe 1,440 calories per day when lactating, far less than a full-sized *H. erectus* mother, who would need about 1,800 calories per day when neither pregnant nor nursing and as much as 2,500 calories per day when nursing. We do not know what sort of cognitive price *H. floresiensis* paid for having such a small brain, but apparently the trade-off was worth it.

What Happened to Archaic Humans?

If you were to travel around the tropics today you'd get the chance to see many different closely related species of primates and appreciate their similarities and differences. For example there are two species of chimpanzees, five species of baboons, and more than a dozen species of macaques. As we have seen, natural selection over the course of the Ice Age led to a similar degree of diversity among the descendents of early *Homo,* including the Neanderthals in Europe, the Denisovans in Asia, the Hobbit in Indonesia, and more. And, of course, there was one additional species: *Homo sapiens.* We evolved at about the same time as the Neanderthals, and if you were to observe the first modern humans about 200,000 years ago you might not have deemed these ancestors to be fundamentally different from their contemporaries. Apart from the Hobbit, modern and archaic humans have generally similar bodies including equally big brains. Yet, obviously, modern humans are unique in some respects, and our species has enjoyed (so far) a very different evolutionary fate. By the time the Ice Age came to an end, all of our close relatives were extinct, leaving just modern humans as the sole surviving species of the human lineage.

Why? Why did other kinds of humans go extinct? What is biologically and behaviorally special about modern humans? What adaptations are unique to modern humans? And how did the legacy of archaic *Homo,* including the ability to use and harness energy in novel ways, set the stage for the next major transformation in the story of the human body?

6

A Very Cultured Species

*How Modern Humans Colonized the World
with a Combination of Brains plus Brawn*

Culture is roughly anything we do and the monkeys don't.

—FITZROY SOMERSET (LORD RAGLAN)

I was eight years old when I first learned that all human beings were once Stone Age hunter-gatherers. I recall being entranced by the grainy images on TV of the Tasaday, a then recently "discovered" tribe of primitive people in the Philippines who had never had any contact with the modern world. There were only twenty-six of them, they were nearly naked, lived in caves, made stone tools, and survived by eating insects, frogs, and wild plants. The discovery electrified the world. Grown-ups, including my teacher at school, were especially excited that the Tasaday had no words for violence or war. If only more people were like the Tasaday . . .

Unfortunately, the Tasaday were a hoax. The tribe's existence was apparently staged by its "discoverer," Manuel Elizalde, who is alleged to have paid a handful of nearby villagers to swap their jeans and T-shirts for orchid leaf loincloths and to eat bugs and frogs instead of rice and pork for the TV cameras. I think the Tasaday fraud fooled the world because Elizalde's orchestrated portrayal of primitive human society was just what many people wanted to see and hear during the Vietnam War. The Tasaday embodied the Rousseauian notion that human beings uncontaminated by civilization are naturally virtuous, peaceful, and healthy. In addition, the Tasaday's laid-back way of life stood in stark contrast to the deeply entrenched assumption that life in the Stone Age was arduous and that human history since the invention of agriculture has been a long process of nearly continuous progress. In the same year that the Tasaday flickered across our TV screens and graced the pages of *National Geographic Magazine,* the anthropologist Marshall Sahlins published his influential book *Stone Age Economics.*[1] Sahlins argued that hunter-gatherers were the "original affluent society" because they had few needs beyond basic sustenance, did not have to work strenuously, ate highly varied and nutritious diets, and had rich social lives with plentiful free time, little marred by violence. According to this way of thinking, still popular, the human condition has been deteriorating ever since we became farmers, starting about six hundred generations ago.

In actual fact, life in the not-too-distant Stone Age was probably neither as dreadful nor as idyllic as some extreme views would have it. Although hunter-gatherers don't have to work as many hours a day as most farmers and they suffer from fewer contagious diseases, it doesn't necessarily follow that hunter-gatherers were leisurely, Paleolithic couch potatoes who barely had to work and were rich only for lack of wanting anything. In actual fact, hunter-gatherers are often hungry, and they manage to get enough food only through a combination of intense cooperation and considerable work, including many hours a day of walking, running, carrying, digging, and more. Yet there is some truth to Sahlin's analysis. Were you a hunter-gatherer, you wouldn't have to work any more than necessary to satisfy your family and group's daily needs. After that, you would benefit from resting and devoting time to social activities such as gossiping and enjoying the company of family and friends. Many contemporary stresses—commuting, the threat of losing one's job, getting into college, saving for retirement—can make one realize that the hunter-gatherer economic system does have certain benefits.

There are no real Tasaday-like tribes left, but a handful of bona fide hunter-gatherer groups persisted until recently and a few still exist, albeit to varying extents as true hunter-gatherers. These people are fascinating and important to study because they are the last human beings whose way of life most closely resembles how our ancestors lived for many thousands of generations. Learning about their diets, activities, and cultures partly helps us appreciate what modern humans are adapted for. However, we cannot figure why humans are the way we are by simply studying contemporary hunter-gatherers because our bodies evolved to do more than just hunt and gather. What's more, none of these populations are pristine Stone Age foragers, and all of them have been interacting for millennia with farmers and herders.

To understand how and why modern human bodies are the way they are, and why we are the last surviving species of human on the planet, we also need to look back in time to consider the final speciation event in our body's history, the origin of *Homo sapiens*. If you were to focus on just the fossil record of this transformation, you might conclude that modern humans originally evolved because of a handful of modest anatomical changes that are mostly evident in our heads, such as smaller faces and more rounded brains and skulls. In actual fact, these shifts, combined with what we can observe from the archaeological record, suggest that what is most profoundly different about modern humans compared to archaic humans is our capacity for cultural change. We have a unique and totally unprecedented ability to innovate and transmit information and ideas from person to person. At first, modern human cultural change accelerated gradually, causing important but incremental shifts in how our ancestors hunted and gathered. Then, starting about 50,000 years ago, a cultural and technological revolution occurred that helped humans colonize the entire planet. Ever since then, cultural evolution has become an increasingly rapid, dominant, and powerful engine of change. Therefore, the best answer to the question of what makes *Homo sapiens* special and why we are the only human species alive is that we evolved a few slight changes in our hardware that helped ignite a software revolution that is still ongoing at an escalating pace.

Who Were the First Homo sapiens?

Every religion has a different explanation for when and where our species, *H. sapiens,* originated. According to the Hebrew Bible, God created Adam from dust in the Garden of Eden and then made Eve from his rib; in other traditions, the first humans were vomited up by gods, fashioned from mud, or birthed by enormous turtles. Science, however, provides a single account of the origin of modern humans. Further, this event has been so well studied and tested using multiple lines of evidence that we can state with a reasonable degree of confidence that modern humans evolved from archaic humans in Africa at least 200,000 years ago.

The ability to pinpoint the time and place of our species' origin comes largely from studying people's genes. By comparing genetic variation among humans from around the globe, geneticists can calculate a family tree of everyone's relationships to one another, and by calibrating that tree, estimate when everyone last shared a common ancestor. Hundreds of such studies using data from thousands of people concur that all living humans can trace their roots to a common ancestral population that lived in Africa about 300,000 to 200,000 years ago, and that a subset of humans dispersed out of Africa starting about 100,000 to 80,000 years ago.[2] In other words, until very recently, all human beings were Africans. These studies also reveal that all living humans are descended from an alarmingly small number of ancestors. According to one calculation, everyone alive today descends from a population of fewer than 14,000 breeding individuals from sub-Saharan Africa, and the initial population that gave rise to all non-Africans was probably fewer than 3,000 people.[3] Our recent divergence from a small population explains another important fact, one that every human ought to know: we are a genetically homogenous species. If you catalog all the genetic variations that exist throughout our species, you'd find that approximately 86 percent are found within any one population.[4] To put this fact into perspective, you could wipe out the entire population of the world except for, say, Fiji or Lithuania and still retain almost every human genetic variation. This pattern contrasts markedly with other apes, like chimpanzees, in which less than 40 percent of the species' total genetic variation exists within any population.[5]

Evidence for our species' recent African origin also comes from fossil DNA. Fragments of DNA can be preserved for many thousands of years in fossil bone when conditions are just right: not too hot, not too acidic, and not too alkaline. Fragments of ancient DNA have been recovered from several early modern humans and more than a dozen archaic humans, mostly Neanderthals. Herculean efforts by Svante Pääbo and colleagues to reassemble and interpret these fragments reveal that the last time the modern human and Neanderthal lineages belonged to the same ancestral population was about 500,000 to 400,000 years ago.[6] Not surprisingly, human and Neanderthal DNA are extremely similar: only one out of every six hundred of your base pairs differs from a Neanderthal's. Much effort is currently being devoted to figuring out which genes these are and what they signify.

Some genealogical surprises also lurk in the DNA of ancient and modern humans. Careful analyses of the differences between Neanderthal and modern human genomes reveal that all non-Africans have a very small percentage, between 2 and 5 percent, of genes that came from Neanderthals. Apparently, a little interbreeding occurred between Neanderthals and modern humans more than 50,000 years ago, probably as modern humans were spreading out of Africa through the Middle East.[7] The descendents of this population then dispersed into Europe and Asia, which explains why Africans lack any Neanderthal genes. Another hybridization event occurred as humans spread into Asia and interbred with the Denisovans. About 3 to 5 percent of the genes among people who live in Oceania and Melanesia come from Denisovans.[8] As more fossil DNA is discovered, we may find traces of additional interbreeding events. Keep in mind that these traces should not be construed as evidence that humans, Neanderthals, and Denisovans are a single species. Closely related species often interbreed very slightly when they come into contact, and humans are evidently no different. I am actually delighted to know that although the Neanderthals are extinct, a little bit of them lives on in me.

Additional, different, and more tangible clues about when and where modern humans first evolved come from fossils. Just as the genetic data predict, the oldest modern human fossils known so far come from Africa, dated to about 195,000 years ago,[9] and a number of other early modern human fossils that are older than 150,000 years also come exclusively from Africa.[10] Ancient bones then trace the initial diaspora of *H. sapiens* around the globe. Modern humans first show up in the Middle East between about 150,000 and 80,000 years ago (these dates are uncertain), and then possibly disappeared for about 30,000 years as Neanderthals moved into the region during the height of a major European glaciation, perhaps displacing humans for a while.[11] Modern humans with new technologies showed up again in the Middle East around 50,000 years ago and then rapidly spread to the north, east, and west. According to the best dates now available, modern humans first appeared in Europe about 40,000 years ago, in Asia about 60,000 years ago, and in New Guinea and Australia by 40,000 years ago.[12] Archaeological sites indicate that humans also managed to cross the Bering Strait and colonize the New World sometime between 30,000 and 15,000 years ago.[13]

The precise chronology of human dispersals will change as more discoveries are made, but the important point is that in a mere 175,000 years after modern humans first evolved in Africa, they colonized every continent except Antarctica. Further, wherever and whenever modern human hunter-gatherers spread, archaic humans soon went extinct. For example, the last known Neanderthals in Europe were found in a cave at the southern tip of Spain, dated to just under 30,000 years ago, about 15,000 to 10,000 years after modern humans first appeared in Europe.[14] The evidence suggests that as modern humans spread rapidly throughout Europe, Neanderthal populations dwindled and ended up confined to isolated refugia before they vanished forever. Why? What was it about *H. sapiens* that made us the only surviving species of human on the planet? How much of our success can we attribute to our bodies and how much to our minds?

What's "Modern" About Modern Humans?

Just as history is written by the victors, prehistory has been written by the survivors (us), and we too often interpret what happened as inevitable. But what if twenty-first-century Neanderthals were writing this book, wondering why *H. sapiens* went extinct many thousands of years ago instead of them? Like us, they would probably start with the fossil and archaeological evidence to ask what is different about our bodies and how we used them.

Paradoxically, the most clear-cut differences to discern between us and archaic humans are anatomical contrasts whose biological relevance is hard to interpret. Most of these differences are evident in the head, and they boil down to two major changes in the way heads are put together, both shown in figure 14. The first is that we have small faces. Archaic humans have voluminous faces that project in front of the braincase, but the modern human face is much less deep and tall and almost completely tucked beneath the forebrain.[15] If you were to poke your finger vertically through a Neanderthal's eye socket, it would likely emerge through the browridge in front of the brain. Your face, in contrast, is more retracted, so a similarly poked finger would almost certainly end up in your brain's frontal lobe. Smaller, retracted faces have several consequences for human facial shape, also evident in figure 14. The most obvious is a smaller browridge. The browridge was once thought to be an adaptation to strengthen the upper part of the face, but it is actually just a shelf of bone that connects the forehead with the top of the eye sockets, hence an architectural by-product of how large the face is and how much it projects in front of the braincase.[16] Shallower faces also cause humans to have smaller, shorter nasal cavities and shorter oral cavities. Vertically smaller faces also give us smaller cheekbones and shorter, squarer eye sockets.

A second distinctive characteristic of the modern human head is its globular shape. When you look at the skull of any archaic human from the side, it is lemon-shaped: long and low, with big ridges of bone above the orbits and at the back of the skull. Modern human skulls, in contrast, are shaped more like oranges: nearly spherical with a high forehead and more rounded contours on the sides and

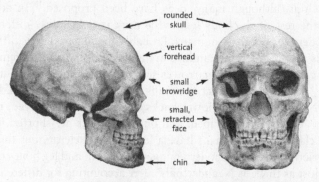

rounded skull

vertical forehead

small browridge

small, retracted face

chin

Early modern human *(Homo sapiens)*

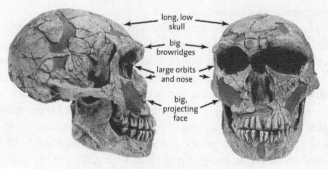

long, low skull

big browridges

large orbits and nose

big, projecting face

Archaic human *(Homo neanderthalensis)*

FIGURE 14. Comparison of an early modern human skull with a Neanderthal skull, illustrating some of the unique characteristics of the modern human head. Many of these features are the result of having a smaller, less projecting face.

back (again, see figure 14). Our more globular heads derive in part from our smaller face, but they also derive from having a rounder brain that sits on a skull base that is much less flat.[17]

Otherwise, there is not much special about human heads. Our brains aren't any larger, our teeth aren't unique, nor are our ears, eyes, or other sense organs. One small but distinctive feature of modern humans is the chin, an upside-down T-shaped projection of bone at the base of the lower jaw. True chins are not found in any archaic humans, and it is unclear why only modern humans have chins, although many ideas have been proposed.[18] In addition, the rest of the body below the neck is only subtly different in modern and archaic humans. Probably the most obvious difference is that modern human hips are a little less flared, and female birth canals are slightly narrower side to side and deeper front to back.[19] In addition, modern humans have less muscular shoulders than Neanderthals, our lower backs are a little more curved, our torsos are less barrel-shaped, and our heel bones are shorter. It is often claimed that humans have a less robust skeleton, but this is not strictly true. Early modern humans have arm and leg bones that were just as thick as Neanderthals' after accounting for differences in body weight and limb length.[20] Overall, the anatomical differences between modern and archaic humans are much more subtle below than above the neck.

Although modern and archaic human bodies are consistently but slightly different, the archaeological record tells a different story. Since stone tools, animal bones, and other artifacts left behind in ancient sites are mostly the products of learned behaviors, it should hardly be surprising that the archaeological evidence for behavioral differences between populations starts out small and then becomes greater over time. In fact, this initial similarity is exactly what you would predict. Neanderthals and modern humans were both big-brained species of hunter-gatherers that diverged from the same last common ancestor more than 400,000 years ago. Consequently, Neanderthals and modern humans inherited the same traditions of toolmaking, known collectively as the Middle Paleolithic (see chapter 5). Further, both species necessarily lived at low population densities, hunted large animals using spears, made fires, and cooked their food. But if you look carefully at the archaeological record in Africa, there are tantalizing traces of something different afoot.[21] A number of African sites older than 70,000 years show that the first modern humans who inhabited Africa at this time were trading over long distances, suggesting large and complex social networks. These early humans were also making new kinds of tools, including small stone points that were used as arrowheads, as well as various new kinds of bone tools, such as harpoons for fishing.[22] Early sites in South Africa also yield evidence for the beginnings of symbolic art, including stained necklace beads and engraved pieces of ocher.[23] Evidence for symbolic behavior among Neanderthals is exceedingly rare.[24] However, the earliest traces of behavioral modernity in Africa are ephemeral. For example, hafted arrowheads appear and then disappear in South Africa between 65,000 to 60,000 years ago and didn't seem to catch on permanently until later.[25] In addition, the very first modern human hunter-gatherers did not create abundant permanent art, build houses, or live at high population densities.

Then, starting about 50,000 years ago, something extraordinary happened: Upper Paleolithic culture was invented. The exact time and place of this revolution is murky, but it may have begun in northern Africa and then spread rapidly northward into Eurasia and southward into the rest of Africa.[26] One very obvious difference about the Upper Paleolithic was how people produced stone tools. In the Middle Paleolithic, complex tools were made in a very laborious and technically demanding way, but Upper Paleolithic toolmakers figured out how to mass manufacture long, thin blades of stone from the edges of prism-shaped cores. This innovation allowed hunter-gatherers to produce lots of thinner and more versatile tools that were easily fashioned into a wide range of specialized shapes. The Upper Paleolithic, however, involved more than just a new way of flaking stone; it was a veritable technological revolution. Unlike their Middle Paleolithic predecessors, the hunter-gatherers of the Upper Paleolithic started to create lots of bone tools, including awls and needles to fabricate clothing and nets, and they made lamps, fishhooks, flutes, and more. They also built more complex camps, sometimes with semipermanent houses. In addition, Upper Paleolithic hunters created much more lethal projectile weapons, such as spear throwers and harpoons.

Thousands of archaeological sites indicate that the Upper Paleolithic involved a revolution in the nature of hunting and gathering. Middle Paleolithic peoples were accomplished hunters who felled mostly large animals, but Upper Paleolithic people added a far broader range of animals to their menu, including fish, shellfish, birds, small mammals, and tortoises.[27] These animals are not only abundant but also can be acquired by women and children with little risk and a high probability of success. We have few remains of the plants consumed during the Paleolithic, but Upper Paleolithic people must have gathered a broad range of plants, which they processed more effectively, not just by roasting but also by boiling and grinding.[28] These and other dietary shifts helped fuel a population explosion. Soon after the Upper Paleolithic appears, the number and density of sites started to increase, even in remote and challenging places like Siberia.

In many ways, the most profound transformation evident in the Upper Paleolithic revolution is cultural: people were somehow *thinking and behaving differently*. The most tangible manifestation of this change is art. A handful of simple artistic objects have been found in Middle Paleolithic sites, but they are rare and pale in comparison with Upper Paleolithic art, which includes spectacular painted scenes in caves and rock shelters, carved figurines, gorgeous ornaments, and elaborate burials with superbly crafted grave goods. To be sure, not all Upper Paleolithic sites and regions have preserved art, but Upper Paleolithic people were the first to regularly express their beliefs or feelings symbolically in permanent media. Another component of the Upper Paleolithic revolution is cultural change. Almost nothing ever changed in the Middle Paleolithic: sites from France, Israel, and Ethiopia are all basically the same regardless of whether they are 200,000, 100,000, or 60,000 years old. But as soon as the Upper Paleolithic begins around 50,000 years ago, one can use artifacts to identify distinctive cultures that have discrete distributions in time and space. Ever since the Upper Paleolithic began, every part of the world has witnessed an endless series of cultural transformations, fueled by endlessly inventive and creative minds. These changes are still going on today at an increasing pace.

In short, if there is anything most different about modern humans compared to our archaic cousins it is our remarkable capacity and proclivity to innovate through culture. Neanderthals and other archaic humans certainly weren't stupid, and a handful of archaeological sites from Europe suggest that after the Neanderthals came into contact with modern humans they tried to create their own version of the Upper Paleolithic.[29] This short-lived response, however, was evidently an imperfect and only partial imitation. Hundreds of archaeological sites testify that the Neanderthals lacked modern humans' tendencies to invent new tools, adopt new behaviors, and express themselves as much using art. Was this lack of cultural flexibility and inventiveness the reason we survived and they went extinct? Or did we simply outbreed them? One way to address these and other related questions is to ask if there is anything special about modern human bodies that made possible or even triggered the Upper Paleolithic and subsequent cultural advances. Obviously, the first place to look is the brain.

Do Modern Humans Have Better Brains?

Brains don't fossilize, and we have yet to find a frozen Neanderthal deep within a glacier. So the only evidence for differences between modern and archaic human brains comes from studying the size and shape of the bones that surround the brain, from comparing human and nonhuman primate brains, and from looking for genes that differ between humans and Neanderthals and that have some effect on the brain in humans. Given our nascent comprehension of how brains work, using these lines of evidence to test if modern human brains function differently from our earlier ancestors' brains is somewhat like trying to figure out how two computers differ from looking at their exteriors and some random components whose function you don't entirely comprehend. Yet try we must, using whatever information is at our disposal.

The most obvious comparison to make is size, and it bears repeating that early modern human and Neanderthal brains were equally voluminous. There is no strong or straightforward relationship between brain size and intelligence (a notoriously difficult variable to measure), but it defies credulity to suppose that the big-brained Neanderthals weren't really smart.[30] That doesn't mean that humans and Neanderthals didn't have some cognitive differences, but it does mean that any differences would have to be in the more subtle, detailed architecture and wiring of the brain. Consequently, there has been much effort to compare the shapes of the bones that house the brain in order to test for differences in underlying brain structure. Although it is not possible to interpret these variations definitively, it turns out that a few key differences in the size of certain components of the brain do contribute to the more globular skull of modern humans.[31] Further, these differences may be relevant to possible cognitive differences between modern and archaic humans.

Of the brain's many structures, the most important to consider are the lobes that make up the largest part of the brain, the cerebrum, shown in figure 15. The outer layer of the cerebrum, the neocortex, is especially expanded in both archaic and modern humans and is responsible for conscious thought, planning, language, and other complex cognitive tasks. The neocortex, moreover, is divided into several lobes with different functions and whose convolutedly folded surface anatomy is partially preserved in fossil crania. The most obvious and significant difference in the neocortex of modern and archaic humans is that the temporal lobes are about 20 percent bigger in just *H. sapiens*.[32] This pair of lobes, which lie behind your temples, performs many functions that use and organize memories. When you listen to someone speaking, you perceive and interpret the sounds in parts of your temporal lobes.[33] The temporal lobe also helps you make sense of sights and smells, such as when you put a name to a face, or recall a memory after hearing or smelling something. In addition, a deep part of the temporal lobes (a structure called the hippocampus) allows you to learn and store information. It is therefore reasonable to hypothesize that enlarged temporal lobes may help modern humans excel at language and memory. A fascinating correlate of these faculties might be spirituality. Brain surgeons have discovered that stimulating the temporal lobe during surgery in alert patients can elicit intensely spiritual emotions even in self-described atheists.[34]

Another part of the human brain that appears to be relatively bigger in modern humans is the parietal lobes.[35] This pair of lobes plays key roles in interpreting and integrating sensory information from different parts of the body. Among its many functions, you use this part of the brain to make a mental map of the world and to figure out where you are, to interpret symbols such as words, to understand how you are manipulating a tool, and to do math.[36] If this part of your brain were to be damaged, you might lose the ability to multitask and do abstract thinking.

Other differences almost surely exist but are harder to measure. One candidate is a portion of the frontal lobe known as the prefrontal cortex. This walnut-sized part of your brain, which lies behind

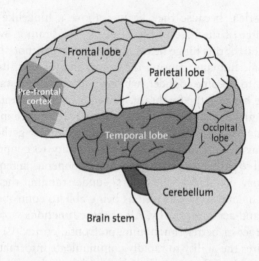

FIGURE 15. Different lobes of the brain. Several regions of the human brain, including the temporal lobes and the prefrontal part of the frontal lobes, are relatively larger in humans than apes. It is possible that some of these regions are larger in modern humans compared to archaic humans.

your brow, is about 6 percent larger in humans than in apes after correcting for size, and it has a more complex structure with greater connectivity.[37] Unfortunately, comparisons of skulls do not reveal just when the prefrontal cortex became relatively bigger in human evolution, so we can only speculate that it is specially enlarged in modern humans. But there is little doubt that its expansion was important, because if the brain were an orchestra, the prefrontal cortex would be its conductor: it helps you to coordinate and plan what other parts of your brain are doing when you speak, think, and interact with others. People with damage to this region have difficulty controlling their impulses, cannot plan or make decisions effectively, and struggle to interpret other people's actions and to regulate their own social behavior.[38] In other words, the prefrontal cortex helps you to cooperate and behave strategically.

One superficial effect of enlarging the temporal and parietal lobes is that these expansions may help make the human head more spherical because they lie just above a hingelike structure in the center of the base of the skull. As the brain grows rapidly just after birth, this hinge flexes about 15 degrees more in modern than archaic humans, causing the brain, and thus its surrounding braincase, to become rounder, while simultaneously rotating more of the face beneath the forebrain.[39] Even more important, evidence for reorganization of the modern human brain may explain some special, adaptive aspects of our cognition. A hunter-gatherer's success is heavily dependent on his or her ability to cooperate with others and to gather and hunt effectively. Cooperating requires having a theory of mind about others—understanding their motivations and mental state—as well as being able to control one's own impulses and act strategically. All of these functions would benefit from a larger or better-functioning prefrontal cortex. Cooperating also requires the ability to rapidly communicate information about emotion and intention, but also ideas and facts. Expansions of the temporal lobe may have also improved these skills, and, together with the parietal lobes, may have helped the first modern humans to reason more effectively as foragers and hunters. These parts of the brain allow us to make mental maps, to interpret sensory clues necessary to track animals, to deduce where resources are located, and to make and use tools. Given evidence for the expansion of these regions in modern humans, it is reasonable to speculate that our rounder brains not only helped us look more modern, they also helped us behave more modernly.

Other aspects of the modern human brain might also be different, but without archaic human brains to study, we can only conjecture. One possibility is that human brains are wired differently. Compared to apes, human brains develop a thicker neocortex with neurons that are larger and more complex and take longer to complete their wiring.[40] As in apes and monkeys, human brains have complex circuits that connect the outer cortical regions of the brain to deeper structures that participate in learning, how the body moves, and other functions. Although these circuits are not wired in a fundamentally different way in human brains, developing humans are apparently able to modify these circuits to a greater extent and with more connections.[41] Perhaps humans uniquely evolved to prolong the body's development in order to provide the brain more time to mature, including during the juvenile and adolescent periods, when many of these complex connections are made and insulated, and when many unused connections (which add noise) are pruned.[42] This hypothesis is admittedly speculative and needs to be tested carefully.[43] However, development did become prolonged at some point in human evolution, and would have been advantageous if it helped hunter-gatherers develop social, emotional, and cognitive skills (including language) that increased their chances of survival and reproduction.[44]

If modern and archaic human brains differ in their structure and function, then there must be genetic differences that underlie them. One might expect that there are genes expressed in the brain that enhance the ability to cooperate and plan that date to about the time modern humans evolved; some scholars propose that such genes evolved more recently, around 50,000 years ago, igniting the Upper Paleolithic.[45] So far, however, no such genes have been identified, but as our understanding of the genetic bases of brain development and function improve we will surely find them and estimate when they evolved. One candidate of much interest is a gene known as *FOXP2*, which plays a critical role in vocalization and other functions such as exploratory behavior.[46] Although the gene differs between humans and apes, it turns out that Neanderthals and humans share the same variant of *FOXP2*.[47] As other genes that differ between humans and Neanderthals are better studied, it will be interesting to find out what, if any, effects they have on human cognition. My guess is that Neanderthals were extremely smart, but that modern humans are more creative and communicative.

The Gift of Gab

How useful is a creative idea or a valuable fact if you can't communicate it? Some of the greatest cultural advances of the last few thousand years occurred thanks to more effective methods of transmitting information, such as writing, the printing press, the telephone, and the Internet. These and other information revolutions, however, all followed from an even earlier, more fundamental great leap forward in communication: modern human speech. Although archaic humans such as the Neanderthals surely had language, the uniquely short and retracted face of modern humans would have made us better at uttering clear, easy-to-interpret speech sounds at a very rapid rate. We are a uniquely silver-tongued species.

Speech sounds are basically a stream of pressurized puffs of air, not unlike those produced by the reed of a musical instrument, such as a clarinet. Just as you alter the volume and pitch of a clarinet by changing the pressure at which you blow on the reed, you vary the volume and pitch of speech sounds by modifying the rate and volume of these puffs as they leave the voice box (larynx) at the top of your windpipe. Once a sound wave leaves your larynx, its quality changes markedly as it passes through your vocal tract. As figure 16 shows, this tract is essentially an r-shaped tube that runs from your larynx to your lips, whose shape you can modify in diverse ways by moving your tongue, lips, and jaw. By changing the vocal tract's shape you alter how much energy is present in different frequencies of those puffs of air as they travel through the tube. The result is an alphabet-like array of sounds. For example, sometimes you constrict the vocal tract at certain locations to add turbulence at specific frequencies (like making the sounds "sss" or "ch"), or sometimes you close and then rapidly open a part of the vocal tract to create a burst of energy at a particular frequency (like "g" or "p").

Most mammals vocalize, but Philip Lieberman has pointed out that the human vocal tract is special for two reasons.[48] One is that our brains are exceptionally skilled at rapidly and precisely controlling the movements of the tongue and other structures that modify its shape. Additionally, the distinctive short and retracted face of modern humans gives our vocal tract a unique configuration with useful acoustic properties. Figure 16, which compares a chimp and human, illustrates this shape change. In both species, the vocal tract has essentially two tubes: a vertical portion behind the tongue and a horizontal portion above the tongue. However, the human vocal tract has different proportions, because a short face causes the oral cavity to be short, and thus requires the tongue to be short and rounded instead of long and flat.[49] Because the larynx is suspended from a tiny floating bone (the hyoid) at the base of the tongue, the low and rounded tongue of a human positions the larynx much lower in the neck than in any other animal. Consequently, the vertical and

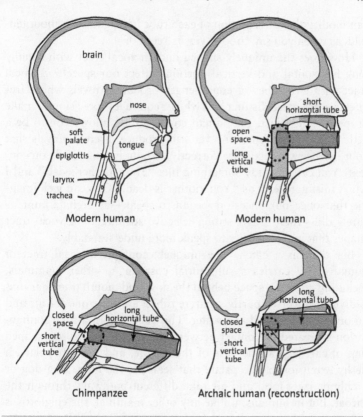

FIGURE 16. Anatomy of speech production. The top left panel (a midsection through a modern human head) shows the low position of the human larynx, the short rounded tongue, and the open space between the epiglottis and the back of the soft palate. This unique configuration causes the vertical and horizontal tubes of the vocal tract to be nearly equal in length and creates an open space between the epiglottis and soft palate (upper right panel). Like other mammals, the chimpanzee has a short vertical tube and a long horizontal tube, with a closed space behind the tongue. Reconstructions of archaic *Homo* suggest that its vocal tract had a more chimplike configuration.

horizontal tubes of the vocal tract are equally long in humans. This configuration differs from all other mammals, including chimps, in which the horizontal portion of the vocal tract is at least twice as long as the vertical portion. A related, important feature of the human vocal tract is that movements of our highly rounded tongue can modify the cross section of each tube independently about ten-fold (as when you say "oooh" versus "eeeh").

How does the uniquely shaped human vocal tract with equally long horizontal and vertical portions affect our speech? A vocal tract with two tubes of equal length produces vowels whose frequencies are more distinct and which require less precision to make properly.[50] In effect, the human configuration allows you to be a little sloppy when speaking yet still produce discrete vowels that your listener will recognize correctly without having to rely on context. You can thus say something like "Your mother's dad" and I won't misinterpret it as "You mother is dead." One can well imagine that once our ancestors began to speak—as archaic humans surely did—there was a strong selective advantage for vocal tract shapes that made it easier to speak more understandably.

But there is a catch. The uniquely configured vocal tract of humans also carries a substantial cost. In all other mammals, including apes, the space behind the nose and mouth (the pharynx) is divided into two partly separate tubes: an inner one for air and an outer one for food and water. This tube-within-a-tube configuration is created by contact between the epiglottis, a gutter-shaped flap of cartilage at the base of the tongue, and the soft palate, a fleshy extension of the palate that seals off the nose. In a dog or a chimpanzee, food and air take different pathways through the throat. But in humans, unlike any other mammal, the epiglottis is a few centimeters too low to contact the soft palate. By dropping the larynx low in the neck, humans lost the tube within a tube and developed a big common space behind the tongue through which food and air both travel to get into either the esophagus or the trachea. As a result, food sometimes gets lodged in the back of the throat, blocking off the airway. Humans are the only species that risks asphyxiation when we swallow something too large or imprecisely. This cause of death is more common than you may think. According to the National Safety Council, choking on food is the fourth leading cause of accidental deaths in the United States, approximately one-tenth the number of deaths caused by motor vehicles. We have paid a heavy price for speaking more clearly.

Next time you have a meal and chat with friends, consider that you are probably doing two unique things: speaking with great clarity and swallowing a little dangerously. Both these activities are special in modern humans, made possible by having an unusually small and retracted face. Certainly, archaic humans also talked with their mouths full at dinner, but their speech was probably a little less clear, and they were probably less likely to choke on their food.

Evolution of Cultural Evolution

Whatever biological traits make us different from archaic humans, they must have been of consequence. The innovations that led to the Upper Paleolithic probably accrued gradually, but once the Upper Paleolithic was in full force, it helped modern humans spread rapidly around the globe, and our archaic cousins vanished whenever and wherever we arrived. The details of this replacement are partly mysterious. Modern humans certainly interacted and sometimes interbred with archaic humans, such as Neanderthals, but no one knows why we, not they, survived.[51] Many theories exist. One possibility is that we simply outbred them, perhaps by weaning our children younger or having lower mortality rates. Very slight differences in birth and death rates have major, sometimes devastating effects on hunter-gatherers, who need to live at low population densities. Calculations show that if both modern humans and Neanderthals were living in the same region but the rate of mortality among Neanderthals was just 1 percent higher than in modern humans, the Neanderthals would have gone extinct in just thirty generations, less than a thousand years.[52] Given evidence that Upper Paleolithic people lived longer than Middle Paleolithic people,[53] the Neanderthals' rate of extinction could have been even faster. Other, nonexclusive hypotheses are that modern humans outcompeted our cousins because we were better at cooperating, that we foraged and hunted for a wider range of resources, including more fish and fowl, and that we had larger, more effective social networks.[54] Archaeologists will continue to debate these and other ideas, but one general conclusion is clear: something about modern human behavior must have been advantageous. In a classic example of circular logic, we define whatever made modern human behavior different to be "behavioral modernity."[55]

However one defines "behavioral modernity," its consequences for our bodies have been profound ever since the Upper Paleolithic began, and they are still of import today, many thousands of generations later. Why? Because whatever biological factors make us cognitively and behaviorally modern, they are primarily manifested through *culture*. Culture is a term with multiple meanings, but it is most essentially a set of learned knowledge, beliefs, and values that cause groups to think and behave differently, sometimes adaptively, sometimes arbitrarily. By this definition, apes, such as chimps, have very simple cultures, and archaic humans, such as *H. erectus* and Neanderthals, had sophisticated cultures. But the archaeological record associated with modern humans indicates unambiguously that we have an extraordinary and special capacity and proclivity to innovate and to transmit new ideas. *H. sapiens* is a fundamentally and exuberantly cultural species. Indeed, culture must be our species' most distinctive characteristic. A visiting alien biologist would surely notice how human bodies differ from other mammals' (we are bipedal, lack fur, and have big brains), but they would be most astonished by the diverse and often arbitrary ways in which we behave, including our clothes, tools, towns, food, art, social organization, and babel of languages.

Human cultural creativity, once unleashed, has been an unstoppable engine of accelerating evolutionary change. Like genes, cultures evolve. But, unlike genes, culture evolves through different processes that make cultural evolution a far more powerful and rapid force than natural selection. This is because cultural traits, known as "memes," differ from genes in several key respects.[56] Whereas new genes arise solely by chance through random mutations, humans often generate cultural variations intentionally. Inventions like farming, computers, and Marxism were created through ingenuity and for a purpose. In addition, memes are transmitted not just from parents to offspring, but from multiple sources. Reading this book is just one of your many horizontal exchanges of information today. Finally, although cultural evolution can occur randomly (think of fashions like tie width or skirt length), cultural change often happens through an agent of change, such as a persuasive leader, television, or a community's collective desire to solve a challenge like hunger, disease, or the threat of Russians on the moon. Together, these differences make cultural evolution a faster and often more potent cause of change than biological evolution.[57]

Culture itself is not a biological trait, but the capacities that enable humans to behave culturally, and to use and modify culture, are basic biological adaptations that appear to be specially derived in modern humans. If Neanderthals or Denisovans were the only species of humans left on the planet, I suspect (but cannot prove) they would still be hunting and gathering in more or less the same way they did 100,000 years ago. This is obviously not the case for *H. sapiens,* and as cultural change has accelerated since the Upper Paleolithic, its effects on our bodies have accelerated too. The most basic interactions between culture and your body's biology are the ways that learned behaviors—the foods you eat, the clothes you wear, the activities you do—alter your body's environment, thus influencing how your body grows and functions. The effects don't cause evolution per se (that would be Lamarckian), but over time some of these interactions do make possible evolutionary change in populations. Sometimes cultural innovations *drive* natural selection on the body. A beautifully studied example is the ability to digest milk sugar as an adult (lactase persistence), which evolved independently in Africa, the Middle East, and Europe among peoples who consumed animal milk.[58] In many other cases, culture ameliorates or negates the effects of the environment on the body, thus *buffering* the body from the effects of natural selection that might otherwise occur. Cultural buffering is so ubiquitous that we are often only aware of its effects when we are deprived of technologies such as clothes, cooking, and antibiotics. Without them, many people alive today would have been removed from the gene pool long ago.

Your body is loaded with features that evolved over hundreds of thousands of years of interactions between culture and biology. Some of these adaptations predate the origin of modern humans. For instance, the invention of stone tools and projectiles made possible selection for increased manual dexterity and the ability to throw with power and accuracy. Teeth were selected to become smaller after stone tools began to be produced in the Lower Paleolithic, and digestive systems changed so much once cooking became prevalent that we are now dependent on cooking to survive.[59] Although it is sometimes assumed that human biology has barely changed since *H. sapiens* evolved 200,000 years ago, our relentless drive to innovate has clearly triggered selection on the human body. Much of this selection has been regional and has contributed to variations that distinguish populations from different parts of the world. As Upper Paleolithic people spread across the globe and encountered new pathogens, unfamiliar foods, and diverse climatic conditions, natural selection adapted these newly isolated populations to their varied environments.

Consider, for instance, how various modern human populations evolved to cope with vastly dissimilar climates. In the hot African environments in which modern humans originated, the biggest problem is to dump heat, but as humans moved into temperate Europe and Asia during the Ice Age, retaining heat became a far more urgent challenge. Remember that these first migrants out of Africa were Africans, and they, like us, would have perished in northern climates during the Ice Age without technologies such as clothing, heating, and dwellings. To a large extent, early modern hunter-gatherers who ventured north devised cultural adaptations to survive in frozen climes. One novel invention of the Upper Paleolithic was bone tools, such as needles, which are entirely absent from the Middle Paleolithic. Apparently, Neanderthal clothing wasn't sewn. Upper Paleolithic people also created warm shelters, lamps, harpoons, and other technologies that facilitated their survival in harsh habitats that, frankly, are unnatural and inhospitable for tropical primates. These cultural innovations, however, did not entirely buffer them from the effects of natural selection but instead made possible selection that would have otherwise not occurred. During bitterly cold Ice Age winters, cultural adaptations enabled people to stay alive enough so that natural selection favored individuals with heritable variations that improved their ability to survive and reproduce. Such selection is evident from changes in body shape. If you want to dump heat in a warm region by sweating, it helps to be tall and skinny with long limbs that maximize your body's surface area, but in order to retain body heat in colder climates, it helps to have shorter limbs along with a wider, more massive frame.[60] As European Upper Paleolithic people endured the extremes of the last major Ice Age, their body shape changed predictably. Like other Africans, the first migrants to Europe were tall and skinny, but over tens of thousands of years they evolved to become shorter and stockier, especially in more northern parts of the continent.[61]

Body shape is just one of many features that vary among populations because of selection that occurred since modern human hunter-gatherers dispersed across the planet to habitats as diverse as deserts, arctic tundras, rain forests, and high mountains. Perhaps no trait has been subject to more misguided attention than skin color. At least six genes cause the skin's outer layer to synthesize pigments that act like a natural sunscreen to block damaging ultraviolet radiation but that also impede the synthesis of vitamin D (which your skin makes in response to sunlight).[62] Consequently, there was strong selection for dark pigmentation near the equator, where ultraviolet radiation is intense year-round, but populations who moved into temperate zones were selected to have less pigmentation to ensure sufficient levels of vitamin D. Studies of human genetic variation have identified hundreds of other genes that bear signatures of strong selection over the last few thousand years (which later chapters will discuss). One caveat to keep in mind is that a large number of the traits that cause people and populations to differ, such as hair texture and eye color, are literally skin deep, and many are just random variations that have nothing to do with natural selection, let alone cultural evolution.

Brains, Brawn, and the Triumph of Modern Humans

By now, it should be evident that the history of the human body provides no single answer to the question chapter 1 posed: "What are humans adapted for?" Our long evolutionary path adapted humans to be upright, to eat a diverse diet, to be hunters, to forage widely, to be endurance athletes, to cook and process our food, to share food, and more. But if there is any special adaptation of modern humans that accounts for our evolutionary success (so far) it must be our ability to be *adaptable* because of our extraordinary capacities to communicate, cooperate, think, and invent. The biological bases for these capabilities are rooted in our bodies, especially our brains, but their effects are manifested primarily through the way we use culture to innovate and to adjust to new and diverse circumstances. After the first modern humans evolved in Africa, they gradually invented more advanced weapons and other new tools, created symbolic art, engaged in more long-distance trading, and behaved in other novel, quintessentially modern ways. It took more than 100,000 years for the Upper Paleolithic way of life to emerge, but that revolution was just one of many cultural leaps forward that are still ongoing at ever faster rates. In the last few hundred generations, modern humans have invented farming, writing, cities, engines, antibiotics, computers, and more. The pace and scope of cultural evolution now vastly exceeds the pace and scope of biological evolution.

It is therefore reasonable to conclude that of all the qualities that make modern humans special, our cultural abilities have been the most transformative and the most responsible for our success. These abilities probably explain why the last Neanderthals went extinct soon after modern humans first set foot in Europe and why, as our species spread through Asia, we probably caused the demise of the Denisovans, the Hobbits of Flores, and any other remaining descendants of *H. erectus*. Many additional cultural innovations enabled modern human hunter-gatherers to inhabit just about every corner of the earth by 15,000 years ago, even inhospitable places like Siberia, the Amazon, the Australian central desert, and Tierra del Fuego.

Viewed in this light, human evolution appears to be, first and foremost, a triumph of brains over brawn. In fact, many narrative accounts of human evolution emphasize this triumph.[63] Despite a lack of strength, speed, natural weapons, and other physical advantages, we have used cultural means to flourish and establish dominion over most of the natural world—from bacteria to lions, from the Arctic to the Antarctic. A large percentage of the billions of humans alive today are enjoying longer and healthier lives than ever before. Thanks to the same powers of inventiveness that sparked the Upper Paleolithic, we can now fly, replace diseased organs, peer into atoms, and travel to the moon and back. Perhaps someday our brains will allow us to understand the fundamental laws of physics that govern the universe, to colonize other planets, and to eliminate poverty.

Much as our remarkable abilities to think, learn, communicate, cooperate, and innovate have made possible our species' recent successes, I think it is not just incorrect but also dangerous to view modern human evolution as solely a triumph of brains over brawn. The Upper Paleolithic and other cultural innovations that helped modern humans colonize the planet and outcompete other species of humans brought many benefits, but they never freed hunter-gatherers from having to work and use their bodies to survive. As we have seen, hunter-gatherers are essentially professional athletes whose livelihood requires them to be physically active. For example, an average male hunter-gatherer from the Hadza tribe in Tanzania weighs 51 kilograms (112 pounds), walks 15 kilometers a day (about 9 miles), and also has to climb trees, dig tubers, carry food, and perform other physical tasks on a daily basis.[64] His total energy expenditure is about 2,600 calories a day. Since 1,100 of those calories sustain his body's basic needs (his basal metabolism), he spends 1,500 calories a day being physically active, amounting to nearly 30 calories per kilogram per day. In contrast, a typical American or European male weighs about 50 percent more and works 75 percent less, expending just 17 calories per kilogram per day on physical activity.[65] In other words, a hunter-gatherer works about twice as much per unit of body weight as the Westerner (which explains a lot about why the Westerner is more likely to be overweight).

Modern human hunter-gatherers therefore flourished with a combination of brains *plus* brawn, and they lead more arduous, physically demanding lives compared to most postindustrial humans. That said, it is important to stress that hunting and gathering, despite its physical demands, is hardly the backbreaking existence of toil and misery that some people imagine it to be. When anthropologists first began to quantify the effort required to be a hunter-gatherer, they were astonished at how much time typical hunter-gatherers actually spend doing "work" even in harsh environments. The Bushmen of the Kalahari, for example, devote an average of six hours per day to activities such as foraging, hunting, making tools, and doing housework.[66] That doesn't mean, however, that the rest of the day is spent relaxing and having fun. Since hunter-gatherers don't produce surpluses of food, they often rest whenever possible to avoid wasting energy, they can never afford to retire when they reach sixty-five years of age, and if they become injured or disabled, others have to work harder to compensate. Because of our species' special cognitive and social skills, modern human hunter-gatherers work fairly hard, but not *that* hard.

Our species' capacities and propensities to use culture to adapt, improvise, and improve also explain another fundamental characteristic of modern human hunter-gatherers: extraordinary variability. As modern hunter-gatherers colonized the planet, they invented a stunning array of technologies and strategies to cope with diverse new conditions.[67] In the frigid open expanses of northern Europe, they learned to hunt mammoths and build huts from their bones. In the Middle East, they harvested fields of wild barley and invented grinding stones to make flour. In China, they created the first pottery, probably to boil food and make soup. Whereas foragers in most tropical places obtain only about 30 percent of their calories from hunting large mammals, hunter-gatherers who colonized temperate and arctic habitats devised ways to survive by obtaining the majority of their calories from animal foods, mostly fish. And while most hunter-gatherers have to move camp regularly to follow seasonal foods, some foragers, such as the native Americans of northwest America, managed to settle in permanent villages. In truth, there is no single hunter-gatherer diet, just as there was no one system of kinship or religion, no one mobility strategy, division of labor, or group size.

The irony of human cultural adaptability is that our species' unique talents for innovating and solving problems not only enabled hunter-gatherers to thrive nearly everywhere on earth but also eventually enabled some of them to cease being hunter-gatherers. Starting around 12,000 years ago, a few groups of people began to settle down in permanent communities, grow plants, and domesticate animals. These shifts were probably gradual at first, but over the next few thousand years they sparked a worldwide agricultural revolution whose effects are still rocking the planet, as well as our bodies. As we will see, farming brought many advantages but also caused many serious problems. Farming enabled humans to have more food, hence more children, but also required new forms of work, transformed diets, and opened a Pandora's box of diseases and social ills. Farming has been around for just a few hundred generations, but it accelerated the pace and scope of cultural change so dramatically that many people today can barely imagine the way we lived before our ancestors invented agriculture, not to mention writing, wheels, metal tools, and engines.

Were these and other recent cultural developments a mistake? Since the human body was molded, bit by bit, over millions of years to be fruit-eating bipeds, then australopiths, and finally big-brained, culturally creative hunter-gatherers, does it not follow that our bodies would be better off if we lived as our evolutionary past adapted us? Has civilization led the human body astray?

PART II

Farming and the Industrial Revolution

7

Progress, Mismatch, and Dysevolution

The Consequences—Good and Bad—of Having
Paleolithic Bodies in a Post-Paleolithic World

> Though we are not so degenerate but that we might
> possibly live in a cave or a wigwam or wear skins
> today, it certainly is better to accept the advantages,
> though so dearly bought, which the invention and
> industry of mankind offer.
>
> —HENRY DAVID THOREAU, *Walden*

Have you ever wanted to abandon it all and seek a simpler life more in tune with your evolutionary legacy? In *Walden*, Henry David Thoreau describes the two years he spent in a hut in the woods by Walden Pond, detached from mid-nineteenth-century American culture, whose growing consumerist and materialist tendencies troubled him. People who have never read *Walden* sometimes mistakenly think that Thoreau spent these years as a hermit. In fact, he was seeking simplicity, self-sufficiency, a greater connection with nature, and only temporary solitude. Thoreau's hut was a several-mile walk from the center of Concord, Massachusetts, which he visited every day or two to gossip and dine with friends, have his clothes laundered, and enjoy other comforts befitting a well-to-do man of letters. Even so, *Walden* has become a sort of bible for primitivists who decry the advances of civilization and yearn for a return to the good old days. According to this line of thinking, modern technology has led to the unfair development of social classes of "haves" and "have-nots," to widespread alienation and violence, and to an erosion of dignity. Some primitivists want to return the human species to an idealized agrarian way of life, and a few even think that the quality of human existence has been going downhill ever since we ceased to be Paleolithic hunter-gatherers.

There is much to be said for a return to more of life's simple pleasures, but a knee-jerk opposition to technology and progress is facile and futile (and was never advocated by Thoreau). By many measures, the human species has thrived since the end of the Paleolithic. The world's population at the start of twenty-first century is at least a thousandfold greater than during the Stone Age. Despite ongoing poverty, war, hunger, and infectious disease in the poorest parts of the world, an unprecedented number of people around the globe not only have enough food but are also enjoying long and healthy lives. As an example, the typical Englishman today is 7 centimeters (nearly 3 inches) taller than his great-grandfather who lived one hundred years ago, his life expectancy is thirty years longer, and his children have about a tenfold greater chance of surviving infancy.[1] In addition, capitalism has allowed average people such as me to take for granted opportunities unimagined by the richest aristocrats a few centuries ago. I have no desire to live permanently as a transcendentalist in the woods, let alone as a caveman without health care, education, and sanitation. I also enjoy the diversity of tasty foods I eat, I love my job, and I get a thrill out of living in a vibrant city full of interesting people, restaurants, museums, and shops. I also take pleasure in recent technologies like air travel, iPods, hot showers, air-conditioning, and 3D movies. Thoreau and others are correct in diagnosing modern life as increasingly consumerist and materialistic, but people's desires haven't changed so much as have their opportunities to satisfy them.

On the other hand, it is equally facile and foolish to ignore the many serious, novel challenges that human beings now confront. What followed the Paleolithic—farming, industrialization, and other forms of "progress"—may have been a boon to the average person, but they promoted new diseases and other problems that were rare or absent during the Paleolithic. Almost every major infectious epidemic, such as smallpox, polio, and the plague, happened after the Agricultural Revolution began. In addition, studies of recent hunter-gatherers show that although they don't enjoy surpluses of food, they rarely suffer from famines or serious malnutrition. Modern lifestyles have also fostered new noncommunicable but widespread illnesses such as heart disease, certain cancers, osteoporosis, type 2 diabetes, and Alzheimer's, as well as scores of other lesser ailments, such as cavities and chronic constipation. There is good reason to believe that modern environments contribute to a sizeable percentage of mental illnesses, such as anxiety and depressive disorders.[2]

The story of progress achieved by the march of civilization since the end of the Stone Age has also been less gradual and continuous than many people suppose. As the next few chapters will show, farming created more food and allowed populations to grow, but for most of the last few thousand years, the average farmer had to work much harder than any hunter-gatherer, experienced worse health, and was more likely to die young. The majority of improvements in human health, such as greater longevity and decreased infant mortality, occurred over just the last hundred years. In fact, from the body's perspective, many developed nations have recently made *too much progress*. For the first time in human history, a large number of people face excesses rather than shortages of food. Two out of three Americans are overweight or obese, and more than a third of their children weigh too much. In addition, a majority of adults in developed nations such as the United States and United Kingdom are physically unfit because our culture has made it easy, hence common, to spend the day without ever raising one's heart rate. Thanks to "progress," I can wake up in my soft and comfortable bed, press a few buttons to get breakfast, drive to work, take an elevator to my office, and then pass the next eight hours sitting in a comfortable chair without breaking a sweat, getting hungry, or being too cold or too hot. Machines now perform for me almost every task that once required physical effort: getting water, washing, acquiring and preparing food, traveling, even brushing my teeth.

In short, the human species has achieved considerable progress over the last few thousand years since we ceased to be hunter-gatherers, but how and why has some of this progress been bad for our bodies? The next few chapters will review how the human body has changed following the Paleolithic, but first let's pause to consider the pros and cons of no longer living in ways for which our bodies were adapted by millions of years of evolution. Are some forms of ill health a necessary consequence of civilization? And, more generally, how have biological and cultural evolution inter-acted after the Paleolithic in ways that affect the human body for the better and the worse?

How Are We Still Evolving?

I have been teaching human evolution to college students for more than twenty years, and for most of that time, I wound up my lectures more or less where chapter 6 just ended, with the origin of modern humans and the dispersal of people across the globe. My reasoning for finishing in the Paleolithic was the general con-sensus that little significant biological evolution has occurred in H. sapiens since then. According to this view, ever since cultural evolution became a more powerful force than natural selection, the human body has barely altered, and whatever changes did take place over the last 10,000 years are more the province of historians and archaeologists than evolutionary biologists.

I now regret the way I used to teach human evolution. For one, it is simply not true that H. sapiens stopped evolving once the Paleolithic ended. In fact, the idea must be wrong because natural selection is the consequence of heritable genetic variation and dif-ferential reproductive success. People continue to pass on genes to their children, and today, as in the Stone Age, some people have more offspring than others. It follows that if there is any heritable basis to differences in people's fertility then natural selection must still be chugging along. What's more, accelerating rates of cultural evolution have rapidly and substantially changed what we eat, the ways we work, the diseases we encounter, and other environmen-tal factors that have created new selective pressures. Evolutionary biologists and anthropologists have shown that cultural evolution hasn't halted natural selection, and it has not only *driven* but sometimes even accelerated selection.[3] As we will see, the Agricultural Revolution has been an especially powerful force for evolutionary change.

One of the reasons we don't think of evolution as being much of a force today is that natural selection is gradual, often requiring hundreds of generations to have a dramatic effect. Since a human generation is typically twenty or more years, one cannot easily detect evolutionary changes in humans of the magnitude that we can observe rapidly in bacteria, yeast, and fruit flies. However, it is possible to measure very recent natural selection in humans over just a few generations with enormous samples and much effort, and a few such studies have managed to find evidence for low levels of selection during the last few hundred years. In Finnish and American populations, for example, there has been selection on the age that women first give birth and the age women start menopause, as well as people's weight, height, cholesterol, and blood sugar levels.[4] If we look over longer time periods, we can detect even more evidence for recent selection. New technologies that rapidly and inexpensively sequence entire genomes have revealed hundreds of genes that have been under strong selection during the last few thousand years within particular populations.[5] As you might expect, many of these genes regulate reproduction or the immune system and were strongly selected because they helped people have more offspring and survive infectious diseases.[6] Others play a role in metabolism and helped certain farming populations adapt to foods such as dairy products and starchy staple crops. A few selected genes are involved in thermoregulation, presumably because they enabled far-flung populations to adapt to a wide range of climates. My colleagues and I, for example, found evidence of strong selection for one gene variant that evolved in Asia near the end of the Ice Age, causing East Asians and native Americans to have thicker hair and more sweat glands.[7] One practical benefit of studying these and other recently evolved genes is to understand better how and why people vary in their susceptibility to certain diseases and how they respond to different medicines.

Although natural selection has not ceased since the Paleolithic, it is nonetheless true that relatively less natural selection has occurred in humans over the last few thousand years compared to the previous few million years. This difference is to be expected because it has been only six hundred generations since the first farmers began to till the soil of the Middle East, and most people's ancestors started farming more recently, probably within the last three hundred generations. For perspective, about the same number of generations of mice have lived in my house over the last century. Although considerable selection can occur in three hundred generations, the strength of selection needs to be very great to cause a beneficial mutation to sweep through a population or a harmful mutation to be eliminated that quickly.[8] In addition, during the last few hundred generations, selection has not always been operating in a consistent direction, which can obscure its traces. For example, as temperatures and food supplies have fluctuated, selection during some periods probably favored people who were bigger, but then during other periods it likely favored people who were smaller. Finally, and most important, there is no question that some cultural developments have buffered untold numbers of humans from natural selection that might have otherwise occurred. Consider how penicillin must have affected selection once the drug became widely available in the 1940s. Millions of people are alive today who would otherwise have been more likely to die from diseases like tuberculosis or pneumonia if they have genes that increase their susceptibility. Consequently, although natural selection has not ceased to act, we know that it has had only limited, regional effects on human biology over the last few thousand years. If you were to raise a Cro-Magnon girl from the Upper Paleolithic in a modern French household, she would still be a typical modern human girl except for some modest biological differences, probably mostly in her immune system and her metabolism. We know this is true because everyone from every corner of the planet shares a last common ancestor from less than 200,000 years ago, and yet different populations are for the most part genetically, anatomically, and physiologically the same.[9]

Regardless of just how much selection has occurred since the Paleolithic, there are other important ways in which humans have evolved over the last few thousands and hundreds of years. Not all evolution occurs through natural selection. An even more powerful and rapid force today is cultural evolution, which has altered many crucial interactions between genes and the environment by altering environments, not genes. Every organ in your body—your muscles, bones, brain, kidneys, and skin—is the product of how your genes were affected by signals from the environment (such as forces, molecules, temperatures) during the period you developed, and their current functions continue to be influenced by aspects of your current environment. Although human genes have changed modestly over the last few thousand years, cultural changes have dramatically transformed our environments, often resulting in a very different, arguably more important kind of evolutionary change than natural selection. For example, toxins in tobacco, certain plastics, and other industrial products can cause cancer, often years after initial exposure. If you grow up chewing soft, highly processed food, your face will be smaller than if you grow up chewing hard, tough food.[10] If you spend your first few years in a hot climate, you develop more working sweat glands than if you were born in a cool environment.[11] These and other changes aren't genetically heritable, but they are *culturally heritable*. Just as you pass on a last name to your children, you also pass on environmental conditions, such as the toxins they encounter, the foods they eat, the temperatures they experience. As cultural evolution is accelerating, environmental changes that affect how our bodies grow and function are also accelerating.

How cultural evolution is changing interactions between the genes we inherited and the environments in which we live is of great consequence. Over the last few hundred generations, the human body has changed in various respects because of cultural change. We mature faster, our teeth have gotten smaller, our jaws are shorter, our bones are thinner, our feet are often flatter, and many of us have more cavities.[12] As future chapters will examine, there is also good reason to believe that today more people sleep less, experience higher levels of stress, anxiety, and depression, and are more likely to be shortsighted. In addition, human bodies these days have to contend with numerous infectious diseases that used to be rare or nonexistent. Each of these changes to the human body has some genetic basis, but what has changed is not so much the genes that play a role in these diseases as the environments with which these genes interact.

Consider type 2 diabetes, a metabolic disease that used to be rare but is now becoming common all across the globe. Some people are genetically more susceptible to type 2 diabetes, which helps explain why the disease is rapidly becoming more prevalent in places like China and India than in Europe and America.[13] However, type 2 diabetes is not booming faster in Asia than in America because of novel genes that are now spreading in the East. Instead, new Western lifestyles are sweeping across the globe and interacting with ancient genes that previously did not have negative effects.

Put differently, not all evolution occurs through natural selection, and interactions between genes and the environment have been changing rapidly, sometimes radically, primarily because of changes in our bodies' environments caused by rapid cultural evolution. You may have genes that predispose you to having flat feet, myopia, and type 2 diabetes, but the distant ancestors from whom you inherited the very same genes likely did not suffer from these problems. We therefore have much to gain by using the lens of evolution to consider shifting gene-environment interactions that have occurred since the Paleolithic ended. How well do the genes and bodies we inherited from our early modern human ancestors fare in the novel environments to which we subject them? And how can an evolutionary perspective on these changes be of practical use?

Why Medicine Needs a Dose of Evolution

Few words cause more terror in a doctor's office, and are less likely to make you think about evolution, than "cancer." If I were to receive a diagnosis of cancer tomorrow, my first concern would be to figure out how to rid myself of the disease. I'd want to know what kind of cells were cancerous, what mutations were causing them to divide out of control, and what medical interventions such as surgery, radiation, and chemotherapy would have the best chance of killing those cells without killing me. Even though I study human evolution, the theory of natural selection would be far from my mind as I confronted the disease. The same would be true if I had a heart attack, a painful tooth cavity, or a torn hamstring. When sick, I see a doctor, not an evolutionary biologist. By the same token, my doctors have studied little if any evolutionary biology as part of their training. And why should they? Evolution, after all, is something that mostly occurred in the past, and today's patients are not hunter-gatherers, let alone Neanderthals. Someone with heart disease needs surgery, drugs, or other medical procedures that require a thorough understanding of fields such as genetics, physiology, anatomy, and biochemistry. Doctors and nurses are therefore not required to take courses on evolutionary biology, and I doubt if they, insurance companies, and others in the health-care industry ever give Darwin or Lucy much thought at their jobs. Just as knowing the history of the Industrial Revolution will not help a mechanic fix your car, why would knowing the Paleolithic history of the human body help a doctor treat your disease?

Considering evolution to be irrelevant to medicine may seem logical at first, but this way of thinking is deeply flawed and shortsighted. Your body was not engineered like a car, but instead evolved through descent with modification. It therefore follows that knowing your body's evolutionary history helps to evaluate *why* your body looks and works as it does, hence *why* you get sick. Although scientific fields such as physiology and biochemistry can help us understand the proximate mechanisms that underlie a disease, the burgeoning field of evolutionary medicine helps us make sense of why the disease occurs in the first place.[14] Cancer, for example, is actually an aberrant evolutionary process going on within a body. Every time a cell divides, its genes have a certain chance of mutating, so cells that divide more frequently (examples include blood and skin cells) or that are more often exposed to chemicals that cause mutations (for example, lung and stomach cells) have a greater chance of accidentally acquiring mutations that cause them to divide out of control, forming tumors. Most tumors, however, are not cancers. To become cancerous, the tumor cells need to gain further mutations that allow them to outcompete other, healthy cells by taking their nutrients and interfering with normal function. In essence, cancerous cells are nothing more than abnormal cells with mutations that enable them to survive and reproduce better than other cells. If we hadn't evolved to evolve, we would never get cancer.[15]

To go a step further, since evolution is an ongoing process that is still occurring, an appreciation of how evolution works can prevent some failures and missed opportunities, as well as improve our ability to ward off and treat many diseases. An especially urgent and obvious example of the need for evolutionary biology in medicine is the way we treat infectious diseases, which are still evolving along with us. By failing to appreciate that humans and diseases such as AIDS, malaria, and tuberculosis remain locked in an evolutionary arms race, we sometimes unwittingly aid or intensify these infectious agents by using drugs ineptly or by rashly disrupting ecological conditions.[16] Preventing and treating the next epidemic will require a Darwinian approach. Evolutionary medicine also yields vital perspectives for improving how we use antibiotics to treat everyday infections. Overusing antibiotics not only promotes the evolution of novel superbugs but also alters the body's ecology in ways that may contribute to new autoimmune illnesses, such as Crohn's disease (see chapter 11). Evolutionary biology even holds some promise in helping us better prevent and treat cancer. We often fight cancer cells by trying to kill them with radiation or toxic chemicals (chemotherapy), but an evolutionary approach to cancer explains why these treatments sometimes backfire. Radiation and chemotherapy not only raise the probability of nonlethal tumors developing mutations that transform them into cancer cells, they also alter the cells' environment in a way that can increase the selective advantage of the new mutations. For this reason, it is hypothesized that less aggressive treatments may sometimes be more beneficial to patients with certain less malignant forms of cancer.[17]

Another application of evolutionary medicine is to recognize that many symptoms are actually adaptations, thus helping doctors and patients rethink the way we treat some illnesses and injuries. How often do you take an over-the-counter medication at the first sign of fever, nausea, diarrhea, or just aches and pains? These discomforts are widely regarded as symptoms to alleviate, but evolutionary perspectives indicate that they can be adaptations to heed and put into service. Fevers help your body fight infections, joint and muscle pains can be signals to cause you to cease doing something harmful like running incorrectly, and nausea and diarrhea assist you in purging harmful bugs and toxins. Moreover, as chapter 1 emphasized, adaptation is a tricky concept. The human body's adaptations evolved long ago solely because they increased how many surviving offspring our ancestors had. Consequently, we sometimes get sick because natural selection generally favors fertility over health, meaning we didn't necessarily evolve to be healthy. For example, because Paleolithic hunter-gatherers faced periodic shortages of food and they had to be very physically active, they were selected to crave energy-rich foods and rest whenever possible, helping them to store fat and devote more energy to reproduction. An evolutionary perspective predicts that most diets and fitness programs will fail, as they do, because we still don't know how to counter once-adaptive primal instincts to eat donuts and take the elevator.[18] Further, because the body is a complex jumble of adaptations, all of which have costs and benefits, and some of which conflict with one another, there is no such thing as a perfect, optimal diet or fitness program. Our bodies are full of compromises.

Finally—and most important for this book—considering and knowing about evolution in general, and human evolution in particular, is indispensable for preventing and treating a class of diseases and other problems known as *evolutionary mismatches*.[19] The idea behind the mismatch hypothesis is extremely simple. Over time, natural selection adapts (matches) organisms to particular environmental conditions. A zebra, for example, is adapted to walk and run on the African savanna, to eat grass, to run from lions, to resist certain diseases, and to cope with a hot, arid climate. If you were to transport a zebra to where I live, New England, it would no longer have to worry about lions, but it would suffer from a variety of other problems as it struggled to find enough grass to eat, to stay warm in the winter, and to resist a new set of diseases. Without help, the transplanted zebra would almost certainly get sick and perish because it is so poorly adapted (mismatched) to the New England environment.

The emerging and important new field of evolutionary medicine proposes that, despite much progress since the Paleolithic, we have become like that zebra in some respects. As innovation has accelerated, especially since farming began, we have devised or adopted a growing list of novel cultural practices that have had conflicting effects on our bodies. On the one hand, many relatively recent developments have been beneficial: farming led to more food, and modern sanitation and scientific medicine led to lower infant mortality and increased longevity. On the other hand, numerous cultural changes have altered interactions between our genes and our environments in ways that contribute to a wide range of health problems. These illnesses are *mismatch diseases,* defined as diseases that result from our Paleolithic bodies being poorly or inadequately adapted to certain modern behaviors and conditions.

I don't think it is possible to overemphasize just how important mismatch diseases are. You are most likely going to die from a mismatch disease. You are most likely to suffer from disabilities caused by mismatch diseases. Mismatch diseases contribute to the bulk of health-care spending throughout the world. What are these diseases? How do we get them? Why don't we do more to prevent them? And how might an evolutionary approach to health and medicine—including a serious consideration of the human body's evolutionary history—help avert and treat mismatch diseases?

Mismatch

Fundamentally, the hypothesis of evolutionary mismatch applies the theory of adaptation to changing interactions between genes and environments. To summarize: everyone in every generation inherits thousands of genes that interact with his or her environment, and most of these genes were selected over the previous few hundreds, thousands, or even millions of generations because they improved their ancestors' ability to survive and reproduce under certain environmental conditions. Therefore, thanks to the genes you inherited, you are adapted to varying extents for certain activities, foods, climatic conditions, and other aspects of your environment. At the same time, because of changes to your environment, you are sometimes (but not always) inadequately or poorly adapted for other activities, foods, climatic conditions, and so on. These maladaptive responses can sometimes (but again, not always) make you sick. For example, since natural selection adapted the human body over the last few million years to consume a diverse diet of fruits, tubers, wild game, seeds, nuts, and other foods that are rich in fiber but low in sugar, it should hardly be surprising that you can develop illnesses such as type 2 diabetes and heart disease from consistently eating foods that are loaded with sugar but depleted of fiber. You would also get sick from eating nothing but fruit. Note, however, that not all novel behaviors and environments react negatively with the bodies we inherited, and sometimes they are even beneficial. For example, human beings did not evolve to drink caffeinated beverages or to brush their teeth, but I know of no evidence that moderate amounts of tea or coffee cause any harm, and brushing your teeth is unquestionably healthy (especially if you eat lots of sugary food). Remember also that not all adaptations promote health. We were adapted to crave salt because it is essential for our bodies, but eating too much salt can make you sick.

There are many mismatch diseases, but all of them are caused by environmental changes that alter how the body functions. The simplest way to classify mismatch diseases is by how a given environmental stimulus has changed. Broadly speaking, most mismatch diseases occur when a common stimulus either increases or decreases beyond levels for which the body is adapted, or when the stimulus is entirely novel and the body is not adapted for it at all. Put simply, mismatches are caused by stimuli that are *too much, too little,* or *too new.* For example, as cultural evolution transforms people's diets, some mismatch diseases occur from eating too much fat, others from eating too little fat, and yet others from eating new kinds of fat that the body cannot digest (such as partially hydrogenated fats).

A complementary way to think about the origins of mismatch diseases is on the basis of different processes that alter environments, changing the degree to which individuals are adapted to their circumstances.[20] By this logic, the simplest cause of mismatch is migration, when people move into new environments for which they are poorly adapted. For example, when northern Europeans move to very sunny places like Australia, they become more likely to get skin cancer because pale skin offers little natural protection against high levels of solar radiation. Mismatches caused by migration are not just a modern problem and must have also occurred during the Paleolithic, when populations dispersed out of Africa and across the globe, encountering new pathogens and new foods.

A key difference between now and then, however, is that population dispersals in the past took place more gradually over longer timescales, allowing plenty of time for natural selection to occur in response to resulting mismatches (as discussed in chapter 6).

Of the processes that alter environments to cause evolutionary mismatches, the most common and powerful occur because of cultural evolution. Technological and economic changes over the last few generations have altered the infectious diseases we contract, the foods we eat, the drugs we take, the work we do, the pollutants we ingest, how much energy we spend and consume, the social stresses we experience, and more. Many of these changes have been beneficial, but as the following chapters will outline we are poorly or insufficiently adapted to handle others, contributing to disease. A common characteristic of these diseases, moreover, is that they occur from interactions whose cause and effect are not immediate or otherwise obvious. It takes many years for pollution to cause some illnesses (most lung cancers develop decades after people begin smoking), and when you've been bitten thousands of times by mosquitoes and fleas, it can be hard to realize that these insects sometimes transmit malaria or plague.

A final and related cause of mismatch has been shifts in life history. As we mature, we go though different developmental stages that affect our susceptibility to disease. For example, living longer may increase how many offspring you have, but it also makes you more likely to build up more damage to your heart and blood vessels and to accumulate more mutations in various cell lines. Aging doesn't directly cause heart disease and cancer, but these diseases do become more prevalent with age, helping account for their increased incidence as life spans have extended. In addition, going through puberty at a younger age can potentially increase the chances of having more offspring, but it also elevates exposure to reproductive hormones that increase the chances of certain diseases. Breast cancer rates, for example, are higher in women who begin having menstrual periods earlier (a more detailed explanation is in chapter 10).[21]

Given the complex causes of mismatch diseases, determining which diseases are evolutionary mismatches is a challenge and can be contentious. One especially thorny problem, emphasized previously, is that there is no straightforward answer to what humans are adapted for. Our species' evolutionary history was not simple, not all features in the body are adaptations, many adaptations involve trade-offs, and the body's jumble of different adaptations sometimes conflict with one another. Consequently, it can be difficult to identify what environmental conditions are adaptive and to what extent. For example, how well are we adapted to eating spicy foods? We are adapted to be physically active, but are we maladapted to be *too active*? It is well known that too much running or other sports can lower a woman's fertility, and it is unclear to what extent extreme endurance events like ultramarathons increase people's risk of injury and disease.

Another problem with identifying mismatch diseases is that we often lack a good enough understanding of many diseases to pinpoint the environmental factors that cause or influence them. Autism, for example, might be a mismatch disease because it used to be rare but is only recently becoming common (not just because of changed diagnostic criteria) and it mostly occurs in developing nations. Yet the genetic and environmental causes of autism are obscure, making it challenging to figure out whether the disease is caused by a mismatch between ancient genes and modern environments.[22] In the absence of better information we can only hypothesize that many diseases such as multiple sclerosis, attention deficit hyperactivity disorder (ADHD), and pancreatic cancer, as well as afflictions such as generalized lower back pain, are cases of evolutionary mismatch.

A final problem with identifying mismatch diseases is that we lack good data on hunter-gatherer health, especially from the Paleolithic. The essence of mismatch diseases is that they are caused by our bodies being poorly adapted to novel environmental conditions. Therefore, diseases that are common in Western populations but rare among hunter-gatherers are good candidates to be evolutionary mismatches. Conversely, if a disease is common among hunter-gatherers who have presumably been well adapted for the environments in which they live, then it is less likely to be a mismatch disease. There have been a number of efforts to identify mismatch diseases. The first comprehensive attempt was by Weston Price (1870–1948), an American dentist who traveled all over the globe before World War II to collect evidence to support his theory that modern Western diets (especially too much flour and sugar) cause cavities, dental crowding, and other health problems.[23] Since then, several other researchers have collected data on the relationship between health and environment among hunter-gatherers and populations who practice subsistence-level agriculture.[24] Unfortunately, these studies are few in number, they sometimes rely on anecdotal or limited data, and they tend to have small sample sizes. One can conclude with reasonable confidence that type 2 diabetes, myopia, and certain forms of heart disease are rare among these populations, but there is very little information about many other diseases such as cancer, depression, and Alzheimer's. Skeptics are correct to point out that absence of evidence isn't always evidence of absence. Moreover, none of the available data from non-Western societies derive from randomized controlled studies, which experimentally test the effect of a given variable such as a food or activity on health while controlling for other potential factors that might affect the results. Finally, there are no pristine hunter-gatherer groups anymore, nor have there been for hundreds if not thousands of years.[25] Most of the hunter-gatherers whose health has been studied smoke cigarettes, drink alcohol, trade for food from farmers, and have long been contending with infectious diseases contracted from outside populations.

With these caveats in mind it is still useful to consider what diseases are or might be evolutionary mismatches. Table 3 is a partial list of diseases and other health problems for which there is some reason to hypothesize that they are either caused or exacerbated by evolutionary mismatches. Stated differently, these diseases may be more prevalent, more severe, or afflict people at younger ages because human beings are not well adapted to novel environmental conditions that play some role in causing them to occur. Please note that table 3 is only a partial list; many of the diseases are only hypothesized mismatches that need to be tested, and I have omitted from the list all infectious diseases that occur from humans coming into contact with new pathogens. Had I included those diseases, the list would be much longer and much scarier.

TABLE 3. Hypothesized Noninfectious Mismatch Diseases

Acid reflux/chronic heartburn	Flat feet
Acne	Glaucoma
Alzheimer's disease	Gout
Anxiety	Hammer toes
Apnea	Hemorrhoids
Asthma	High blood pressure (hypertension)
Athlete's foot	Iodine deficiency (goiter/cretinism)
Attention deficit hyperactivity disorder	Impacted wisdom teeth
Bunions	Insomnia (chronic)
Cancers (only certain ones)	Irritable bowel syndrome
Carpal tunnel syndrome	Lactose intolerance
Cavities	Lower back pain
Chronic fatigue syndrome	Malocclusion
Cirrhosis	Metabolic syndrome
Constipation (chronic)	Multiple sclerosis
Coronary heart disease	Myopia
Crohn's disease	Obsessive-compulsive disorder
Depression	Osteoporosis
Diabetes (type 2)	Plantar fasciitis
Diaper rash	Polycystic ovarian syndrome
Eating disorders	Preeclampsia
Emphysema	Rickets
Endometriosis	Scurvy
Fatty liver syndrome	Stomach ulcers
Fibromyalgia	

If table 3—which is only a partial list—astounds and alarms you, it should! It is important to emphasize that not every illness listed is always caused by mismatch, and many of them are just hypothesized mismatches for which more data are necessary to test if they are really caused or exacerbated by novel gene-environment interactions. In spite of these caveats, it should be clear that most of the diseases that are likely to afflict you are triggered or intensified by environmental factors that have mostly become common since farming and industrialization. For most of human evolution, people did not have the opportunity to get sick or become disabled from diseases such as type 2 diabetes and myopia. It therefore follows that a large percentage of the medical conditions that afflict human beings today are evolutionary mismatches because they are caused or aggravated by modern lifestyles that are out of sync with our bodies' ancient biology. In fact, given that heart disease and cancer are responsible for more deaths in developed nations than any other diseases, you are most likely to die from a mismatch disease. Further, the disabilities that are most likely to lower your quality of life as you age are also likely to be caused by evolutionary mismatches. And, again, please remember that table 3 is only a partial list because it excludes many deadly infectious diseases such as tuberculosis, smallpox, influenza, and measles that spread widely after the origins of agriculture, mostly because we came into contact with farm animals and started to live in large groups with high densities and poor sanitation.

The Vicious Circle of Dysevolution

Before we resume the story of the human body and consider how cultural evolution since the end of the Paleolithic altered environments in ways that sometimes *cause* mismatch diseases, there is one additional evolutionary dynamic to consider: how cultural evolution sometimes *responds* to these diseases. This is hardly a trivial issue because the nature of the response helps explain why some mismatch diseases like smallpox and goiter are now extinct or rare, while others like type 2 diabetes, heart disease, and flat feet remain prevalent or are becoming more common.

To explore this dynamic, let's compare two common mismatch diseases whose evolutionary origins we will learn more about in chapter 8: scurvy and cavities. Scurvy is caused by insufficient vitamin C and used to be common in sailors, soldiers, and others whose diets lacked fresh fruit and vegetables, the primary natural sources of this vitamin.[26] Modern science did not figure out the underlying cause of scurvy until 1932, but many societies figured out how to prevent the disease by eating certain plants that are rich in this vitamin.[27] Today, scurvy is seldom seen because it is easily prevented—even among people who don't consume fresh fruit or vegetables—by adding vitamin C to processed foods. Scurvy is therefore a mismatch disease of the past because we now effectively prevent its causes.

In contrast, consider cavities. Cavities are the work of bacteria that adhere to teeth in a thin film of plaque. Most bacteria in your mouth are natural and harmless, but a few species create problems when they feed off starches and sugars in the food we chew and then release acids that dissolve the underlying tooth, creating a pit.[28] Untreated, a cavity can expand and worm its way deep into the tooth, causing excruciating pain as well as serious infection. Unfortunately, humans have little natural defense against cavity-causing microbes other than saliva, presumably because we did not evolve to eat copious quantities of starchy, sugary foods. Cavities occur at low frequencies in apes, they are rare among hunter-gatherers, they started to become rampant following the origin of agriculture, and they spiked in the nineteenth and twentieth centuries.[29] Today cavities afflict nearly 2.5 billion people worldwide.[30]

Although cavities are evolutionary mismatches whose causal mechanisms are as well understood as scurvy, they remain extremely common today because we don't effectively prevent their root causes. Instead, cultural evolution has devised successful treatments to cure cavities once they occur by having a dentist drill them out and replace them with fillings. In addition, we have developed some partially effective ways to prevent cavities from being more common through brushing, flossing, sealing teeth, and having a hygienist scrape plaque off our teeth once or twice a year. Without these preventive measures, there would be many billions more cavities than the billions that already exist, but if we really wanted to prevent them, we would have to reduce our consumption of sugar and starch drastically. However, ever since farming, most of the world's population has been dependent on cereals and grains for most of their calories, making a truly cavity-preventing diet impossible for all but a few. In effect, cavities are the price we pay for cheap calories. Like most parents, I let my daughter eat cavity-causing foods, encouraged her to brush her teeth, and sent her to the dentist, knowing full well that she'd probably get a few cavities. I hope she forgives me.

Unlike scurvy, cavities are therefore a kind of mismatch disease that is still prevalent because of a feedback loop—a vicious circle—caused by interactions between cultural evolution and biology. The circle begins when we get sick or injured from an evolutionary mismatch that results from being inadequately adapted to a change in the body's environment, either from too much, too little, or too novel a stimulus. Although we often treat the disease's symptoms with varying degrees of success, we fail to or choose not to prevent the disease's causes. When we pass on those environmental conditions to our children, we set in motion a feedback loop that allows the disease itself to persist and perhaps increase in prevalence and intensity from one generation to the next. In the case of cavities, I didn't pass on my cavities to my daughter, but I did pass on a diet that causes them, and she is likely to do the same to her children.

The drawbacks of not treating a disease's causes have been discussed and debated for centuries, often in the context of a patient's illness. According to the *Oxford English Dictionary*, the original meaning of the word "palliative" (first used in the fifteenth century) was to refer to care that "relieves the symptoms of a disease or condition without dealing with the underlying cause."[31] In addition, many evolutionary biologists and anthropologists have elucidated how culture and biology interact with each other over long periods of time not just to stimulate biological change but also to stimulate cultural change.[32] For example, the migration of Paleolithic people to temperate climates spurred the invention of new forms of clothing and housing. The same processes also apply to mismatch diseases. However, we lack a good term for the deleterious feedback loop that occurs over multiple generations when we don't treat the causes of a mismatch disease but instead pass on whatever environmental factors cause the disease, keeping the disease prevalent and sometimes making it worse. I am generally averse to neologisms, but I think "dysevolution" is a useful and fitting new word because, from the body's perspective, the process is a harmful (*dys*) form of change over time (*evolution*). To reiterate, dysevolution is not a form of biological evolution, because we don't pass on mismatch diseases directly from one generation to the next. Instead, it is a form of cultural evolution, because we pass on the behaviors and environments that promote mismatch diseases.

Cavities, unfortunately, are just the tip of the iceberg for mismatch diseases of dysevolution. In fact, I suspect that a large percentage of the mismatch conditions listed in table 3 are subject to this pernicious feedback loop. Consider high blood pressure (hypertension), which afflicts more than one billion people and which is a leading risk factor for strokes, heart attacks, kidney disease, and other illnesses.[33] Like almost all conditions, high blood pressure is caused by interactions between genes and the environment, and since arteries naturally stiffen with age, it is also a by-product of old age. But the major causes of high blood pressure among young and middle-aged people are diets that promote obesity, as well as very high salt intake, low levels of physical activity, and overconsumption of alcohol. Many medications are available for treating hypertension, but the best treatment is also the best form of prevention: good old-fashioned diet and exercise.[34] Therefore, like cavities, high blood pressure is a common case of dysevolution because even though we know how to lessen its prevalence, our culture creates and passes on the environmental factors that cause the condition and keep it common. As chapters 10 through 12 will explore, similar feedback loops help explain the incidence of type 2 diabetes, heart disease, some forms of cancer, malocclusions, myopia, flat feet, and many other common mismatch diseases.

Although dysevolution is caused by not treating a mismatch disease's causes, it is possible we sometimes aggravate the process by how we treat symptoms. Symptoms are, by definition, departures from normal health such as fever, pain, nausea, and rashes that signal the presence of a disease condition. Symptoms don't instigate disease, but they cause suffering and so they are what we notice and care about when we get sick. When you have a cold, you don't complain about the viruses in your nose and throat, you complain about the fever, cough, and sore throat that make you miserable. Similarly, a patient with diabetes probably doesn't think about her pancreas but instead is bothered by the toxic effects of too much blood sugar. As I argued above, symptoms are often evolved adaptations that prompt action. In many cases, treating symptoms aids the healing process. For some diseases (like common colds), we have no alternative but to treat symptoms. It is properly humane to relieve suffering, and it is often beneficial, even life-saving, to treat symptoms. However, it is possible that we are sometimes so effective at treating a mismatch disease's symptoms that we reduce the urgency of treating its causes. I suspect this is the case for cavities, and later chapters will explore the effects of treating symptoms for other novel diseases.

I believe that how we respond to mismatch diseases through dysevolution is an important ongoing process worth considering as we explore how the human body changed in the last 10,000 years since we started to farm, eat new foods, use machines to do work, and sit in chairs all day. To be sure, not all mismatches lead to dysevolution, but plenty do, and they share several common, predictable characteristics. First, and most obviously, they tend to be chronic, noninfectious diseases whose causes are difficult to treat or prevent. Since the advent of modern scientific medicine, we have become adept at treating or preventing many infectious diseases by identifying and killing the pathogens that cause them. Diseases caused by insufficient food or malnutrition are effectively preventable by alleviating poverty or providing dietary supplements. In contrast, chronic noninfectious diseases remain challenging to prevent or cure because they typically have many interacting causes and involve complex trade-offs. For example, we evolved adaptations to crave sugar, gain weight, and take it easy, and myriad factors, biological and cultural, conspire to make it difficult for overweight people to shed pounds (more on this in chapter 10). Other new illnesses, like Crohn's disease, are probably mismatch diseases but their causes remain elusive. There will never be any Pasteur for these afflictions.

A second characteristic of dysevolution is that one expects the process to apply mostly to mismatch diseases that have a low or negligible effect on reproductive fitness. Diseases like cavities, myopia, or flat feet are treated so effectively they don't hamper one's ability to find a mate and have children. Others, like type 2 diabetes, osteoporosis, or cancer, tend not to occur until people are grandparents. Such illnesses of middle to late age might have had strong negative selective consequences in the Paleolithic because hunter-gatherer grandparents play a critical role in provisioning their children and grandchildren.[35] But the economic role of being a grandparent in the twenty-first century is very different, and it is doubtful that being infirm or dying in your fifties or sixties today has much if any negative effect on how many children or grandchildren you have.

A final characteristic of mismatch diseases that are common or becoming more prevalent because of dysevolution is that their causes have other cultural benefits, often social or economic. The causes of many mismatch diseases, such as smoking cigarettes or drinking too much soda, are popular because they provide immediate pleasures that override concerns about or rational valuations of their long-term consequences. In addition, there is a strong incentive for manufacturers and advertisers to cater to our evolved desires and sell us products that increase our convenience, comfort, efficiency, and pleasure—or that carry the illusion of being advantageous. Junk food is popular for a reason. If you are like me, you use commercial products nearly twenty-four hours a day, even when you are asleep. Many of these products, like the chair I am sitting on, make me feel good, but not all of them are healthy for my body. The hypothesis of dysevolution predicts that as long as we accept or cope with the symptoms of the problems these products create, often thanks to other products, and as long as the benefits exceed the costs, then we will continue to buy and use them and pass them on to our children, keeping the cycle going long after we are gone.

The staggering burden of the mismatch diseases from which humans suffer, and the feedback loop of dysevolution that keeps them common, raises many questions. How do we know they really are mismatch diseases? What aspects of modern environments cause these diseases? How does cultural evolution perpetuate them? And what should we do about them? Are heart attacks, cancers, and flat feet necessary by-products of civilization, or can we effectively prevent them without having to give up bread, cars, and shoes?

Chapters 10 through 12 will explore the biological bases of different kinds of mismatch diseases and why some (but not all) are not inevitable consequences of progress. I will also consider how an evolutionary perspective might help us prevent mismatch diseases by focusing more effectively on their environmental causes. But first let's look more closely at what happened to the human body after the Paleolithic ended. How did the Agricultural and Industrial Revolutions change how our bodies grow and function in ways both good and bad?

8

Paradise Lost?

The Fruits and Follies of Becoming Farmers

> With the introduction of agriculture mankind
> entered upon a long period of meanness, misery,
> and madness, from which they are only now being
> freed by the beneficent operation of the machine.
>
> —BERTRAND RUSSELL, *The Conquest of Happiness*

In *Paradise Lost* (book 4), Milton imagines how paradise looked to Satan before the fall of man, when all was perfect in Eden. Paradise, it turns out, is a landscaped, perfumed parkland, teeming with luscious fruit and herds of munching herbivores: "A happy rural seat of various view; groves whose rich trees wept odorous gums and balm, others whose fruit burnished with golden rind hung . . . of delicious taste; betwixt them lawns, or level downs, and flocks grazing the tender herb were interposed."

Paradise may seem attractive to you, but Satan reacts jealously to all this pastoral bliss: "O hell! What do mine eyes with grief behold?" I imagine him as a worldly urbanite condemned to living in pastoral exile far from the comforts of civilization. In addition to having to watch Adam and Eve cavort about naked, he might have been wondering where he could get a decent espresso. Torture! Not so for Adam and Eve, who, tempted into eating fruit from the tree of knowledge of good and evil, are evicted from paradise and condemned to labor for their sins as farmers in the cruel outside world. In the Bible, God delivers their judgment as a curse that encapsulates the enduring miserable essence of the human condition:

> Cursed is the ground for your sake; in toil you shall eat of it all the days of your life. Both thorns and thistles it shall bring forth for you, and you shall eat the herb of the field. In the sweat of your face you shall eat bread till you return to the ground, for out of it you were taken; for dust you are, and to dust you shall return. (Genesis 3:17–19, King James Bible)

It is difficult to read God's verdict without recognizing that the expulsion of Adam and Eve from the Garden of Eden is an allegory for the first really big cause of mismatch: the end of the hunter-gatherer way of life. Ever since this transition, which began about six hundred generations ago, the human species' punishment has been to toil miserably as farmers, growing our daily bread rather than plucking luscious fruits just there for the taking. In a rare instance of accord, creationists and evolutionary biologists agree that it has been downhill for humans ever since. According to Jared Diamond, farming was the "worst mistake in the history of the human race."[1] In spite of having more food, hence more children, than hunter-gatherers, farmers generally have to work harder; they eat a lower-quality diet; they more often confront starvation because their crops occasionally fail from floods, droughts, and other disasters; and they live at higher population densities, which promote infectious diseases and social stress. Farming may have led to civilization and other types of "progress," but it also led to misery and death on a grand scale. Most of the mismatch diseases from which we currently suffer stem from the transition from hunting and gathering to farming.

If farming was such a colossal mistake, why did we start doing it? What is the consequence of having a body adapted by millions of years of evolution for hunting and gathering but then eating only plants that were grown and animals that were grazed? In what ways have human bodies benefited from farming, and what kinds of mismatch diseases have this transition caused? And how have we responded?

The Very First Farmers

Farming is often viewed as an old-fashioned way of life, but from an evolutionary perspective, it is a recent, unique, and comparatively bizarre way to live. What's more, farming originated independently in several different locations, from Asia to the Andes, within a few thousand years of the end of the Ice Age. A first question to ask before considering how farming affected the human body is why did farming develop in so many places and in such a short span of time after millions of years of hunting and gathering?

There is no single answer to this question, but one factor might have been global climate change. The Ice Age ended 11,700 years ago, ushering in the Holocene epoch, which has not only been warmer than the Ice Age, but also more stable, with fewer extreme fluctuations in temperature and rainfall.[2] During the Ice Age, hunter-gatherers sometimes attempted to cultivate plants through trial and error, but their experiments didn't take root, perhaps because they were snuffed out by extreme and rapid climate change. Experiments with cultivation had a greater chance of being successful during the Holocene, when regional rainfall and temperature patterns persisted reliably with little change from year to year and from decade to decade. Predictable, consistent weather may be helpful for hunter-gatherers, but it is essential for farmers.

A far more important factor that spurred on the origin of farming in different parts of the globe was population stress.[3] Archaeological surveys show that campsites—places people lived—became more numerous and larger once the last major glaciation started to end around 18,000 years ago.[4] As the polar ice caps receded and the earth began to warm, hunter-gatherers experienced a population boom. Having more children may seem a blessing, but they are also a source of great stress to hunter-gatherer communities who cannot survive at high population densities. Even when climatic conditions were relatively benevolent, feeding additional mouths would have put foragers under considerable pressure to supplement their typical gathering efforts by cultivating edible plants. However, once begun, such cultivation set up a vicious circle because the incentive to cultivate is amplified when larger families need to be fed. It is not hard to imagine farming developing over many decades or centuries in much the same way that a hobby can turn into a profession. At first, growing food through casual cultivation was a supplemental activity that helped provision big families, but the combination of more offspring to feed plus benign environmental conditions increased the benefits of growing plants relative to the costs. Over generations, cultivated plants evolved into domesticated crops, and occasional gardens turned into farms. Food became more predictable.

Whatever factors tipped the scales to turn hunter-gatherers into full-time farmers, the origin of farming set in motion several major transformations wherever and whenever it occurred. Hunter-gatherers tend to be highly migratory, but incipient farmers benefit from settling down into permanent villages to tend and defend their crops, fields, and herds year-round. Pioneer farmers also domesticated certain plant species by selecting—either consciously or unconsciously—plants that were larger and more nutritious as well as easier to grow, harvest, and process. Within generations, such selection transformed the plants, making them dependent on humans to reproduce. For example, the wild progenitor of corn, teosinte, has just a few, loosely held seed kernels that easily detach from the plant when ripe. As humans selected cobs with bigger, more numerous, and less detachable seeds, the corn plants become reliant on humans to remove and plant the seeds by hand.[5] Farmers also started to domesticate certain animals, such as sheep, pigs, cattle, and chickens, primarily by selecting for qualities that made these creatures more docile. Less aggressive animals were more likely to be bred, leading to more tractable offspring. Farmers also selected for other useful qualities such as rapid growth, more milk, and better tolerance to drought. In most cases, the animals became as dependent on humans as we have come to depend on them.

These processes happened somewhat differently at least seven times in diverse places including southwestern Asia, China, Meso-america, the Andes, the southeastern United States, sub-Saharan Africa, and the highlands of New Guinea. The best-studied center of agricultural innovation is Southwest Asia, where nearly a century of intensive research has revealed a detailed picture of how hunter-gatherers invented farming, spurred on by a combination of climatic and ecological pressures.

The story begins at the end of the Ice Age, when Upper Paleo-lithic foragers were flourishing along the eastern side of the Medi-terranean Sea, taking advantage of the region's natural abundance of wild cereals, legumes, nuts, and fruits, plus animals such as gazelle, deer, wild goats, and sheep. One of the best-preserved sites from this period is Ohalo II, a seasonal camp at the edge of the Sea of Galilee, where at least a half dozen families of foragers, twenty to forty people, lived in makeshift huts.[6] The site contains many seeds of wild barley and other plants that these foragers gathered, as well as the grinding stones they used to make flour, the sickles they made for cutting wild cereals, and the arrowheads they made for hunting. Life for the people who lived in Ohalo II probably differed little from what anthropologists have documented among recent hunter-gatherers in Africa, Australia, and the New World.

The end of the Ice Age, however, brought much change to Ohalo II's descendants. As the Mediterranean region's climate started to warm and become wetter starting 18,000 years ago, archaeological sites become more numerous and widespread, creeping into areas now occupied by the desert. The culmination of this population boom was a period called the Natufian, dated to between 14,700 and 11,600 years ago.[7] The early Natufian was a sort of golden era of hunting and gathering. Thanks to a benevolent climate and many natural resources, the Natufians were fabulously wealthy by the standards of most hunter-gatherers. They lived by harvest-ing the abundant wild cereals that naturally grow in this region, and they also hunted animals, especially gazelle. The Natufians evidently had so much to eat that they were able to settle perma-nently in large villages, with as many as 100 to 150 people, building small houses with stone foundations. They also made beautiful art objects, such as bead necklaces and bracelets and carved figurines, they exchanged with distant groups for exotic shells, and they buried their dead in elaborate graves. If there ever was a Garden of Eden for hunter-gatherers, this must have been it.

But then crisis struck 12,800 years ago. All of a sudden, the world's climate deteriorated abruptly, perhaps because an enormous glacial lake in North America emptied suddenly into the Atlantic, temporarily disrupting the Gulf Stream and wreaking havoc with global weather patterns.[8] This event, called the Younger Dryas,[9] effectively plunged the world back into Ice Age conditions for hundreds of years. Imagine how profoundly stressful this shift was for the Natufians, who were living at high population densities in permanent villages but who still relied on hunting and gathering. Within a decade or less, their entire region became severely colder and drier, causing food supplies to dwindle. Some groups responded to this crisis by returning to a simpler, nomadic lifestyle.[10] Other Natufians, however, evidently dug in their heels and intensified their efforts to maintain their settled way of life. In this case, necessity appears to have been the mother of invention, because some of them experimented successfully with cultivation, creating the first agricultural economy somewhere in the area now encompassing Turkey, Syria, Israel, and Jordan. Within a thousand years, people had domesticated figs, barley, wheat, chickpeas, and lentils, and their culture changed enough to warrant a new name, the Pre-Pottery Neolithic A (PPNA). These farming pioneers lived in large settlements that were sometimes as large as 30,000 square meters (about 7.4 acres, roughly the size of one and a half blocks in New York City), with mud brick houses that had plaster-lined walls and floors. The oldest levels of the ancient town of Jericho (famous for its walls) had about fifty houses and supported a population of five hundred people. PPNA farmers also made elaborate ground stone tools for grinding and pounding food, created exquisite figurines, and plastered the heads of their dead.[11]

And the change kept on coming. At first, PPNA farmers supplemented their diet by hunting, mostly for gazelle, but within a thousand years, they had domesticated sheep, goats, pigs, and cattle. Soon thereafter, these farmers invented pottery. As these and other innovations continued to accrue, their new, Neolithic way of life flourished and expanded rapidly throughout the Middle East and into Europe, Asia, and Africa. It's almost certain you ate something today that these people first domesticated, and if your ancestors came from Europe or the Mediterranean, there's a good chance you have some of their genes.

Farming also evolved in other parts of the world following the end of the Ice Age, but the circumstances were different in each region.[12] In East Asia, rice and millet were first domesticated in the Yangtze and Yellow River valleys about 9,000 years ago. Asian farming, however, began more than 10,000 years after hunter-gatherers started to make pottery, an invention that helped these foragers boil and store food.[13] In Mesoamerica, squash plants were first domesticated about 10,000 years ago, and then corn (maize) was domesticated around 6,500 years ago. As farming took hold gradually in Mexico, farmers began to domesticate other plants, such as beans and tomatoes. Maize agriculture spread slowly and inexorably throughout the New World. Other centers of agricultural invention in the New World are the Andes, where potatoes were domesticated more than 7,000 years ago, and the southeastern United States, where seed plants were domesticated by 5,000 years ago. In Africa, cereals such as pearl millet, African rice, and sorghum were domesticated south of the Sahara starting about 6,500 years ago. Finally, it seems likely that yams and taro (a starchy root) were first domesticated in highland New Guinea between 10,000 and 6,500 years ago.

Just as cultivated crops took the place of gathered plants, domesticated animals took the place of hunted ones.[14] One hotspot of domestication was Southwest Asia. Sheep and goats were first domesticated in the Middle East about 10,500 years ago, cattle were domesticated in the Indus River valley around 10,600 years ago, and pigs were domesticated from wild boar independently in Europe and Asia between 10,000 and 9,000 years ago. Other animals were domesticated more recently around the globe, among them llamas in the Andes about 5,000 years ago and chickens in southern Asia about 8,000 years ago. Man's best friend, the dog, was actually the first domesticated species. We bred dogs from wolves more than 12,000 years ago, but there is much debate over when, where, and how this domestication occurred (and to what extent dogs actually domesticated us).

How and Why Did Farming Spread?

All humans used to be hunter-gatherers, but just a few thousand years later, only a handful of isolated foraging groups remain. Much of this replacement occurred soon after farming began, because regardless of how farming originated, it then spread like contagion. A major reason for this rapid spread was population growth. Recall from previous chapters that modern human hunter-gatherer mothers typically wean their children at age three, they have children every three to four years, and their infant and juvenile mortality rate can be as high as 40 to 50 percent. Thus an average healthy hunter-gatherer mother might give birth over her life span to six or seven children, of which three might survive to become adults. Because of other causes of mortality such as accidents and illness, hunter-gatherer populations, if unchecked, typically grow at an extremely slow pace (approximately 0.015 percent per year).[15] At this rate, a population would double in about 5,000 years and quadruple in 10,000 years.[16] In contrast, a mother who is a subsistence farmer can wean her children between one and two years—at half the age that hunter-gatherer children are weaned—because she usually has enough food to feed many children at once, including cereal, animal milk, and other easily digestible foods. Therefore, if infant mortality rates were as high among farmers as they were among foragers, early farming populations would have had twice the rate of population growth. Even at this modest rate of increase, populations would approximately double every 2,000 years and grow by thirty-two times in 10,000 years. In actual fact, the rate of population growth fluctuated after farming began, and was sometimes even higher, but there is no question that it launched the first major population explosion in human history.[17]

As early farming populations grew and expanded, they inevitably came into contact with hunter-gatherers. Sometimes they fought, but often they coexisted, traded, interbred, and thus exchanged both genes and cultures.[18] The patchwork of languages and cultures across the globe today is largely a leftover of the way farmers spread and interacted with hunter-gatherers. According to some estimates, the world probably had more than a thousand different languages by the end of the Neolithic.[19]

If farming was "the biggest mistake in human history," which triggered lots of evolutionary mismatch diseases, then why did it spread so rapidly and thoroughly? The biggest reason is that farmers pump out babies much faster than hunter-gatherers. In today's economy, a higher reproductive rate often entails ominous connotations of expense: more mouths to feed, more college tuition bills to pay. Too many children can be a source of poverty. But to farmers, more offspring yield more wealth because children are a useful, fantastic labor force. After a few years of care, a farmer's children can work in the fields and in the home, helping to take care of crops, herd animals, mind younger children, and process food. In fact, a large part of the success of farming is that farmers breed their own labor force more effectively than hunter-gatherers, which pumps energy back into the system, driving up fertility rates.[20] Farming therefore leads to exponential population growth, causing farming to spread.

Another factor that encouraged the spread of agriculture is the way farmers alter the ecology around their farms in ways that hinder if not prevent any more hunting and gathering. Occasionally hunter-gatherers are able to live in permanent or semipermanent villages, but most hunter-gatherers move camp about a half dozen times a year because at some point it is less work for a group to break camp, carry their few belongings several dozen miles, and build a new camp than it is to stay put and travel farther every day to get enough food. In contrast, farmers are tied to their fields and cannot migrate as hunter-gatherers do. Fields, crops, and stored harvests must be regularly tended and defended. After they settle permanently, farmers alter the ecology around their settlement by clearing brush, burning fields, and grazing animals like cows and goats, which destroy natural habitats by eating young plants, thus promoting the growth of weeds instead of trees or bushes. Once people become farmers, they have a hard time returning to hunting and gathering. Such reversals do happen, but mostly in exceptional circumstances. When Maori horticulturalists arrived in New Zealand eight hundred years ago, they found it easier to collect shellfish and hunt giant flightless birds (moas) than to plant crops as they did elsewhere in the Pacific. Eventually, however, the Maori depleted these resources (they hunted the moas to extinction) and returned to farming.[21]

A final factor that helped agriculture take off was that early farming was not as laborious and miserable as it later became. The very first farmers certainly had to work hard, but we know from archaeological sites that they still hunted animals, did some gathering, and initially practiced cultivation on a modest scale. Farming pioneers surely had challenging lives, but the popular image of the incessant drudgery, filth, and misery of being a farmer probably applies more to later peasants in feudal systems than to early Neolithic farmers. A girl born to a French farmer in 1789 had a life expectancy of just twenty-eight years, she probably suffered from frequent bouts of starvation, and she was more likely than not to die from diseases such as measles, smallpox, typhoid fever, and typhus.[22] No wonder they had a revolution. The very first farmers of the Neolithic had demanding lives, but they were not yet beset by plagues, such as smallpox or the Black Death, and they were not oppressed by a heartless feudal system in which a handful of powerful aristocrats owned their land and appropriated a large percentage of their harvest. To be sure, these and other miseries were to come, but not until it was much too late to turn back the clock and return to a hunter-gatherer way of life.

In other words, your distant ancestors who gave up hunting and gathering weren't so crazy after all. Faced with the same circumstances, you and I would probably make the same choice. But, generations later, farming did begin to generate a series of mismatch diseases and other problems because millions of years of adaptations for Paleolithic life did not fully prepare the human body to be farmers. To explore these problems, many of which we still confront, let's consider how farming diets, workloads, population sizes, and settlement systems affected human biology in ways both good and bad.

The Farmer's Diet: A Mixed Blessing

My family celebrates Thanksgiving every November, ostensibly to commemorate the Pilgrims' first harvest, an achievement that was largely made possible by help from the local Wampanoag Indians (whose land the Pilgrims subsequently stole). Like other Americans, we make a big deal of Thanksgiving, roasting a turkey and preparing staggering quantities of cranberry sauce, sweet potatoes, and other supposedly local foods. Thanksgiving, however, is far from unique because farmers in just about every corner of the world celebrate the harvest's success with a feast of locally grown foods. Such banquets serve many functions, not the least of which is to remind us to be grateful for our good fortune in being blessed with an abundance of food. And rightly so. Can you imagine what a Paleolithic hunter-gatherer would think if he or she were to be transported to a typical supermarket?

Thanks to today's supermarkets, any day can be Thanksgiving, but the bounty available to modern shoppers is hardly representative of the way most farmers have eaten for the last few thousand years. Before the era of food transport, refrigerators, and supermarkets, almost all farmers suffered from a dreadfully monotonous diet. A typical farmer's diet in Neolithic Europe consisted mostly of bread made from wheat or other grains like rye and barley. The calories from these cereals were supplemented with peas and lentils, dairy products like milk and cheese, occasional meat, and fruits when in season.[23] That was it, day after day, year after year, century after century. The principal benefit of growing just a few staple foods is the ability to produce greater quantities. A typical adult hunter-gatherer female will manage to collect about 2,000 calories a day, and a male can hunt and gather between 3,000 and 6,000 calories in a day.[24] A hunter-gatherer group's combined efforts yield just about enough food to feed small families. In contrast, a household of early Neolithic farmers from Europe using solely manual labor before the invention of the plow could produce an average of 12,800 calories per day over the course of a year, enough food to feed families of six.[25] In other words, the first farmers could double their family size.

More food is good, but agricultural diets can provoke mismatch diseases. One of the biggest problems is a loss of nutritional variety and quality. Hunter-gatherers survive because they eat just about anything and everything that is edible. Hunter-gatherers therefore necessarily consume an extremely diverse diet, typically including many dozens of plant species in any given season.[26] In contrast, farmers sacrifice quality and diversity for quantity by focusing their efforts on just a few staple crops with high yields. It is likely that more than 50 percent of the calories you consume today derived from rice, corn, wheat, or potatoes. Other crops that have sometimes served as staples for farmers include grains like millet, barley, and rye and starchy roots such as taro and cassava. Staple crops can be grown easily in massive quantities, they are rich in calories, and they can be stored for long periods of time after harvest. One of their chief drawbacks, however, is that they tend to be much less rich in vitamins and minerals than most of the wild plants consumed by hunter-gatherers and other primates.[27] Farmers who rely too much on staple crops without supplemental foods such as meat, fruits, and other vegetables (especially legumes) risk nutritional deficiencies. Unlike hunter-gatherers, farmers are susceptible to diseases such as scurvy (from insufficient vitamin C), pellagra (from insufficient vitamin B_3), beriberi (from insufficient vitamin B_1), goiter (from insufficient iodine), and anemia (from insufficient iron).[28]

Relying heavily on a few crops—sometimes just one crop—has other serious disadvantages, the biggest being the potential for periodic food shortages and famine. Humans, like other animals, can cope with seasonal food shortages by burning fat and losing weight as long lean seasons are balanced by seasons of bounty in which the weight can be regained. In general, the body mass of subsistence farmers fluctuates several pounds between seasons as food availability and workload shift. These seasonal variations, however, can sometimes be extreme. Farmers in the Gambia, for example, typically lose 4 to 5 kilograms (9 to 11 pounds) during the wet season, when they have to work intensively to plant and weed crops at a time of food shortage and more disease; if all goes well, they regain the weight during the dry season when they harvest their crops and rest.[29] However, when the harvest is poor, farmers in the Gambia and elsewhere suffer from serious malnutrition, and death rates soar, especially among children. Hunter-gatherers also have cycles of losing and gaining weight, but when climatic variation disrupts normal growing cycles, the consequences are less extreme because foragers are not tied to staple crops and simply shift to alternative foods. In other words, farmers can bring in many more calories than hunter-gatherers, but they are more vulnerable to disasters such as drought, floods, blights, and war that regularly wipe out entire crops, sometimes in a trice. Farmers can and do survive bad years by storing enough food during years of surplus (as Joseph advised the pharaoh in Genesis). But multiple years of crop failures in a row cause disastrous famines that have been an occasional and recurring cause of death ever since farming was invented.

Consider the Irish Potato Famine. Potatoes were imported to Ireland in the seventeenth century from South America, and the plant was so well suited to the island's ecology that it became a staple crop in the eighteenth century (encouraged by a system of tenant farms too small to yield enough food from growing a variety of crops). Potatoes provided the bulk of calories for an average Irish farmer (especially in the winter), helping to fuel a population boom. But then the blight, a funguslike microorganism, spread throughout the potato fields in 1845, wiping out more than 75 percent of the harvest for four years in a row and causing more than a million deaths.[30] Sadly, the Irish Potato Famine is just one of thousands of famines that have claimed untold lives since the origin of agriculture.[31] More likely than not, a famine is occurring in some part of the world while you read these words. Although some hunter-gatherers no doubt have died from lack of food over the course of many millions of years of human evolution, a hunter-gatherer's chances of dying from starvation must be orders of magnitude lower than any farmer's.

Another set of mismatch diseases that can be caused by farming diets are nutrient deficiencies. Many of the molecules that make grains like rice and wheat nutritious, healthful, and sustaining are the oils, vitamins, and minerals present in the outer bran and germ layers that surround the mostly starchy central part of the seed. Unfortunately, these nutrient-rich parts of the plant also spoil rapidly. Since farmers must store staple foods for months or years, they eventually figured out how to refine cereals by removing the outer layers, transforming rice or wheat from "brown" into "white." These technologies were not available to the earliest farmers, but once refining became common the process removed a large percentage of the plant's nutritional value. For instance, a cup of brown and white rice have nearly the same caloric content, but the brown rice has three to six times as much B vitamins, plus other minerals and nutrients such as vitamin E, magnesium, potassium, and phosphorus. Refined cereals and domesticated plants like corn are also lower in fiber (the indigestible part of the plant). Fiber speeds the rate of food and waste passage through the intestines, and it plays a vital role in slowing the rate of digestion and absorption (more in chapter 10). Another risk of long-term food storage is contamination. Aflatoxins, for example, are harmful compounds produced by funguses that thrive on cereals, nuts, and oilseeds and that can cause liver damage, cancer, and neurological problems.[32] Since hunter-gatherers don't store foods for more than a day or two, they rarely if ever encounter these toxins.

An additional and very significant health problem caused by farmers' diets is due to lots of starch. Hunter-gatherers eat plenty of complex carbohydrates, but farmers grow and then process cereals, roots, and other plants that are rich in simple carbohydrates, also known as starch. Starch tastes great, but too much can cause a raft of mismatch diseases. The most common of these maladies is rotten teeth. After a meal, starches and sugars stick to your teeth and attract bacteria that multiply and combine with proteins in your mouth to form plaque, a whitish film surrounding the tooth. As the bacteria digest sugars they excrete acid, which is trapped by the plaque and then dissolves the enamel crown, causing cavities. Cavities are rare among hunter-gatherers but extremely common in early farmers.[33] In the Near East, the percentage of individuals with cavities jumped from about 2 percent before agriculture to about 13 percent in the early Neolithic and became even higher in later periods.[34] Figure 17 shows some painful-looking examples. Cavities, I should add, were hardly a trivial concern before the invention of antibiotics and modern dental care. A cavity that penetrates below the crown into the dentine is not only excruciatingly painful but also can cause a severe, possibly fatal infection that starts in the jaw and moves into the rest of the head.

Meals that are high in simple carbohydrates can also challenge the body's metabolism. Starchy foods, especially those that are processed to remove fiber, are rapidly and readily turned into sugar, causing blood sugar levels to spike quickly (a focus of chapter 10). Our digestive systems are simply not able to cope effectively with

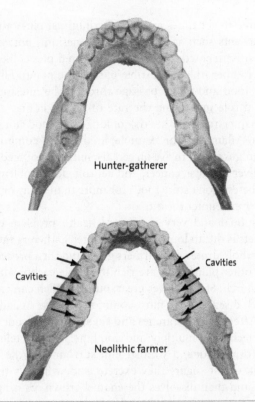

FIGURE 17. Cavities became more common after the origin of farming, as illustrated by these two jaws, one from a hunter-gatherer, the other from an early Neolithic farmer. Images courtesy of the Peabody Museum, Harvard University.

too much sugar too fast, and over time diets that are high in simple starch can contribute to type 2 diabetes and other problems. Early farmer diets, however, were not nearly as refined and starchy as modern, highly processed industrial diets, and the negative effects of rapid rises in blood sugar levels are countered by regular, vigorous physical activity. Therefore, adult onset diabetes was rare until recently. Nonetheless, surges in blood sugar levels from consuming lots of simple carbohydrates apparently affected early farmers because there is evidence that over several millennia, some farming populations evolved a number of adaptations to increase insulin production and to decrease insulin resistance.[35] We will return later to these adaptations and their relationship to mismatch conditions such as diabetes and heart disease.

Of course, diets vary substantially among farmers: peasants in China, Europe, and Mesoamerica grew and ate totally different foods. The development of farming in all these different locales, however, led to similar trade-offs between caloric quantity and nutritional quality. Farmers—even Neolithic pioneers who lacked fertilizers, irrigation, and plows—can grow much more food than hunter-gatherers can acquire, but a farmer's diet is generally less healthy and more risky. Farmers consume foods that are starchier and contain less fiber, less protein, and fewer vitamins and minerals. Farmers are also more susceptible to eating contaminated food, and they risk famine more regularly and intensely than hunter-gatherers. In terms of diet, humans have paid a high price for the pleasure of enjoying a yearly harvest feast.

Farm Labor

How did farming change how much physical activity we do and how we use our bodies to do the work? Although hunting and gathering is not easy, nonfarming populations like the Bushmen or the Hadza generally work only five to six hours a day.[36] Contrast this with a typical subsistence farmer's life. For any given crop, a farmer has to clear a field (perhaps by burning vegetation, clearing brush, removing rocks), prepare the soil by digging or plowing and perhaps fertilizing, sow the seeds, and then weed and protect the growing plants from animals such as birds and rodents. If all goes well and nature provides enough rain, then comes harvesting, threshing, winnowing, drying, and finally storing the seeds. As if that were not enough, farmers also have to tend animals, process and cook large batches of foods (for example by curing meat and making cheese), make clothing, build and repair homes and barns, and defend their land and stored harvests. Farming involves endless physical toil, sometimes from dawn to dusk. As George Sand put it, "It is sad, no doubt, to exhaust one's strength and one's days in cleaving the bosom of this jealous earth, which compels us to wring from it the treasures of its fertility, when a bit of the blackest and coarsest bread is, at the end of the day's work, the sole recompense and the sole profit attaching to so arduous a toil."[37]

There is no question that farmers, especially those oppressed by feudal landlords or trying to survive famines, have to work extremely hard, but the empirical evidence is that farming was not always as miserable as Sand's hyperbole suggests. One very simple way to compare the workloads of farmers, hunter-gatherers, and modern postindustrial people is to measure physical activity levels (PALs). A PAL score measures the number of calories spent per day (total energy expenditure) divided by the minimum number of calories necessary for the body to function (the basal metabolic rate, BMR). In practical terms, a PAL is the ratio of how much energy one spends relative to how much one would need to sleep all day at a comfortable temperature of about 25 degrees Celsius (78 degrees Fahrenheit). Your PAL is probably about 1.6 if you are a sedentary office worker, but it could be as a low as 1.2 if you spent the day in a hospital on bed rest, and it could be 2.5 or higher if you were training for a marathon or the Tour de France. Various studies have found that PAL scores for subsistence farmers from Africa, Asia, and South America average 2.1 for males and 1.9 for females (range: 1.6 to 2.4), which is just slightly higher than PAL scores for most hunter-gatherers, which average 1.9 for males and 1.8 for females (range: 1.6 to 2.2).[38] These averages don't reflect the considerable variation—daily, seasonal, and annual—within and between groups, but they underscore that most subsistence farmers work as hard if not a little harder than hunter-gatherers and that both ways of life require what people today would consider a moderate workload.

Evidence that subsistence farming involves amounts of overall physical labor similar to or slightly higher than hunting and gathering should not be surprising if one considers the kinds of physical activities that farmers did before the invention of mechanized machines such as tractors. Like hunter-gatherers, farmers generally have to walk many miles a day, but they also do many activities that require considerable upper body strength such as digging, carrying, and lifting. Farmers probably require more power and less endurance than hunter-gatherers, but their activities vary considerably (as is true for hunter-gatherers). In any case, the biggest work-load difference between these economic systems is not in terms of adult labor, but child labor. According to the anthropologist Karen Kramer, children in most hunter-gatherer societies work just an hour or two per day, mostly foraging, hunting, fishing, collecting firewood, and helping with domestic tasks such as food processing.[39] In contrast, a subsistence farmer's children work on average between four to six hours a day (the range is from two to nine hours) doing gardening, tending animals, hauling water, collecting firewood, processing food, and doing other domestic tasks. In other words, child labor has an ancient agricultural history because children are needed for their substantial contributions to a family's economic success, especially on a farm. Child labor also helps teach youngsters the skills they will need as adults. Today we have substituted school for manual labor, but to accomplish many of the same ultimate goals.

Populations, Pests, and Plagues

Of all the advantages of farming, the most fundamental and consequential is that more calories allow people to have bigger families, leading to population growth. But larger populations and their effects on human settlement patterns also fostered new kinds of infectious diseases. Without a doubt, these diseases have been and remain the most devastating of the evolutionary mismatches caused by the Agricultural Revolution.

One prerequisite of plagues is large populations, which didn't happen until farming. The earliest farming villages were small by today's standards, but as the Reverend Malthus famously pointed out in 1798, even modest increases in a population's birthrate will cause rapid increases in overall population size over just a few generations.[40] An initial village of farmers will grow exponentially faster than an equivalent-sized band of hunter-gatherers solely by weaning their children at eighteen months instead of three years, even with the same rate of infant mortality. We lack accurate data on the world's population before modern censuses, but educated guesses summarized in figure 18 suggest that the number of humans living

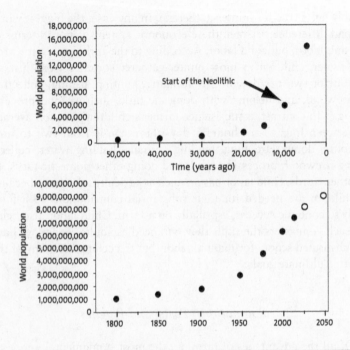

FIGURE 18. World population growth. The top panel plots approximate guesses of how many people lived at the end of the Paleolithic and how populations increased rapidly after the Neolithic began about 10,000 years ago. The bottom panel graphs more recent population growth since the start of the Industrial Revolution. For more information, see J. Hawks et al. (2007). Recent acceleration of human adaptive evolution. *Proceedings of the National Academy of Sciences USA* 104: 20753–58; C. Haub (2011). *How Many People Have Ever Lived on Earth?* Population Reference Bureau, http://www.prb.org/Articles/2002/How ManyPeopleHaveEverLivedonEarth.aspx.

had multiplied at least a hundredfold from just 5 or 6 million people at 12,000 years ago to 600 million people at the time of Jesus's birth; by the start of the nineteenth century, the world probably had approximately 1 billion people.[41]

Another prerequisite of plagues are permanent settlements with high population densities. Farmers live primarily in villages, which allows them to share common resources such as mills and irrigation ditches, to trade more easily, and to benefit from economies of scale. These economic and social benefits, combined with rapid population growth, led to a steady expansion of settlement sizes once farming took off. Over the course of a few thousand years in the Middle East, villages grew from tiny hamlets of ten houses in the Natufian, to Neolithic villages of fifty houses, to small towns with well over one thousand inhabitants by 7,000 years ago. By 5,000 years ago, a few towns ballooned into early cities such as Ur and Mohenjo Daro with tens of thousands inhabitants. As populations grew in size, population densities soared. Hunter-gatherers necessarily live at low population densities, well below one person per square kilometer (a little more than a third of a square mile), but farmers live at population densities many orders of magnitude higher, between one to ten people per square kilometer in simple agrarian economies, and upward of fifty people per square kilometer in towns.[42]

Living in larger, denser communities is socially stimulating and economically profitable, but such communities also pose life-threatening health hazards. The biggest peril is contagion. There are many kinds of infectious disease, but all of them are caused by organisms that make a living by invading hosts, feeding off their bodies, reproducing, and then being transmitted to new hosts to keep the cycle going. A disease's survival therefore depends on how many hosts are available in a population for it to infect, the disease's ability to spread from one host to another, and the rate at which its hosts survive the infection.[43] By aggregating many potential hosts in close contact with one another, villages and towns become ideal places for infectious diseases to thrive, hence dangerous places for human hosts. Another boon to the spread of infectious diseases is commerce. Because they have surpluses, farmers regularly trade goods, and in so doing they also exchange microbes, allowing infectious organisms to jump rapidly from one community to another. Not surprisingly, farming ushered in an era of epidemics, including tuberculosis, leprosy, syphilis, plague, small-pox, and influenza.[44] This is not to say that hunter-gatherers did not get sick, but before farming, humans primarily suffered from parasites such as lice, pinworms they acquired from contaminated food, and viruses or bacteria, such as herpes simplex, which they got from contact with other mammals.[45] Diseases such as malaria and yaws (the nonvenereal precursor of syphilis) were probably also present among hunter-gatherers, but at much lower rates than in farmers. In fact, epidemics could not exist prior to the Neolithic because hunter-gatherer population densities are below one person per square kilometer, which is beneath the threshold necessary for virulent diseases to spread. Smallpox, for example, is an ancient viral disease that humans apparently acquired from monkeys or rodents (the disease's origins are unresolved) that was unable to spread appreciably until the growth of large, dense settlements.[46]

Another insalubrious by-product of farming that promoted infectious mismatch diseases is poor sanitation. Hunter-gatherers who live in small temporary camps simply head off into the bushes to defecate, and they produce only modest amounts of refuse. As soon as people settle down permanently, they inevitably accumulate lots of waste and foul their nests. Permanent latrines contaminate drinking water and soil with human fecal matter, refuse piles up and rots, and dwellings create an ideal environment for small animals such as mice, rats, and sparrows that feed off food and trash, and that benefit from the safe haven that humans provide from their natural predators, such as owls or snakes. In fact, the house mouse (*Mus musculus*) first evolved in the permanent villages of Southwest Asia at the dawn of agriculture, and rats evolved so effectively to take advantage of human settlements that most cities have far more rats than humans.[47] These pests sometimes reciprocate our hospitality by being vectors of disease. Rodents carry lethal viruses such as Lassa fever and host fleas that harbor the plague and typhus. Sparrows and pigeons carry salmonella, bedbugs, and mites, which in turn carry diseases like encephalitis. Until people began to construct closed sewers, septic fields, and other forms of public sanitation, the transition to village life was the source of much sickness.

The evolution of farming and the growth of villages and cities also created windfall ecological conditions for many insects that transmit deadly diseases. Most egregiously, when farmers clear vegetation and irrigate crops, they create ideal habitats for mosquitoes, which lay their eggs in pools of stagnant water. Mosquitoes, which don't like heat or sun, also hide in cool houses and nearby bushes, placing them in ideal proximity to humans, whose blood they crave. Although malaria is a very ancient disease, the combination of ideal breeding grounds and plentiful human hosts increased its prevalence dramatically during the Neolithic.[48] Other mosquito-borne diseases that have thrived since the origin of agriculture include yellow fever, dengue, filariasis, and encephalitis. In addition, slow-moving water from irrigation ditches promoted the spread of the parasitic disease schistosomiasis (bilharzia), which is caused by worms whose life cycle begins in freshwater snails and then continues once the worms burrow into the legs of a wading human. Another boon for some diseases has been clothing, which creates hospitable environments for mites, fleas, and lice. Hunter-gatherers, especially those who inhabit temperate climates, have clothing, but there are a lot more farmers and they have a lot more clothing. Adam and Eve supposedly donned fig leaves when they were expelled from the Garden of Eden, but their descendants' filthy clothes became a godsend for millions of future generations of nasty, tiny pests.

Finally, humans have unleashed upon ourselves a frightening array of horrid diseases—more than fifty—that we acquired by living in close contact with animals.[49] These diseases are some of the scariest, nastiest pathogens that pose a serious risk to humans, including tuberculosis, measles, and diphtheria (from cattle); leprosy (from water buffalo); influenza (from pigs and ducks); and plague, typhus, and possibly smallpox (from rats and mice). Influenzas, for example, are a type of constantly mutating virus that come from waterfowl and then jump to barnyard animals, such as pigs and horses, where they further evolve and reassort into new forms, some of which are especially infectious to humans. When contracted, the virus causes an inflammatory response in the cells that line your nose, throat, and lungs, causing you to then cough and sneeze, thereby spreading millions of copies of the virus to fellow humans.[50] Most influenza strains are mild, but a few become lethal, usually when they trigger pneumonia or other respiratory infections. The great influenza epidemic that swept the globe in 1918 at the close of World War I killed between forty and fifty million people,[51] three times more than the number of civilians and soldiers who perished in the war itself. One alarming characteristic of this pandemic was that it was especially lethal among healthy young adults, rather than the elderly, perhaps because young adults had naïve immune systems with fewer antibodies to influenza, making them more susceptible to pneumonia, which was often the actual cause of death.

All told, there are probably more than one hundred infectious mismatch diseases that were caused or exacerbated by the origin of agriculture. Fortunately, in the last few generations modern medicine and public health have made great strides in preventing and combating many of these diseases. For the first time in millennia, people in developed nations rarely, if ever, worry about epidemics or succumbing to contagion. Perhaps this complacency is misguided. In spite of many new technologies that help us avert, track, and treat infectious diseases, human populations are bigger and denser than ever before, keeping us vulnerable to new epidemics.[52]

Was Farming Worth It?

Despite all the famines, increased work, and diseases caused by farming, how well did humans and their bodies actually do over the course of the momentous transition from hunting and gathering to agriculture? Were the mismatch diseases caused by the Agricultural Revolution worth it?

As is often the case, one's perspective is colored by the criteria one uses to measure success or failure. If, like most people, you think that agriculture was the biggest step toward progress ever taken by humans, then you have some justification for being glad your ancestors adopted this way of life many hundreds of generations ago. The very first farmers benefited from having more food, and this surplus was quickly invested in producing more children, which in turn increased their reliance on farming rather than foraging. So if hunter-gatherers switched to agriculture because of population stress, the benefits surely outweighed the costs, especially from an evolutionary perspective, in which the primary currency of success is how many children you have. Farming not only allowed people to have bigger families but also to settle down in villages, towns, and cities, causing a massive, still ongoing shift in human settlement patterns. Farming was also a precursor to surpluses, which made possible art, literature, science, and many other human achievements. In effect, farming made civilization possible. The other side of the coin, however, is that farming surpluses also made possible social stratification, hence oppression, slavery, war, famine, and other evils unknown to hunter-gatherer societies. Farming also ushered in many mismatch diseases that range from cavities to cholera. Hundreds of millions of people have died from plagues, malnutrition, and starvation—deaths that would not have occurred had we remained hunter-gatherers. Yet, despite these many deaths, there are nearly six billion more people alive today than would be the case had the Agricultural Revolution never begun.

Although farming has been a boon for the human species as a whole, it was a mixed blessing to the human body. One useful proxy for assessing the success of farming on human health is to look at changes in height. In general, a person's maximum height is strongly influenced by genes, but actual height is highly constrained by environment: people who suffer from poor nutrition, disease, or other physiological stresses don't grow to their full genetic potential. This is because a growing child usually has a finite amount of energy, which can be used to maintain the body, to fight infections, to do things, or to grow. If a child needs to devote lots of limited energy to fight infections or to work intensively, then less energy will be available for growth. So studying changes in height is a good overall measure of documenting changes in how well people are fed and how much they suffered from disease and other kinds of stress. Analyses of people's height suggest that the initial stages of farming were initially beneficial for people's health in many, though not all, parts of the world. Not surprisingly, one success story is the Middle East, where farming first started. Careful studies show that as the Neolithic began around 11,600 years ago and then progressed over its first few thousand years, people's stature initially increased by about 4 centimeters (1.5 inches) in males and a little less in females. However, stature then started to decline beginning around 7,500 years ago at the same time that skeletal markers of disease and nutritional stress also become more common.[53] Similar patterns of initial progress then reversal are also evident in other parts of the world, including the Americas. For example, as maize agriculture was gradually incorporated into the diet in eastern Tennessee between 1,000 and 500 years ago, stature increased by 2.2

centimeters in men (0.86 inches) and by approximately 6 centimeters (2.4 inches) in women.[54] Judging by height, many (though not all) populations of early farmers initially benefited from their new way of life.

However, if one steps back from comparing populations immediately before and after the Agricultural Revolution and considers changes in height over longer periods of time, the effects of farming lifestyles are generally less salubrious.[55] With few exceptions, people shrank as agricultural economies intensified. For example, the height of farmers in early Neolithic China and Japan decreased by 8 centimeters (3.1 inches) over several thousand years as rice farming progressed,[56] and as agriculture took hold in Mesoamerica, height decreased by 5.5 cm (2.2 inches) in men and 8 centimeters (3.1 inches) in women.[57] In other words, the unfortunate irony of agricultural intensification is that even though farmers produced more food overall, the energy available for each child to grow diminished, probably because they were spending relatively more energy fighting infections, coping with occasional shortages of food, and toiling long hours in the fields.

Other types of data confirm that the transition to farming generally challenged people's health. Acute stress from infection or starvation leaves deep, permanent furrows in teeth; anemia from lack of dietary iron causes skeletal lesions; and infections such as syphilis leave traces of inflammation on bones. Researchers who have tabulated the incidence of these and other pathologies dating to before and after the transition to agriculture repeatedly find that the skeletons of the descendants of pioneer farmers have more signs of disease, malnutrition, and dental problems, regardless of whether one looks in South America, North America, Africa, Europe, or elsewhere.[58] Simply put, over time, farming life generally became nastier, more brutish, shorter, and more painful.

Mismatch and Evolution Since Farming

Although the first farmers reaped some benefits from switching to agricultural economies, this new way of life also led to many mismatch diseases and other problems. What sort of evolutionary developments did these changes, especially the mismatch diseases, trigger? To what extent did farming drive natural selection and cultural evolution or simply lead to mismatch diseases, hence more misery and death?

Let's first consider how farming led to natural selection. It bears repeating that the very first farmers lived about 600 to 500 generations ago, and in most parts of the world, farming has been practiced for less than 300 generations. From an evolutionary perspective, this is not much time for lots of major evolutionary change, such as a new species evolving, but it is enough time for genes with strong effects on survival and reproduction to change their frequencies appreciably within populations. In fact, because farming so profoundly altered people's diets, the pathogens they encountered, the work they did, and the number of children they could have, the origins of agriculture probably *intensified* selection on certain genes.[59] Consider also that natural selection can operate only on existing heritable variations. In this respect, farming clearly boosted rates of evolution, because, as populations have exploded in size (by more than a thousandfold), each generation has made available many new mutations upon which selection can act. Efforts to measure this surge in diversity have identified more than one million new genetic variations that arose in various populations across the planet in the last few hundred generations.[60] The existence of so many recent mutations is sobering because many of them are harmful.

Most of the mutations that have arisen in the last few hundred generations have not been subject to much selection, especially positive selection, and in fact more than 86 percent of the new mutations that have arisen probably have negative effects.[61] But with so many new mutations, it should hardly be surprising that studies have identified more than one hundred genes that have been favored by recent natural selection, many because of farming.[62] It will take years of research to study all these genes carefully, but as you might predict a large percentage of them help the immune system cope with some of the most lethal pathogens that have afflicted humans since the origins of agriculture: bubonic plague, leprosy, typhoid fever, Lassa fever, malaria, measles, and tuberculosis. Among the best-studied cases are genes that help provide immunity from malaria. Malaria is an ancient disease caused by mosquito-borne parasites. The prevalence of malaria thus increased as farming spread because of higher population densities and farming practices that promoted mosquito breeding. Because the malaria parasite feeds on hemoglobin, the iron-containing protein that transports oxygen in blood, several mutations that affect hemoglobin have been selected in populations plagued by malaria.[63] One of these mutations causes sickle cell anemia, in which blood cells have an abnormal semicircular shape; other mutations decrease the blood cell's ability to make energy following an infection or retard the formation of hemoglobin molecules.[64] In these and other cases, partial immunity comes from carrying just one copy of the mutated gene, but having two copies causes serious, sometimes fatal anemias. That genes with such life-threatening effects ever evolved makes sense only in the context of natural selection to provide immunity against a disease with even more disastrous effects. In other words, the benefit of providing partial immunity to farmers in malaria-affected areas outweighed the terrible costs of some of their relatives dying from anemia.

Other genes that underwent recent positive selection because of agriculture play important roles in helping humans adapt to domesticated foods. There are several examples, but the best studied are genes that help adults digest milk. Milk contains a special form of sugar, lactose, which is broken down by the enzyme lactase. Preagricultural humans never had to digest milk after they stopped nursing, and as most humans mature, their digestive system naturally stops producing lactase by the time they are five or six years old. But after people domesticated mammals like goats and cows that supply milk, the ability to digest lactose after infancy became an advantage, promoting selection for genes that permit lactase production in adults. In fact, several such mutations evolved independently among East Africans, northern Indians, Arabs, and inhabitants of Southwest Asia and Europe.[65] Other adaptations evolved to help farmers cope with spikes in blood sugar caused by eating lots of carbohydrates. For example, the *TCF7L2* gene, which promotes insulin secretion after a meal, has several variants that evolved separately in Europe, East Asia, and West Africa about the time of the Neolithic.[66] These and other gene variants help protect the descendants of these farmers today from type 2 diabetes.

Natural selection is a never-ending process that must still be acting now, aided by the recent bonanza of new genetic variation. Yet even though the Agricultural Revolution led to selection that helped struggling farmers cope with novel diets and infectious diseases, it would be wrong to conclude that natural selection has been the dominant engine of evolutionary change over the last few thousand years. By any yardstick, recent genetic adaptations that have evolved independently in different parts of the New and Old Worlds are modest compared to the scale and degree of cultural innovation that humans have cooked up over the same time frame. Many of these cultural innovations—the wheel, plows, tractors, writing—have improved economic productivity, but quite a few were responses to mismatch diseases caused by the farming way of life. Stated more precisely, many of these innovations have acted as *cultural buffers* that have insulated or even protected farmers from the dangers and drawbacks of agriculture, which would otherwise have resulted in even stronger selection than we can detect.

Consider malnutrition, a problem that confronts farmers more than hunter-gatherers, because the way farmers rely on a few staple foods decreases their diets' nutritional diversity and quality. One example is pellagra, a horrible illness from insufficient vitamin B_3 (niacin), which causes diarrhea, dementia, skin rashes, and eventually death if untreated. Pellagra is common among farmers who eat mostly corn (maize) because vitamin B_3 in corn is bound to other proteins, making it unavailable to the human digestive system. Native American farmers never evolved genes that would give them resistance to pellagra, but they did learn long ago to make a special kind of corn flour, called masa flour, by soaking the corn in an alkali solution before grinding. This process (termed nixtamalization) not only liberates vitamin B_3 for digestion but also increases the corn's calcium content.[67]

Making masa flour is one of thousands of cultural evolutionary responses to the changes wrought by agriculture. These cultural innovations—which include primitive sanitation, dentistry, pottery, domesticated cats, and cheese—have obviated or mitigated many mismatch diseases that have appeared or intensified since we ceased to be hunter-gatherers. Some of these inventions, like making masa flour and cheese, were brilliant solutions to problems that arose from farming, but that then buffered humans from natural selection. But others are not so much solutions as Band-Aids that treat only the symptoms of mismatch conditions. Such palliative responses can create a problem, because treating the symptoms rather than causes of mismatch diseases sometimes provokes a pernicious feedback loop, which I term dysevolution, that allows the disease to persist or even intensify. However, before we consider this vicious circle, we first need to consider the next major chapter in the history of the human body: the industrial era.

9

Modern Times, Modern Bodies

The Paradox of Human Health in the Industrial Era

> A clattering of clogs upon the pavement; a rapid
> ringing of bells; and all the melancholy mad
> elephants, polished and oiled up for the day's
> monotony, were at their heavy exercise again.
>
> —CHARLES DICKENS, *Hard Times*

Human existence has undergone many profound changes over the last few million years, but never has so much change occurred so rapidly as in the last 250 years. My grandfather's life exemplifies this transformation. He was born in about 1900 in Bessarabia, a poor, rural region along the border between Russia and Romania. Like many parts of Eastern Europe at the time, Bessarabia was an agrarian economy, barely touched by the Industrial Revolution. In the village where he was born, no one had electricity, gas, or indoor plumbing. All work was done by humans and farm animals. As a boy, however, my grandfather fled with his family to America because of the pogroms. In America, he had the opportunity to go to public school; he then fought in World War I, and thanks to veteran's benefits he was able to attend medical school and become a doctor in New York City. Many of us have seen considerable change in our lifetimes, but my grandfather essentially traversed the entire Industrial Revolution in a few short years as a youngster and then experienced most of the changes of the twentieth century.

And, boy, did he love the change. Far from being a Luddite opposed to technological progress,[1] my grandfather embraced the many benefits of science, industrialization, and capitalism. Perhaps because he was born a peasant, my grandfather especially enjoyed having a swank bathroom, a big car, air-conditioning, and central heating. He was also intensely proud of the progress that occurred within his profession, pediatrics. At the time he was born, about 15 to 20 percent of American infants died in their first year of life, but infant mortality plunged to less than 1 percent over the course of his career.[2] This impressive decline in mortality was largely attributable to antibiotics and to other new medicines for treating infants stricken with respiratory illnesses, infectious diseases, and diarrhea. Infant mortality rates also declined dramatically during the twentieth century because of preventive health measures such as improved sanitation, better nutrition, and more access to doctors. Unlike many physicians, who tend to see their adult patients only when they are sick, pediatricians regularly and frequently see their young patients when they are healthy to keep them from getting ill. The dramatic successes of pediatrics during the twentieth century prove that preventive medicine really is the best medicine.

My grandfather died in the early 1980s, but I am sure he would despair at the state of preventive medical care for children today in the United States. The majority of American children still get regular checkups, inoculations, and dental care, but 10 percent of them don't because of poverty and poor access to health care. The percentage of low birth weight babies, now 8.2 percent, has not declined in decades and in fact has been rising recently, even though low birth weight substantially increases a child's risk of dozens of short-term and long-term health problems.[3] In 1900, Americans were, on average, the tallest people in the world, but today they tend to be shorter than most Europeans.[4] Finally, Americans and others are failing shamefully at preventing childhood obesity. Since 1980, the percentage of children who are obese has more than tripled in the United States, from 5.5 percent to approximately 17 percent, and a similar trend is occurring worldwide.[5] So far, the combined efforts of doctors, parents, public health professionals, educators, and others to reverse this growing problem have been mostly ineffectual. More and more children (and their parents) are getting fatter, and overweight children are so prevalent they are now perceived by some as normal.

If you view the current status of the human body as a whole, many countries, like the United States, now confront a novel paradox. On the one hand, more wealth and impressive advances in health care, sanitation, and education since the Industrial Revolution have dramatically improved billions of people's health, especially in developed nations. Children born today are far less likely to die from infectious mismatch diseases caused by the Agricultural Revolution and they are much more likely to live longer, grow taller, and be generally healthier than children born in my grandfather's generation. As a consequence, the world's population tripled over the course of the twentieth century. But on the other hand, our bodies face new problems that were barely on anyone's radar screen a few generations ago. People today are much more likely to get sick from new mismatch diseases such as type 2 diabetes, heart disease, osteoporosis, and colon cancer, which were either absent or much less common for most of human evolutionary history, including most of the agricultural era.

To understand how and why all this happened—and how to address these new problems—requires considering the industrial era through the lens of evolution. How did the Industrial Revolution along with the growth of capitalism, medical science, and public health affect the way our bodies grow and function? In what ways did the momentous social and technological changes of the last few hundred years ameliorate or solve the many mismatch diseases created by the development of farming yet simultaneously cause new mismatch diseases?

What Was the Industrial Revolution?

Most fundamentally, the Industrial Revolution was an economic and technological revolution in which humans started to use fossil fuels to generate power for machines to manufacture and transport things in massive quantities. Factories first appeared in the late eighteenth century in England, and methods of industrial production quickly spread to France, Germany, and the United States. Within one hundred years, the Industrial Revolution spread to Eastern Europe and the Pacific Rim, including Japan. As you read this, a wave of industrialization is sweeping through India, Asia, South America, and parts of Africa.

Some historians object to the term "Industrial *Revolution*." Compared with political revolutions, which can happen in a few days or years, the transition from agrarian to industrial economies occurred over many hundreds of years; some parts of the globe, such as rural China, are only just beginning to industrialize. Yet from the perspective of evolutionary biology, the term "revolution" is totally appropriate because within less than a dozen generations, humans altered their framework of existence, not to mention the earth's environment, more rapidly and profoundly than any previous cultural transformation. Before the Industrial Revolution began, the world's population was less than one billion, mostly consisting of rural farmers who did all their work using manual labor or domesticated animals. Now there are seven billion people, more than half of us live in cities, and we use machines to do the majority of our work. Before the Industrial Revolution, people's work on the farm required a wide range of skills and activities, such as growing plants, tending animals, and doing carpentry. Now many of us work in factories or offices, and people's jobs often require them to specialize in doing just a few things, such as adding numbers, putting the doors on cars, or staring at computer screens. Before the Industrial Revolution, scientific inventions had little effect on the daily life of the average person, people traveled little, and they ate only minimally processed food that was grown locally. Today, technology permeates everything we do, we think nothing of flying or driving hundreds or thousands of miles, and much of the world's food is grown, processed, and cooked in factories far from where it is consumed. We have also changed the structure of our families and communities, the way we are governed, how we educate our children, how we entertain ourselves, how we get information, and how we perform vital functions like sleep and defecation. We have even industrialized exercise: more people get pleasure from watching professional athletes compete in televised sports than by participating in sports themselves.[6]

So much change in so little time is impressive. For some, like my grandfather, the changes unleashed by the Industrial Revolution were liberating and exhilarating, and there is little doubt that humans in Western economies today are generally healthier and more prosperous than they were for hundreds of generations. But for some, the changes brought by the Industrial Revolution have been confusing, unsettling, or disastrous. Regardless of whether you think the industrial era has been good or bad, three profoundly fundamental shifts underlie this revolution. The first is that industrialists harnessed new sources of energy, primarily to produce things. Preindustrial people occasionally used wind or water to generate power, but they mostly relied on muscles—human and animal—to generate force. Industrial pioneers such as James Watt (who invented the modern steam engine) figured out how to transform energy from fossil fuels such as coal, oil, and gas into steam, electricity, and other kinds of power to run machines. The first of these machines were designed to make textiles, but within decades others were invented to make iron, mill wood, plow fields, transport things, and do just about everything else one can manufacture and sell (including beer)[7].

A second major component of the Industrial Revolution was a reorganization of economies and social institutions. As industrialization gathered steam, capitalism, in which individuals compete to produce goods and services for profit, became the world's dominant economic system, spurring the development of further industrialization and social change. As workers changed their locus of activity from the farm to factories and companies, more people had to work together even as they needed to perform more specialized activities. Factories required more coordination and regulation. In addition, new private companies and government institutions had to be created to transport, sell, and advertise goods, to finance investments, and to accommodate as well as manage the hordes of people who relocated to massive cities that sprang up around factories. As women and children entered the workforce (child labor was common during the early part of the Industrial Revolution), families and neighborhoods reconfigured, as did work hours, eating habits, and social classes. As the middle class expanded, a combination of government services and private industries evolved to cater to their needs, to educate them, to provide basic resources and amenities like roads and sanitation, to disseminate information, and to entertain. The Industrial Revolution created not just blue-collar but also white-collar jobs.

Finally, the Industrial Revolution coincided with a transformation of science from a pleasant but nonessential branch of philosophy into a vibrant profession that helped people make money. Many heroes of the early Industrial Revolution were chemists and engineers, often amateurs such as Michael Faraday and James Watt who lacked formal degrees or academic appointments. Like many young Victorians excited by the winds of change, Charles Darwin and his elder brother Erasmus dreamed as boys of becoming chemists.[8] Other fields of science, such as biology and medicine, also made profound contributions to the Industrial Revolution, often by promoting public health. Louis Pasteur began his career as a chemist working on the structure of tartaric acid, which was used in wine production. But in the process of studying fermentation he discovered microbes, invented methods to sterilize food, and created the first vaccines. Without Pasteur and other pioneers in microbiology and public health, the Industrial Revolution would not have progressed so far and so fast.

In short, the Industrial Revolution was actually a combination of technological, economic, scientific, and social transformations that rapidly and radically altered the course of history and reconfigured the face of the planet in less than ten generations—a true blink of an eye by the standards of evolutionary time. Over the same period, the Industrial Revolution also changed everyone's bodies. It changed what we eat, how we chew, how we work, and how we walk and run, as well as how we keep cool and warm, give birth, get sick, mature, reproduce, grow old, and socialize. Many of these changes have been beneficial, but some have had negative effects on the human body, which has yet to evolve to cope with this new environment. Because harnessing energy to run machines was the foundation of the Industrial Revolution, the first place to look at how this revolution caused many mismatch conditions is how much and what kind of work we now do.

Physical Activity

In the 1936 movie *Modern Times,* Charlie Chaplin arrives at the factory in his overalls and dutifully gets to work on an assembly line with a pair of wrenches, tightening an endless stream of nuts. As the conveyer belt speeds up, Chaplin comically emphasizes something that every factory worker knows: working on an assembly line can be hard, intense work. Even though the Industrial Revolution mostly replaced muscles with engines as the source of mechanical force used to make and move things, factory workers often do demanding, arduous labor. In a typical nineteenth-century factory, employees were required to arrive and be ready for work when the factory's whistle blew or forfeit half a day's wages. They were then expected to work steadily and rapidly for twelve or more hours under the supervision of foremen whose job was to ensure that production continued efficiently and effectively. Eighty-plus-hour weeks, low wages, and dangerous working conditions were so prevalent that eventually labor unions and governments began to enact reforms to make industrial work safer and less inhumane. After the English Factories Act of 1802, child workers below the age of thirteen were no longer allowed to work more than eight hours a day, and adolescents between the ages of thirteen and eighteen could work no more than twelve hours a day (child labor was not banned in the United Kingdom until 1901).[9] Since then, labor agreements in some countries have continued to improve working conditions: the average factory worker in the United States today works a forty-hour week, about 50 percent fewer hours than during the nineteenth century.[10] However, many factory jobs in less developed nations, such as China, still exceed ninety hours a week.[11] In short, industrial jobs until recently involved as much or even more work time than agricultural jobs, and in some places still involve punishing hours.

From the body's perspective, a key measure of work is how much actual physical activity the labor requires. Despite depictions of relentlessly strenuous work on the factory floor in movies such as *Modern Times* or *Metropolis*, industrial jobs have always varied enormously in their energetic costs. Table 4 summarizes measurements of the calories workers spend per hour doing a variety of activities. Many of these activities are typical of the labor done in factories and offices, others are more typical of farming, and I included the cost of walking and running for comparison. As you would expect, the most strenuous jobs are those such as mining or loading in which one operates heavy machinery or uses one's own physical strength. These industrial jobs are about as energetically costly as farming, if not more so. A second, more moderate class of industrial jobs requires workers to stand and do things with the help of tools and machines. These jobs, which include working on an assembly line or doing laboratory work, tend to be as energetically costly as walking at a comfortable speed. A final class of industrial jobs, which have become increasingly common as robots and other machines replace or alter human labor, mostly involve sitting and doing things with one's hands. Tasks such as typing, sewing, or doing general work at a desk are only slightly more costly than just sitting still. In a typical day, a receptionist or bank clerk who spends an eight-hour day seated in front of a computer expends about 775 calories while doing her job, a worker at an automobile factory spends about 1,400 calories, and a really hardworking coal miner could spend a whopping 3,400 calories. Measured by donuts, a receptionist spends about as much energy each day to do his job as he gets from eating three glazed donuts, but the coal miner would need to eat fifteen donuts on the job to stay in energy balance.

In other words, the industrial era was initially very demanding energetically, but changes in technology have made many (but not all) workers' jobs less strenuous in terms of physical activity. These differences are consequential because even minor changes in expenditure add up over many long hours. Consider sewing, a common type of industrial labor. A person operating an electric sewing machine typically expends about 73 calories per hour, about the same energy cost as just sitting; operating an old-fashioned pedal-powered sewing machine, however, is 30 percent more expensive, costing 98 calories per hour.[12] Over the course of a year, the operator of the electric machine will spend approximately 52,000 fewer calories, enough energy to run about eighteen marathons![13] Consider also that those differences are modest compared to the differences in energetic demands of workers who do their job while sitting or standing. It costs about 7 to 8 percent more calories to stand than sit, and even more calories if you move about. Over the course of a 260-day work year of eight-hour days, a blue-collar worker on the floor of a car assembly plant will expend approximately 175,000 more calories than a white-collar worker in an office, enough to run nearly sixty-two marathons. Nothing over the last few million years of human history has changed human energetics as much as the low cost of working at a desk using machines run by electric power.

One of the ironies of industrialization is that its spread across the globe has required more people to spend more time sitting. This is because, paradoxically, greater industrialization eventually decreases the percentage of manufacturing jobs and increases the number of workers employed in service, information, or research jobs. In developed countries such as the United States, only 11 percent of workers actually work in factories. Several factors underlie the trend from jobs that produce goods to jobs that provide services. One is that manufacturing creates more wealth, thus creating the need for bankers, lawyers, secretaries, and accountants. In addition, more wealth increases the cost of labor, giving manufacturers a strong incentive to ship jobs to less developed countries with lower labor costs. The service sector is the largest and fastest-growing part of most developed economies such as those of the United States and Western Europe. More people than ever make their living simply by typing, reading a computer screen, talking on the phone, and occasionally walking to and from meetings within a building.

And it's not just your job. The Industrial Revolution profoundly altered how much physical activity people do not just at work, but also for the rest of the day. Many of the most successful products invented and manufactured since the start of the Industrial Revolution have been labor-saving devices. Cars, bicycles, airplanes, subways, escalators, and elevators reduce the energy cost of traveling. Recall that over the last few million years the average hunter-gatherer walked 9 to 15 kilometers every day (roughly 5 to 9 miles), but today a typical American walks less than a half a kilometer per day (a third of a mile) while commuting an average of 51 kilometers (32 miles) by car.[14] Less than 3 percent of shoppers in an American mall voluntarily take the stairs when an escalator is available to make their journey easier (the percentage doubles with signs that encourage stair use).[15] Food processors, dishwashers, vacuum cleaners, and clothes-washing machines have substantially lessened the physical activity required to cook and clean.[16] Air conditioners and central heating have decreased how much energy our bodies spend to maintain a stable body temperature. Countless other devices, such as electric can openers, remote controls, electric razors, and suitcases on wheels, have reduced, calorie by calorie, the amount of energy we expend to exist.

In short, over the course of just a few generations the Industrial Revolution drastically reduced how much physical activity we do. If you are like me, you can easily spend your days mostly sitting and never have to exert yourself beyond taking a few steps and pressing various buttons. If you exercise by going to the gym or jogging a few miles, you do so because you want to, not because you have to.

How much less physical activity do our bodies actually perform now than before the Industrial Revolution? As chapter 8 discussed, a simple measure of overall energy expenditure is the physical activity level (PAL), the ratio of the energy you spend per day relative to the energy you would spend by resting in bed and doing absolutely nothing. PALs for male adults with clerical or administrative jobs that involve sitting all day long average 1.56 in developed countries and 1.61 in less developed countries; in contrast, PALs for workers involved in manufacturing or farming average 1.78 in developed countries and 1.86 in less developed countries.[17] Hunter-gatherer PALs average 1.85, about the same as those of farmers or other people whose job requires them to be active.[18] Therefore, the amount of energy a typical office worker spends being active on an average day has decreased by roughly 15 percent for many people in the last generation or two. Such a reduction is not trivial. If an average-sized male farmer or carpenter who spends approximately 3,000 calories per day suddenly switches to a sedentary lifestyle by retiring, his energy expenditure will decline by about 450 calories a day. Unless he compensates by eating a lot less or exercising more intensively, he'll grow obese.

TABLE 4. Energetic Cost of Different Tasks

Task	Cost (calories/hour) [19]
Knitting	70.7
Operating electric sewing machine	73.1
Working at desk while sitting	92.4
Operating foot-driven sewing machine	97.7
Typing while seated	96.9
Standing at rest	107.0
Standing, light work (washing)	140.0
Working on car assembly line	176.5
Forging metal	187.9
Walking on level, 3–4 km/hr	181.8
Household chores (general)	196.5
Lab work (general)	205.6
Gardening	322.7
Hoeing	347.3
Coal mining	425.3
Loading a truck	435.9
Running (endurance speeds)	600–1,500

Data from W. P. T. James and E. C. Schofield (1990). *Human Energy Requirements: A Manual for Planners and Nutritionists.* Oxford: Oxford University Press. Note that the values are actually kilocalories per hour.

Industrial Diets

According to science fiction shows like *Star Trek,* food in the future will be produced by replicators. All you have to do is walk up to a machine that looks like a microwave and command it to produce something you desire like "tea, Earl Grey, hot" or "macaroni and cheese" and, voilà, the atoms necessary to make the dish will be assembled in just the right way. This fantasy of future food is actually not that far off from the way many people sustain themselves today and makes the differences between Paleolithic and agricultural era diets seem fairly trivial. Even though farmers neither hunt nor gather, they do at least grow and process their food. How about you? Did you grow or raise anything you ate today? In fact, did you even have to process it? The average American or European consumes about a third of all meals outside the house, and when we cook, we mostly unwrap, combine, and heat different ingredients.

I love to cook, but the most intensive work I usually do is to peel a carrot, dice an onion, or grind stuff in a food processor.

From a physiological perspective, the Industrial Revolution changed our diets as much if not more than the Agricultural Revolution. As chapter 8 reviewed, by switching from hunting and gathering to herding and cultivating, the first farmers increased the amount of food they could obtain, but at a cost. Farmers not only must work hard, but the food they produce is less diverse, less nutritious, and less certain than what a hunter-gatherer eats. By using machines to produce, transport, and store food in the same way we do textiles and cars, the Industrial Revolution abated some of these trade-offs and magnified others. These shifts began in the nineteenth century, but they intensified after World War II, especially in the 1970s, as gigantic industrial corporations took over the business of making and producing food from small-scale farmers.[19] In much of the developed world, the food we eat is now as industrial as the cars we drive and the clothes we wear.

The biggest change brought about by the industrial food revolution is that food producers (one cannot really call them farmers) have figured out how to grow and manufacture as cheaply and efficiently as possible exactly what people have desired for millions of years: fat, starch, sugar, and salt. The result of their ingenuity is a superabundance of inexpensive calorie-dense food. Consider sugar. The only really sweet food a hunter-gatherer can eat is honey, which usually requires walking many miles to find a hive, climbing the tree, smoking out the bees, and then bringing the honeycomb back. Sugarcane became a crop in the Middle Ages, and its cultivation accelerated during the eighteenth century, largely by using slaves to produce massive quantities in plantations.[20] With the end of slavery in the late nineteenth century, industrial methods were applied to sugar production, and modern farmers now use specialized tractors to plant enormous fields of domesticated sugarcane and sugar beets, which have been bred to be as sweet as possible. Other machines are used to irrigate the plants and to make and spread fertilizers and pesticides, which increase yields and minimize crop losses. Once grown, these supersweet plants are harvested and processed by yet more machines to extract the sugar, which is then packaged and shipped all over the world by ships, trains, and trucks. The availability of sugar increased even more dramatically in the 1970s when chemists devised a method to transform cornstarch into a sugary syrup (high fructose corn syrup). About half the sugar Americans consume now derives from corn. After adjusting for inflation, a pound of sugar today costs one-fifth what it did one hundred years ago.[21] Sugar has become so superabundant and so cheap that the average American consumes more than 100 pounds (45 kilograms) a year![22] Perversely, some people now pay extra money to buy foods made with less sugar.

Unless you have a garden or go to farmers' markets, the chances are that most of what you eat—including free-range eggs and organic lettuce—was grown industrially, often with the support of government subsidies to keep quantities abundant and prices low. Between 1985 and 2000, when the purchasing power of a U.S. dollar decreased by 59 percent, the price of fruits and vegetables doubled, fish increased by 30 percent, and dairy remained about the same; in contrast, sugars and sweets became about 25 percent less expensive, fats and oils declined in price by 40 percent, and soda became 66 percent less expensive.[23] At the same time, portion sizes ballooned. If you were to walk into an American fast-food restaurant in 1955 and order a hamburger and fries, you'd consume about 412 calories, but today for the same price (in inflation-adjusted dollars) the same order would have double the amount of food, totaling 920 calories.[24] Soda consumption in the United States has more than doubled since 1970, now averaging more than 150 liters (about 40 gallons) per year.[25] According to U.S. government estimates, bigger and more calorie-dense portions have caused the average American to consume about 250 more calories per day in 2000 than in 1970, a 14 percent increase.[26]

Industrial food may be inexpensive, but its production exacts a significant toll on the environment and on the health of workers. For every calorie of industrial food you eat, approximately 10 calories of fossil fuel were spent to plant, fertilize, harvest, ship, and process the food before it got to your plate.[27] Further, unless the food was organic, massive quantities of pesticides and inorganic fertilizers were used, polluting water supplies and sometimes poisoning workers. The most extreme and disturbing type of industrial food is meat. Because humans have craved meat perhaps more than anything else (except possibly honey) for millions of years, there is a strong incentive to produce cheap, plentiful meat, especially beef, pork, chicken, and turkey. Satisfying this craving, however, was a challenge until recently, keeping meat consumption modest. Despite having domesticated animals, early farmers generally ate less meat than hunter-gatherers, because animals are more valuable alive for their milk than dead for their flesh and because farm animals require lots of land and labor, especially if one has to make and store hay to feed them during the winter. Food industrialization altered this equation dramatically by employing new technologies and economies of scale. Most of the meat Americans and Europeans eat is grown in giant facilities called concentrated animal feeding operations (CAFOs). CAFOs are huge fields or barns where hundreds to thousands of animals are fed grain (usually corn) in crowded conditions. The animals respond just as we do to being fed an abundance of starch without exercising: they get fat. They also have high rates of disease because concentrated animal wastes and high animal densities promote infectious diseases, and because species such as cows have digestive systems adapted for grass rather than grain. As a result, the animals require endless administrations of antibiotics and other medicines to keep their chronic diarrhea under control and to prevent them from dying (the antibiotics also increase weight gain). CAFOs also generate copious quantities of pollution. Do the economic benefits of industrially producing so much low-quality inexpensive meat outweigh the costs to human health and the environment?

The other major shift in human diet since the industrial food revolution is how foods are increasingly modified and processed to increase their desirability, convenience, and storability. Millions of years of struggling to get enough food probably explains why people consistently prefer processed foods with low fiber and high concentrations of sugar, fat, and salt.[28] In turn, manufacturers, parents, schools, and anyone else who sells or provides food are happy to give us what we want, and an entire new profession of food engineers has been created to design new processed foods that are appealing, inexpensive, and have a long shelf life.[29] If your supermarket is anything like mine, more than half the foods for sale are substantially processed and more ready to eat than most "real food." I spent years as a parent trying to limit people's efforts to serve these processed foods to my daughter. Instead of giving her an apple, she'd be given a fruit roll, a fruit-flavored candy ludicrously marketed as a substitute for fruit that has the same number of calories and vitamin C, but without the fiber or any other nutrients.

Processing food by grinding it into tiny particles, removing fiber, and increasing its starch and sugar content changes how our digestive systems function. When you eat something, you have to expend some energy to digest it, to break the molecules down and transport the nutrients from your gut to the rest of your body. (You can feel and measure the energy cost of digestion by how much your body temperature rises after eating a meal.) This cost is significantly reduced—by more than 10 percent—when you eat more highly processed foods that have smaller particle sizes.[30] If you grind up a steak into hamburger or a handful of peanuts into peanut butter, your body will extract more calories per gram of food with less cost. Your gut digests the food using enzymes, proteins that bind to the surface of food particles and break them down. Small particles have more surface per unit mass, so smaller particles are digested more efficiently. In addition, processed foods with less fiber, such as white flour and white rice, require fewer steps and less time to digest, causing blood sugar levels to rise more quickly. Such foods (termed high glycemic foods) are quickly and easily broken down, but our digestive systems are not well adapted to the rapid swings in blood sugar levels they cause. When the pancreas tries to produce enough insulin that fast, it often overshoots, causing elevated levels of insulin, which then causes blood sugar levels to plunge below normal, making you hungry. Such foods promote obesity and type 2 diabetes (more on this in chapter 10).

So just how much has industrialization changed what individuals eat? One should mistrust simplistic characterizations of diets, both today and in the past, because there was no single diet eaten by hunter-gatherers or farmers, just as there is no single modern Western diet. Even so, table 5 compares reasonable approximations of a typical, generalized hunter-gatherer diet with estimates of what a typical modern American eats, and with U.S. government recommended daily allowances (RDAs). Compared to foragers, people who eat industrial diets consume a relatively high percentage of carbohydrates, especially sugars and refined starches. Industrial diets are also comparatively low in protein, high in saturated fats, and exceedingly low in fiber. Finally, despite manufacturers' abilities to load foods with calories, industrial diets contain low quantities of most vitamins and minerals, with the obvious exception of salt.

In short, the invention of agriculture caused the human food supply to increase in quantity and deteriorate in quality, but food industrialization multiplied this effect. Over the last hundred years, people have developed many technologies to produce orders of magnitude more food that is usually nutrient poor but calorie rich. Since the Industrial Revolution began about twelve generations ago, these changes have enabled us to feed more than an order of magnitude more people and to feed them more. Although approximately 800 million people today still face shortages of food, more than 1.6 billion people are overweight or obese.

TABLE 5. Comparison of Standard Hunter-Gatherer and American Diets, and the U.S. Government Recommended Daily Allowances (U.S. RDA)

Data are averaged for males and females.

Item	Hunter-gatherer	Average American	U.S. RDA
Carbohydrate (% daily energy)	35–40%	52%	45–65%
Simple sugars (% daily energy)	2%	15–30%	<10%
Fat (% daily energy)	20–35%	33%	20–35%
Saturated fat (% daily energy)	8–12%	12–16%	<10%
Unsaturated fat (% daily energy)	13–23%	16–22%	10–15%
Protein (% daily energy)	15–30%	10–20%	10–35%
Fiber (g/day)	100 g	10–20 g	25–38 g
Cholesterol (mg/day)	>500 mg	225–307 mg	<300 mg
Vitamin C (mg/day)	500 mg	30–100 mg	75–95 mg
Vitamin D (IU/day)	4,000 IU	200 IU	1,000 IU
Calcium (mg/day)	1,000–1,500 mg	500–1,000 mg	1,000 mg
Sodium (mg/day)	<1,000 mg	3,375 mg	1,500 mg
Potassium (mg/day)	7,000 mg	1,328 mg	580 mg

Modern American dietary data is from http://www.cdc.gov/nchs/data/ad/ad334.pdf, and the hunter-gatherer diet estimates are based on M. Konner and S. B. Eaton (2010). Paleolithic nutrition: 25 years later. *Nutrition in Clinical Practice* 25: 594–602.

Industrial Medicine and Sanitation

Until the Industrial Revolution, medical progress (if one can use the term) largely consisted of replacing ignorant ideas with quackery. To be sure, people still employed folk remedies—some of which probably date back to the Paleolithic—but they had little useful knowledge about how to deal with diseases of civilization like plagues, anemia, vitamin deficiencies, and gout, which began following the Agricultural Revolution and which hunter-gatherers rarely if ever have to contend with. In Europe and America, popular but ineffectual remedies for illness included copious bloodletting, immersing oneself in mud, or ingesting small amounts of poison, such as mercury. Anesthesia did not exist, and hygienic practices like washing your hands before pulling a tooth or delivering a baby were rarely considered and sometimes ridiculed. Not surprisingly, sensible people avoided doctors, who mostly believed that people got sick from an imbalance in the four basic humors: yellow bile, black bile, phlegm, and blood.[31]

This abysmal state of medical knowledge was matched by appallingly unsanitary conditions that frequently made people sicken and die. Hunter-gatherers never reside in any camp long enough or in sufficiently high numbers to accumulate much filth, and they generally stay fairly clean. As soon as people settled down in villages, life got more squalid, and as populations swelled and aggregated into towns and cities, living conditions became increasingly unsanitary and malodorous. Cities and towns reeked like pigsties. European cities were full of cesspools, giant underground caverns into which people dumped feces and other waste. One major problem with cesspools is that they leak liquid fecal matter (euphemistically termed "black water"), contaminating local streams and rivers, and thus people's drinking water. Sewers, when they existed, were rare or ineffective. Toilets were a luxury for the wealthy, and sewage treatment was usually nonexistent. Soap was an extravagance, few people had access to regular showers and baths, and clothing and bedding were rarely washed. To top it all off, sterilization and refrigeration had yet to be invented. For thousands of years after the origin of agriculture, life stank, diarrhea was common, and cholera epidemics were regular occurrences.

In spite of being unsanitary death traps, cities became magnets as the agricultural economy progressed. People flocked to cities because urban areas generally had more wealth, more jobs, and more economic opportunities than impoverished rural areas. Prior to 1900, death rates were actually higher in large English cities such as London than in rural areas, requiring a regular influx of rural immigrants to maintain urban population sizes.[32] However, as the Industrial Revolution progressed, urban conditions started to improve significantly, thanks to the rise of modern medicine, sanitation, and government. In fact, the economic transformations of the Industrial Revolution were inextricably linked with contemporary revolutions in medicine, sanitation, and public health. These different revolutions shared similar roots in the Enlightenment, and it is hard to imagine the Industrial Revolution succeeding without necessary improvements in medicine and hygiene, which in turn provided more impetus for goods and services. Factories need workers both to make and buy their goods. In addition, industrialization supplied the technical ability and financial capital necessary to engineer sewers, manufacture soaps, and produce inexpensive drugs. These lifesaving advances helped populations explode, increasing the demand for economic output.

If there was one advance in medicine that most revolutionized human health, it was the discovery of microbes and the ensuing knowledge of how to combat them. Antonie van Leeuwenhoek, who made substantial improvements to the microscope, published the first descriptions of bacteria and other microbes in the 1670s, but he and his contemporaries did not realize these "animalcules," as he called them, could be pathogens. However, people had long known or suspected that invisible agents of contagion existed and that contact with infected individuals was somehow dangerous. Leviticus, for example, is filled with tips to diagnose leprosy and rules about burning lepers' clothes, cleaning their houses, and quarantining them: "And the leper in whom the plague is, his clothes shall be rent, and his head bare, and he shall put a covering upon his upper lip, and shall cry, Unclean, unclean!"[33] Some cultures knew that pus from smallpox victims could infect but also sometimes inoculate people (the Chinese made it into a medicinal snuff). In 1796, Edward Jenner famously invented and tested the process of vaccination by scratching the arm of an eight-year-old lad with pus from a cowpox-infected farmer's daughter. A few weeks later, he brazenly scratched the lad's arm again with smallpox pus from a human, eliciting no infection.

In spite of this knowledge, the fact that microbes caused infections was not proved until 1856, when Louis Pasteur, a chemist, was commissioned by the French wine industry to help them prevent their valuable wine from mysteriously turning into vinegar. Pasteur not only discovered that airborne bacteria contaminated the wine but also that heating the wine to 60 degrees Celsius (140 degrees Fahrenheit) was sufficient to kill the troublesome microbes. Pasteurization, the simple process of heating wine, milk, and other substances, instantly improved winemakers' profits and subsequently prevented billions of infections and millions of deaths. Pasteur quickly recognized the broader implications of his discovery and turned his attention to other microbial villains, discovering the streptococcus and staphylococcus bacteria and developing vaccines against anthrax, chicken cholera, and rabies. Pasteur also saved the French silk industry by figuring out the source of a plague that was killing their silkworms.[34]

Pasteur's discoveries electrified the scientific world, creating the new field of microbiology and unleashing an avalanche of further discoveries over the next few decades as newly minted microbiologists feverishly hunted down and identified the bacteria that cause other diseases such as anthrax, cholera, gonorrhea, leprosy, typhoid, diphtheria, and plague. The tiny *Plasmodium* protozoan, which causes malaria, was discovered in 1880, and viruses were discovered in 1915. Of equal importance was the discovery that many infectious diseases were transmitted by mosquitoes, lice, fleas, rats, and other vermin. Then came drugs. Although Pasteur and other early pioneering microbiologists had observed that certain bacteria or funguses could inhibit the growth of lethal bacteria, such as anthrax, the first drugs that effectively killed bacteria were developed by Paul Ehrlich in Germany in the 1880s. The first sulfur-based antibiotics were synthesized in the 1930s. Penicillin was discovered accidentally in 1928, but its significance was not immediately recognized, and this first true wonder drug was not mass manufactured until World War II. The number of lives that have been saved by penicillin is too great to count, but it must be in the hundreds of millions.

The desire and means to improve people's health, combined with the profitability of the new health-care industry, led to many other great medical advances during the first hundred years or so of the Industrial Revolution. Important, lucrative steps forward include the discovery of vitamins, the discovery of diagnostic tools such as X-rays, the development of anesthesia, and the invention of the rubber condom. The invention of anesthesia aptly illustrates the interplay between profits and progress during the industrial era.[35] William Morton, a dentist, conducted the first successful public surgery using ether as an anesthetic in September 1846 at the Massachusetts General Hospital in Boston and then promptly patented the anesthetic. Patenting medical discoveries seems unremarkable now, but Morton's action caused outrage among the medical establishment, who disapproved of his attempt to control and profit from a substance that could alleviate human suffering. Morton spent the rest of his life engaged in lawsuits, even though his discovery was quickly eclipsed by chloroform, which was cheaper, safer, and more effective. Of course, the desire for profit also helped inspire—and still inspires—plenty of bad medical ideas. People who are sick or worried about becoming sick spend fortunes on various forms of quackery and willingly suspend their disbelief about the efficacy of their chosen treatment. For example, during the nineteenth century, regular enemas were frequently marketed as a magic bullet to promote good health. Entrepreneurs such as John Harvey Kellogg built luxury "sanitariums," resorts where rich people paid handsomely to have their colons irrigated daily while enjoying plenty of exercise, a fibrous diet of whole-grain foods, and other treatments.[36]

The other major success of the industrial era's battle against disease was preventing infections in the first place through better sanitation and hygiene. These innovations received much of their impetus from the discovery of germs and were also aided by new methods of building and manufacturing. Necessity is also the mother of invention, and better sanitation and hygiene became an urgent concern because rapidly growing cities simply couldn't cope with so many humans excreting so much. Early cities such as Rome had networks of moderately effective sewers, many of which were constructed by covering streams that carried away waste. But many cities relied on giant, stinky, leaking cesspools. London's thousands of overflowing cesspools become so intolerable that the city foolishly permitted them to be emptied into the Thames River in 1815, thus dumping even more excrement into London's major source of drinking water.[37] Londoners somehow endured these conditions and the frequent cholera epidemics they caused until the unusually hot summer of 1858, the Great Stink, when the city became so foul that Parliament (whose building borders the Thames) finally acted to construct a new sewer system. Queen Victoria was so excited by the sewer that she had an underground railroad built through a section of the sewer that traversed the Thames in order to dedicate its construction. Sewers, major feats of engineering, were also constructed in cities around the world, to the great relief and pride of their residents. The city of Paris still operates a delightful though somewhat malodorous museum (Le Musée des Égouts de Paris), which permits you to see and smell the sewers of Paris and learn their glorious history.

Advances in indoor plumbing and personal hygiene complemented the construction of sewers. You probably take for granted the use of a flush toilet, but until the late nineteenth century, clean places to defecate were luxuries, and the technology to keep human waste away from drinking water was primitive and ineffectual. Although Thomas Crapper did not invent the toilet, he was a pioneer in its mass manufacture, allowing anyone and everyone to safely eliminate their eliminations into newly constructed sewers. During the first part of the twentieth century, the magnate John D. Rockefeller helped construct outhouses throughout the American South to combat hookworm infections, which were transmitted by human feces.[38] You probably also wash your hands with soap after using the toilet, but the ability to cleanse yourself easily, inexpensively, and effectively was substantially boosted by nineteenth-century advancements in indoor plumbing and soap manufacture. Clothes and bedding, too, were hard to wash before laundry soap and easy-to-wash cotton clothing became affordable and common during the Industrial Revolution. In fact, few people recognized the health benefits of washing before the nineteenth century. When Ignaz Semmelweis in Hungary and Oliver Wendell Holmes Sr. in the United States independently suggested in the 1840s that doctors and nurses could drastically cut the incidence of puerperal (childbed) fever by washing their hands, they were greeted with derision. Fortunately, Pasteur's discovery of microbes combined with evidence that basic hygiene saved lives ultimately convinced their skeptics. Another major advancement in the war against germs was the discovery by Joseph Lister in 1864 of how to use carbolic acid to kill microbes, leading to the development of antiseptics, and later to aseptic techniques. Lister was accorded the singular honor of operating on Queen Victoria's armpit in 1871.[39]

Finally, industrialists transformed food safety. Hunter-gatherers don't store food for more than a few days, but farmers cannot survive without storing their harvests for months, if not years. Before the industrial era, salt was the most common and effective food preservative. Canned foods were first invented in 1810 by the French army at the behest of Napoleon Bonaparte, who believed that an army marches on its stomach. Early pioneers in canning quickly figured out that canned foods had to be heated to prevent them from spoiling, but after Pasteur invented pasteurization, food manufacturers rapidly devised ways to store a wide range of foods such as milk, jam, and oil safely and economically in cans, bottles, and other kinds of airtight packages. Another major advance was refrigeration and freezing. People had long kept food cool in cellars, and rich people sometimes had access to ice in the summer, but many foods had to be eaten after they molded or went rancid. Effective refrigeration was developed in the United States starting in the 1830s, primarily using new technologies to manufacture ice, and within decades refrigerated railroad cars were transporting all sorts of foods over long distances for sale.

Advances in medicine, sanitation, and food storage show how the industrial and scientific revolutions did not occur independently but instead spurred each other on by rewarding and inspiring discoveries and inventions that made money and saved untold numbers of lives. Many changes wrought by the industrial era, however, have not necessarily benefited how our bodies grow and function. We have already discussed some of the negative effects of industrialization on the food we eat and the work we do. Since we spend about one-third of our lives asleep, I would be remiss not to consider how we have changed the way we get our forty winks.

Industrial Sleep

Did you get enough sleep last night? A typical American spends an average of 7.5 hours in bed every night but sleeps for only 6.1 hours, 1 hour less than the national average from 1970, and between 2 and 3 hours less than 1900.[40] In addition, only a third of Americans take naps. Most people sleep alone or with a single partner in soft, warm beds raised several feet off the floor, and we often force our babies and children to sleep like adults in an isolated or nearly isolated state in their own rooms with as little sensory stimulus as possible: little light, no sounds, no smells, and no social activity.

You may prefer such sleeping habits, but they are modern and comparatively bizarre. A compilation of reports on the sleeping customs of hunter-gatherers, pastoralists, and subsistence farmers suggests that, until recently, humans rarely slept in solitary, isolated conditions, not sharing their beds with children and other family members; people usually napped every day; and they normally got more sleep than we do.[41] A typical Hadza hunter-gatherer wakes up every morning at dawn (always between 6:30 and 7:00 a.m. at the equator), enjoys a one- to two-hour nap at midday, and goes to bed around 9:00 p.m.[42] People also didn't usually sleep in a single bout but considered it normal to wake in the middle of the night before having a "second sleep."[43] In traditional cultures, beds are usually hard, and bedding is negligible to minimize fleas, bedbugs, and other parasites. People also slept in much more complex sensory environments, usually with a fire nearby, listening to the sounds of outside world, and tolerating one another's noises, movements, and occasional sexual activities.

Many factors account for how and why we sleep so differently than we used to. One is that the Industrial Revolution transformed time and provided us with bright lights, radio, television shows, and other fun things to entertain and stimulate us well beyond an evolutionarily normal bedtime hour.[44] For the first time in millions of years, much of the world can now stay up late, encouraging sleep deprivation. On top of that, many people today suffer from insomnia because they experience more stress from some mix of physical and psychological factors, such as too much alcohol, poor diet, lack of exercise, anxiety, depression, and various worries.[45] It is also possible that the unusual, stimulus-free environments in which we now like to sleep further promote insomnia.[46] Falling asleep is a gradual process in which the body goes through several stages of light sleep and the brain becomes progressively less aware of outside stimuli before entering a deep stage of sleep in which one is unaware of the outside world. For most of human evolution, this slow process may have been an adaptation to help avoid falling into deep sleep in dangerous circumstances, such as when lions are prowling nearby. Having a first and second sleep during the night may have also been adaptive. Perhaps insomnia sometimes occurs because by isolating ourselves in insulated bedrooms we don't hear evolutionarily normal sounds such as the hearth crackling, people snoring, and hyenas barking far in the distance, reassuring subconscious parts of the brain that everything is okay.

Whatever the causes, we sleep less and less well than we used to, and at least 10 percent of the population in developed countries regularly experiences serious insomnia.[47] Lack of sleep rarely kills you, but chronic sleep deprivation prevents your brain from working properly and whittles away at your health. When you are short of sleep for extended periods of time your body's hormonal system responds in several ways that used to be adaptive only during brief periods of stress. Normally when you sleep the body secretes a pulse of growth hormone, which stimulates general growth, cell repair, and immune function, but sleep deprivation curtails this surge and instead induces the body to produce more of the hormone cortisol.[48] High cortisol levels switch your body's metabolism from a state of growth and investment into a state of fright and flight by raising alertness and shuttling sugar into the bloodstream. This shift is useful to get you out of bed in the morning or to help you flee from a lion, but chronically high cortisol levels depress immunity, interfere with growth, and increase the risk of type 2 diabetes. Chronically insufficient sleep also promotes obesity. During normal sleep, the body is at rest, causing levels of one hormone, leptin, to rise, and another hormone, ghrelin, to fall. Leptin suppresses appetite and ghrelin stimulates appetite, so this cycle helps you avoid getting hungry in your sleep. However, when you consistently get too little sleep, your leptin levels fall and your ghrelin levels rise, effectively signaling a state of famine to your brain, regardless of how nourished you may be.[49] Sleep-deprived people thus have more cravings, especially for carbohydrate-rich foods.

The cruelest irony about sleep in the industrial era is that sleeping well is a privilege of wealth. People with higher incomes get more sleep because they sleep more efficiently (they spend less time in bed unable to fall asleep).[50] The likely explanation is that wealthier people are less stressed and thus fall asleep more easily. For people struggling to make ends meet, daily stress and insufficient sleep lead to a vicious circle because stress inhibits sleep and insufficient sleep elevates stress.

The Good News: Taller, Longer-Lived, and Healthier Bodies

The last 150 years have profoundly transformed how we eat, work, travel, fight disease, keep clean, and even sleep. It is as if the human species had a total makeover: our daily lives would be barely comprehensible to our ancestors from just a few generations ago, but we are essentially identical genetically, anatomically, and physiologically. The change has been so rapid that too little time has elapsed for more than a modicum of natural selection to have occurred.[51]

Was it worth it? From the perspective of the human body, the answer to this question must be "very much so—but not much at first." When the first factories were built in Europe and the United States, workers toiled hard for brutally long hours in dangerous conditions, and they swarmed into big, polluted cities that were full of contagion. Working in an urban factory might have been better than starving in the countryside, but for many, the price of early progress was, and still is, misery. Yet the average person's health did begin to improve in developed industrialized nations like the United States, England, and Japan as wealth rapidly accumulated and medical advances accelerated. Sewers, soap, and vaccinations stemmed the relentless outbreaks of infectious diseases unleashed by the Agricultural Revolution many thousands of years before. New methods of food production, storage, and transport increased the quantity and quality of food available to most people. To be sure, war, poverty, and other ills still caused much suffering and death, but ultimately the Industrial Revolution made more people better off than they were a few hundred years before. You are more likely to have been born, less likely to get sick or die prematurely, and are probably taller and heavier.

If there is any one variable that underlies the changes caused by industrialization and medicine it must be energy. As I discussed in chapter 5, human beings, like every other organism, use energy to accomplish three basic functions: we grow, we maintain our bodies, and we reproduce. Before agriculture, the amount of energy hunter-gatherers acquired was only marginally greater than the energy they needed to grow, maintain their bodies, and reproduce at a rate of replacement. Daily physical activity levels and energy returns were moderate, childhood mortality was high, and population growth was slow. Agriculture changed this equation by increasing substantially how much energy was available, enabling reproductive rates to nearly double. For millennia, farmers had to be very physically active, and they suffered from the burden of many mismatch diseases. But then the invention of industrialization suddenly made available seemingly limitless supplies of energy from fossil fuels, and technologies such as engines and mechanical looms transformed that energy into doing work, thus producing exponentially more wealth, including food. At the same time, modern sanitation and medicine substantially decreased not just mortality but also how much energy people spent combating illnesses. If you expend less energy staying healthy, then you will inevitably channel more energy into growth and reproduction. It follows that the three most predictable consequences of the Industrial Revolution on the human body are bigger bodies, more babies, and greater longevity.

Let's first consider body size as measured by stature. Height is affected by both genetic and environmental factors during the period you are growing: good health essentially allows you to grow as high as your genes will let you (but no more); bad health and poor nutrition stunt your growth. As our energy-balance model predicts, human bodies indeed have become bigger since the Industrial Revolution. But if you look carefully at stature over the last few hundred years, most of the change has been recent. As an example, figure 19 graphs how male height has changed since 1800 in France.[52] During the early part of the Industrial Revolution, stature increased moderately (it actually declined in poorer countries, such as the Netherlands). Increases in stature accelerated slightly in the 1860s but then took off during the last fifty years. Ironically, if we consider how height has changed over a longer time scale, the last 40,000 years (as shown in figure 19), it is apparent that recent advances have allowed Europeans to get back to and then slightly exceed where they started in the Paleolithic.[53] Stature in Europe decreased at the end of the Ice Age, perhaps in part because of genetic changes as Europeans adapted to warmer climates, but then they got even shorter during the challenging millennia of the early Neolithic. Agricultural advances started to reverse the trend over the last millennium, and it wasn't until the twentieth century that Europeans were the same height as cavemen. In fact, stature data suggest that Europeans are now taller than anyone else on the planet. In 1850, Dutch males were on average 4.8 centimeters (2 inches) shorter than American males. Since then, stature has increased almost 20 centimeters (8 inches) among Dutch males but only 10 centimeters (4 inches) among American males, making the Dutch now the tallest people in the world.[54]

What about weight? We'll consider expanding waistlines and obesity more in chapter 10, but long-term data from various countries suggest that the extra energy now available to so many people has predictably increased weight relative to height. This relationship is often measured using the body mass index (BMI), a person's weight (in kilograms) divided by height (in meters) squared. Figure 20 plots measurements of the BMIs of American men between the ages of forty and fifty-nine over the last 100 years from a monumental study by Roderick Floud and colleagues.[55] The graph shows that a typical adult American male in 1900 had a healthy BMI of about 23, but since then BMI has steadily increased, albeit with a slight dip after World War II. The average American male of today is overweight (defined as a BMI above 25).

Sadly, increases in adult height and weight over the last hundred years or so have not translated into lowering the percentage

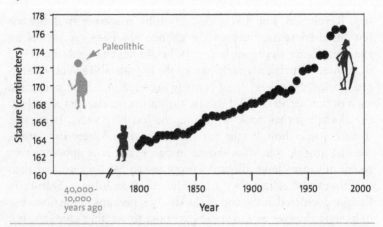

FIGURE 19. Change in stature among French men since 1800 (and compared to Paleolithic Europeans). Data from R. Floud et al. (2011). *The Changing Body: Health, Nutrition, and Human Development in the Western World Since 1700.* Cambridge: Cambridge University Press; T. J. Hatton and B. E. Bray (2010). Long-run trends in the heights of European men, 19th-20th centuries. *Economics and Human Biology* 8: 405–13; V. Formicola and M. Giannecchini (1999). Evolutionary trends of stature in upper Paleolithic and Mesolithic Europe. *Journal of Human Evolution* 36: 319–33.

of babies who are born too small. Infant size at birth is a major health concern because babies born with a low birth weight— clinically defined as less than 2.5 kilograms (5.5 pounds)—are at much greater risk of death or suffering from poor health as children and as adults. Floud and colleagues' data show that mean birth weight in the United States is significantly lower in blacks than whites, but in both groups the proportion of low birth weight babies has barely changed since 1900 (about 11 percent among blacks and 5.5 percent among whites). This disparity is primarily the consequence of socioeconomic differences because birth weight is a direct reflection of how much energy a mother is able to invest in her offspring.[56] Countries such as the Netherlands that provide all residents with access to good health care have lower percentages of low birth weight babies (about 4 percent).

The other obvious prediction from the energy model is that the combination of more calories from plentiful energy-dense food, less physical activity, and less illness will change the demographic char-

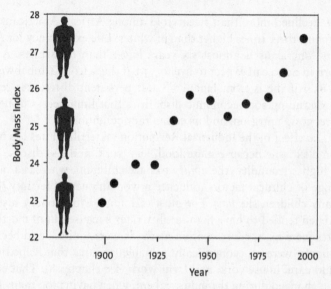

FIGURE 20. Changes in the body mass index (BMI) of American men between the ages of forty and fifty-nine since 1900 (some values are extrapolated). Modified from R. Floud et al. (2011). *The Changing Body: Health, Nutrition, and Human Development in the Western World Since 1700.* Cambridge: Cambridge University Press.

acteristics of human populations. In addition to growing taller and wider, people with a positive energy balance live longer, they can have more children, and their children are more likely to survive. In fact, if there is any one universally accepted measure of progress, it is low infant mortality rates. By that measure, the Industrial Revolution has been an outstanding success. Infant mortality among white Americans decreased thirty-six-fold between 1850 and 2000, from 21.7 percent to 0.6 percent.[57] Lower infant mortality, combined with other advances, also doubled life expectancy. If you were born in 1850, the chances were you'd live to be forty years old, and your cause of death would mostly likely be an infectious disease. An American baby born in the year 2000 can expect to live seventy-seven years and will most likely die from cardiovascular disease or cancer. Amid these heartening statistics, however, are sobering reminders that changes over the last few hundred years have not benefited everyone equally. Since 1850, infant mortality has declined more than twentyfold among African Americans but remains three times higher than in whites. Life expectancy for African Americans is almost six years lower than for whites. A girl born in 2010 can expect to live to 55.1 if she is from Zimbabwe but to 85.9 if she is from Japan.[58] These persistent differences reflect longstanding socioeconomic disparities that limit access to health care, good nutrition, and more sanitary conditions.

The effect of the Industrial Revolution on fertility rates is a more complex issue because more food, less work, and less disease lead to higher fecundity (the *ability* to have children), whereas a broad range of cultural factors influence a woman's actual fertility (how many children she *has*). For most of human evolutionary history, women tended to have high fertility rates because infant mortality rates were high, contraception methods were limited, and because children were an economically valuable resource that helped with child care, housework, and farm work (see chapter 8). That equation changed during the industrial era, when having too many children switched to becoming an economic burden. Families started to limit their fertility, aided by new methods of contraception. In 1929, the American demographer Warren Thompson proposed that as populations went through the Industrial Revolution, they went through a "demographic transition," depicted in figure 21. Thompson's basic observation was that, following industrialization, mortality rates decline because of better conditions, and then families react by lowering their fertility rates. As a consequence, population growth rates are typically high during the early phases of industrialization, but they subsequently level off and sometimes even decline. Thompson's demographic transition model has been controversial because it does not apply to all countries. For example, in France, birth rates actually declined before death rates dropped, and in many countries throughout the developing world in the Middle East, South Asia, Latin America, and Africa, birthrates have remained elevated despite substantial decreases in mortality.[59] These countries have very high rates of population growth. It should hardly be surprising that economic development influences but does not determine family size.

In short, the combined effects of lower infant mortality, higher longevity, and increased fertility have fueled an explosion in the

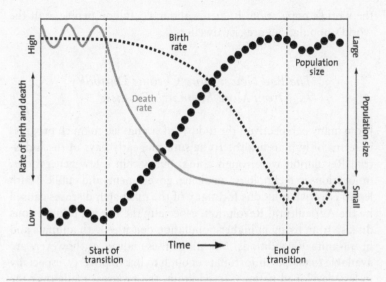

FIGURE 21. The demographic transition model. Following economic development, death rates tend to fall before birth rates decrease, resulting in an initial population boom that eventually levels off. This controversial model, however, only applies to some countries.

world's population, as figure 18 graphs. Since population growth is intrinsically exponential, even small increases in fertility or decreases in mortality spark rapid population growth. If an initial population of 1 million people grows at 3.5 percent per year, then it will roughly double every generation, growing to 2 million in twenty years, 4 million in forty years, and so on, reaching 32 million in a hundred years. In actual fact, the global growth rate peaked in 1963 at 2.2 percent per year and has since declined to about 1.1 percent per year,[60] which translates into a doubling rate of every sixty-four years. In the fifty years between 1960 and 2010, the world's population more than doubled, from 3 to 6.9 billion people. At current rates of growth, we can expect 14 billion people at the end of this century.

One major by-product of population growth plus the concentration of wealth in cities has been a shift to more urbanization. In 1800, only 25 million people lived in cities, about 3 percent of the world's population. In 2010, about 3.3 billion people, half the world's population, are city dwellers.

The Bad News: More Chronic Disability from More Mismatch Diseases

From many perspectives, the industrial era has led to much progress in terms of human health. To be sure, the early days of the Industrial Revolution were rough going, but within a few generations, innovations in technology, medicine, government, and public health led to effective solutions for many of the mismatch diseases caused by the Agricultural Revolution, especially the burden of infectious disease from living at higher population densities with animals and in unsanitary conditions. Not all of these advances, however, are available to people unfortunate enough to live in poverty, especially in less developed nations. In addition, the progress made over the last 150 years has also come with some consequential drawbacks for people's health. Most essentially, there has been an epidemiological transition. As fewer people succumb to diseases from malnutrition and infections, especially when they are young, more people are developing other kinds of noncommunicable diseases as they age. This transition is still ongoing: in the forty years between 1970 and 2010, the percentage of deaths worldwide from infectious disease and malnutrition fell by 17 percent and life expectancy increased by eleven years, while the percentage of deaths from noncommunicable diseases rose by 30 percent.[61] As more people live longer, more of them are suffering from disability. In technical terms, lower rates of mortality have been accompanied by higher rates of morbidity (defined as a state of ill health from any form of disease).

To put this epidemiological transition into perspective, compare how senior citizens live today in the United States with the way their grandparents or great-grandparents experienced old age. When Franklin D. Roosevelt signed the Social Security Act in 1935, old age was defined as sixty-five years, yet estimated life expectancy in the United States at the time was sixty-one years for males and sixty-four years for females.[62] A senior citizen today, however, can expect to live eighteen to twenty years longer. The downside is that he or she also should expect to die more slowly. The two most common causes of death in 1935 America were respiratory diseases (pneumonia and influenza) and infectious diarrhea, both of which kill rapidly. In contrast, the two most common causes of death in 2007 America were heart disease and cancer (each accounted for about 25 percent of total deaths). Some heart attack victims die within minutes or hours, but most elderly people with heart disease survive for years while coping with complications such as high blood pressure, congestive heart failure, general weakness, and peripheral vascular disease. Many cancer patients also remain alive for several years following their diagnosis because of chemotherapy, radiation, surgery, and other treatments. In addition, many of the other leading causes of death today are chronic illnesses such as asthma, Alzheimer's, type 2 diabetes, and kidney disease, and there has been an upsurge in the occurrence of nonfatal but chronic illnesses such as osteoarthritis, gout, dementia, and hearing loss.[63] Altogether, the growing prevalence of chronic illness among middle-aged and elderly individuals is contributing to a health-care crisis because the children born during the post–World War II baby boom are now entering old age, and an unprecedented percentage of them are suffering from lingering, disabling, and costly diseases. The term epidemiologists coined for this phenomenon is the "extension of morbidity."[64]

One way to quantify the extension of morbidity currently occurring is a metric known as disability-adjusted life years (DALYs), which measures a disease's overall burden as the number of years lost to ill health plus death.[65] According to an impressive recent analysis of medical data worldwide from between 1990 and 2010, the burden of disability caused by communicable and nutrition-related diseases has plunged by more than 40 percent, while the burden of disability caused by noncommunicable diseases has risen, especially in developed nations. As examples, DALYs have risen by 30 percent for type 2 diabetes, by 17 percent for neurological disorders, such as Alzheimer's, by 17 percent for chronic kidney disease, by 12 percent for musculoskeletal disorders, such as arthritis and back pain, by 5 percent for breast cancer, and by 12 percent for liver cancer.[66] Even after factoring in population growth, more people are experiencing more chronic disability that results from noncommunicable diseases. For the diseases just mentioned, the number of years a person can expect to live with cancer has increased by 36 percent, with heart and circulatory diseases by 18 percent, with neurological diseases by 12 percent, with diabetes by 13 percent, and with musculoskeletal diseases by 11 percent.[67] To many, old age is now equated with various disabilities (and high medical bills).

Is the Epidemiological Transition the Price of Progress?

How much is the paradox of human health trends today—that more people are living to be older but also suffering more often and for longer from chronic, costly diseases—simply the price of progress? After all, you have to die of something. Since communicable diseases are killing fewer younger people, then it makes sense to expect more diseases like cancer and type 2 diabetes that tend to strike older people. As your body ages, your organs and cells function less effectively, your joints wear out, mutations accumulate, and you encounter more toxins and other harmful agents. According to this logic, if you are less likely to die as a youth from malnutrition, influenza, or cholera, you should consider yourself fortunate to die at an older age from heart disease or osteoporosis. The same logic would also have you view nonlethal but nonetheless annoying common conditions such as irritable bowel syndrome, myopia, and cavities as necessary, collateral consequences of civilization.

Has the industrial era caused a trade-off between lower mortality and an extension of morbidity? To some extent, the answer is unquestionably yes. Because of more food, better sanitation, and better work conditions, fewer people, especially children, contract infectious diseases and suffer from insufficient food, and so they live longer. It is also inevitable that, with age, the chance of cancer-causing mutations increases, arteries harden, bones lose mass, and other functions deteriorate. Many health problems correlate strongly with age, which makes them more prevalent as more populations grow and a larger percentage of them are middle-aged and elderly. According to some estimates, the number of years people live with disabilities has increased worldwide by 28 percent simply from population growth, and by nearly 15 percent because there are now more old people.[68] Yet for every year of added life that has been achieved since 1990, only 10 months is healthy.[69] By the year 2015, there will be more people over the age of sixty-five than under the age of five, yet nearly half of those above the age of fifty will be in some state of pain, disability, or incapacity that requires medical care.

When examined from an evolutionary perspective, however, the epidemiological transition cannot be explained solely as a trade-off between mortality and morbidity. Almost every published analysis of changing health trends considers shifts in human mortality and morbidity from only the last hundred or so years using data on just people from industrial or subsistence agricultural economies. Yet without considering data on hunter-gatherer health, these assessments of changes in global heath are like trying to figure out who won a soccer game using only the goals scored during the last few minutes. Further, although it makes sense for doctors and public health officials to categorize diseases based on whether they are caused by infections, malnutrition, tumors, and so on, an evolutionary perspective suggests that we should also look at the extent to which diseases are caused by evolutionary mismatches between the environmental conditions (including diet, physical activity, sleep, and other factors) for which we evolved and the environmental conditions that we now experience.

If we reconsider the current epidemiological transition—the trade-off between dying young from infectious diseases and extending morbidity from noncommunicable diseases—with an evolutionary perspective, a somewhat different picture emerges. Viewed in this light, it is evident that as populations grow and people live longer, more people are getting sick from mismatch diseases that used to be uncommon or nonexistent and that are not necessarily or entirely the inevitable by-products of progress.

A key line of evidence to support this view comes from what we know about hunter-gatherer health from the few groups who still remain to study. Recall that hunter-gatherers live in small populations because mothers have babies infrequently and their children suffer from high rates of infant and childhood mortality. Even so, recent hunter-gatherer lives are not necessarily nasty, brutish, and short, as is often assumed. Hunter-gatherers who survive childhood typically live to be old: their most common age of death is between sixty-eight and seventy-two, and most become grandparents or even great-grandparents.[70] They most likely die from gastrointestinal or respiratory infections, diseases such as malaria or tuberculosis, or from violence and accidents.[71] Health surveys also indicate that most of the noninfectious diseases that kill or disable older people in developed nations are rare or unknown among middle-aged and elderly hunter-gatherers.[72] These admittedly limited studies have found that hunter-gatherers rarely if ever get type 2 diabetes, coronary heart disease, hypertension, osteoporosis, breast cancer, asthma, and liver disease. They also don't appear to suffer much from gout, myopia, cavities, hearing loss, collapsed arches, and other common ailments. To be sure, hunter-gatherers don't live in perpetually perfect health, especially since tobacco and alcohol have become increasingly available to them, but the evidence suggests that they are healthy compared to many older Americans today despite never having received any medical care.

In short, if you were to compare contemporary health data from people around the world with equivalent data from hunter-gatherers, you would not conclude that rising rates of common mismatch diseases such as heart disease and type 2 diabetes are straightforward, inevitable by-products of economic progress and increased longevity. Moreover, if you look carefully, some of the epidemiological data used to support the inevitability of a trade-off between dying young from infectious disease and dying at an older age from heart disease or certain cancers fails to hold up to scrutiny. Consider, for example, recent breast cancer trends. In the United Kingdom, the incidence of breast cancer among fifty- to fifty-four-year-old women nearly doubled between 1971 and 2004, but there was no doubling of the population of women in their early fifties (instead life expectancy increased by only five years over the same period).[73] In addition, metabolic diseases such as type 2 diabetes and hardening of the arteries aren't just cropping up because people are living longer, they are actually becoming more prevalent at younger ages as the incidence of obesity also rises among young people.[74] To be sure, some diseases such as prostate cancer are now easier to diagnose, hence they appear more common, but doctors in developed nations now have to treat many diseases that used to be extremely rare and that seldom appear in the nonindustrial world. One example is Crohn's disease, in which the body's immune system attacks the gut, causing horrid symptoms that include cramps, rashes, vomiting, and even arthritis. Rates of Crohn's disease are rising throughout the world, most especially among people in their teens and twenties.[75]

Another important line of evidence that the epidemiological transition does not result from an unavoidable trade-off caused by progress comes from examining the causes of changing trends in mortality and morbidity. This is a tricky task because it is impossible to tease apart precisely which factors cause most chronic noncommunicable diseases and by how much. Even so, several studies consistently rank the following factors as especially important causes of morbidity among people in developed nations (in rough order): high blood pressure, tobacco smoking, overuse of alcohol, pollution, a diet low in fruit, high body-mass index, high fasting levels of blood glucose, physical inactivity, high sodium, diets low in nuts and seeds, and high cholesterol.[76] Note that many of these factors are not independent. Smoking, poor diet, and physical inac-

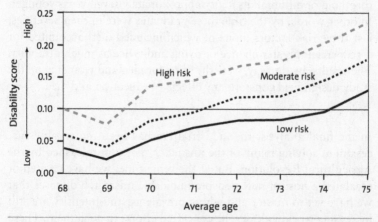

FIGURE 22. Compression of morbidity among graduates of the University of Pennsylvania. The subjects were divided into different risk categories based on BMI, smoking, and exercise habits. Individuals with higher risk factors had more disability at younger ages. Modified from A. J. Vita, et al. (1998). Aging, health risks, and cumulative disability. *New England Journal of Medicine* 338: 1035–41.

tivity are each well known to cause high blood pressure, obesity, high levels of blood sugar, and bad cholesterol profiles. Regardless, none of these risk factors were common before the Agricultural and Industrial Revolutions.

Last but not least, there is some evidence to question or at least temper the assumption that an extension of morbidity necessarily accompanies greater longevity. A seminal test of this hypothesis by James Fries and colleagues analyzed data from 1,741 people who attended the University of Pennsylvania in 1939 and 1940 and were then surveyed repeatedly for more than fifty years.[77] Data were collected on three key risk factors (BMI, smoking habits, and how much they exercised), the chronic illnesses from which they suffered, and their degree of disability (quantified on the basis of how well they performed eight basic daily activities: dressing, rising, eating, walking, grooming, reaching, gripping, and performing errands). Those classified as high risk because they were overweight, smoked, and didn't exercise much had a 50 percent higher mortality rate than those who were low risk. Additionally, as figure 22 illustrates, these high risk individuals had disability scores that were 100 percent greater than those who were low risk and they crossed the threshold of minimal disability approximately seven years younger. In other words, by the time these graduates were in their seventies, just three risk factors (none of which included diet) accounted for a 50 percent greater chance of dying and twice as much disability. The results, by the way, were the same for men and women, and the study design held constant any effects of education and race.

In the final analysis, the industrial era has been remarkably successful at solving many of the mismatch diseases unleashed by the Agricultural Revolution. But at the same time, we have created or escalated a host of new noncommunicable mismatch diseases that we have yet to master and whose prevalence and intensity are still increasing worldwide despite concerted efforts to quell them. These diseases and the extension of morbidity that has accompanied the ongoing epidemiological transition are not simple, inevitable byproducts of greater longevity and less infectious disease. There is no ineluctable trade-off behind the correlation between greater longevity and higher morbidity. Instead, the evidence confirms the commonsense notion that it is possible to live a long and healthy life without being condemned to contracting chronic noninfectious diseases that cause years of disability. Yet, sadly, not enough people age so well. In order to try to understand these trends, let's now use the lens of evolution to look more deeply at the causes of the mismatch diseases that have arisen since the Agricultural and Industrial Revolutions. An equally important issue is how our failure to treat the causes of these diseases sometimes fosters dysevolution, the pernicious feedback loop that allows them to stay prevalent or increase in frequency.

Of the various mismatch diseases we confront, some of the most worrying are those caused by too much of a formerly rare stimulus. And of these diseases, the most quintessential and widespread are related to obesity, which is caused by having too much energy.

PART III

The Present, the Future

PART ONE

The Present, the Future

10

The Vicious Circle of Too Much

Why Too Much Energy Can Make Us Sick

My exit is the result of too many entrées.

—RICHARD MONCKTON MILNES

I was raised to fear fat—both eating it and being it. On the assumption that you are what you eat, my mother considered cheese, butter, and anything else with lots of fat to be forms of poison to be avoided as much as possible. Eggs were giant poison pills. She was not entirely correct about what foods make you fat, but she was right to worry about obesity. Of the many health problems that the human species confronts today, obesity has become the biggest, both literally and figuratively. Although obesity itself is not a disease, it arises from having too much of a formerly rare stimulus: energy. In turn, too much energy, including too much bodily fat (especially in the abdomen), can cause many mismatch diseases that are rapidly becoming more prevalent because of the environments we have created and because we fail to prevent their causes effectively.

Obesity is such a widespread, conspicuous problem and the subject of so much discussion that many people are becoming fed up with reading, talking, or thinking about it. How often do you need to be reminded that two-thirds of adults in countries like the United

States are overweight or obese, that one-third of their children are too heavy, and that the percentage of obese people has doubled since the 1970s? How many advertisements can we digest for plus-size clothing and new diet plans? If there is one thing everyone knows about obesity, it is that trying to shed pounds is extremely difficult and sometimes impossible. Further, what's wrong with being fat in the first place? If Venus figurines—carved statues of faceless women with large breasts, ample thighs, and swollen bellies—are any indication, we used to venerate lots of body fat during the Stone Age.[1]

I do not wish to sugarcoat an important topic, but widespread confusion, debate, anger, and angst over the obesity epidemic testify that we desperately need to understand better when and why obesity is a problem. Why do humans so readily become fat? Why does obesity predispose people to certain diseases if humans are also adapted to store fat? Why is the incidence and intensity of obesity-related diseases increasing now? Why do some overweight people get sick but others don't? To address these and other why questions requires looking through the lens of evolution. An evolutionary perspective confirms that humans are exquisitely adapted to gain weight and that storing a relatively large quantity of body fat is normal. An evolutionary perspective highlights why we are inadequately adapted not so much to surplus fat in our butts, legs, and chins, but to excess fat in our bellies. An evolutionary perspective helps call attention to the ultimate root causes of the problem. Chief among these is that what matters is not just how much we eat, but also what we eat and that our bodies are inadequately adapted to cope with relentless supplies of excess energy, contributing to many of the most serious mismatch diseases we now confront, like type 2 diabetes, hardening of the arteries, and some cancers. Finally, an evolutionary perspective reveals that the way we treat these mismatch diseases of affluence sometimes creates a feedback loop that compounds the problem.

How the Body Stores, Uses, and Converts Energy

Obesity and its related diseases of affluence such as type 2 diabetes and heart disease are types of mismatch caused by what you eat and by how much energy you consume relative to how much you use. Although it is intuitively obvious that too much ice cream is bad for you, how can too much of a good thing like energy be harmful? A first step toward making sense of this problem is to get a handle on how the body converts different kinds of food into energy, and how that energy is burned or stored. I'll do my best to explain these complex processes as simply as possible.

Whenever you do anything, such as grow, walk, digest, sleep, or read these words, you spend energy. Almost all of the energy that your body uses to fuel activities is stored in a tiny ubiquitous molecule called ATP (adenosine triphosphate). ATPs are like minuscule batteries that circulate in your body's cells, giving off energy when needed. In turn, your body synthesizes and recharges ATP molecules by burning fuels, mostly carbohydrates and fats. You eat not just to replenish these energy stores but also to create an energy reserve so you never run out of ATP, even for an instant. ATP thus functions in your body like money that you acquire, use, and save. Just as your bank balance is a function of the difference between how much money you earn versus how much you spend, your *energy balance* is the difference between how much energy you take in versus how much you expend over a given period of time. Measured over the short term, you are rarely in energy balance: when you eat or digest you are usually in positive energy balance, and for the rest of the day (and night) you tend to be in slight negative energy balance. However, over long periods of time, such as days, weeks, and months, your energy balance is at steady state if you are neither gaining nor losing weight. Simplistically, weight gain or loss is caused by extended periods of being in positive or negative energy balance. Because weeks or months of negative energy balance are bad for reproductive success, most organisms, including humans, are well adapted to avoid this state.

One way to avoid being in negative energy balance is by regulating how much energy you expend. Just as you spend or squander your salary on goods and services such as food, rent, and entertainment, your body spends energy on diverse functions. A large component of your body's budget, your resting metabolism, goes toward taking care of essential needs such as feeding your brain, circulating blood, breathing, repairing tissues, and maintaining your immune system.

A typical adult's resting metabolism requires about 1,300 to 1,600 calories a day, but this cost varies widely, largely because of variations in fat-free body mass (bigger bodies consume more energy).[2] The remainder of your energy budget is spent doing things, primarily being physically active, but also digesting and keeping a stable body temperature. If you lounge about in bed all day you can stay in energy balance by ingesting just a fraction more than your resting metabolic demands. If, however, you decide to run a marathon, you'll need an additional 2,000 to 3,000 calories.

The other way we regulate energy balance is by eating food, which contains energy in the form of chemical bonds. Much as my brain enjoyed the delicious meal I just consumed, my digestive system is now treating it as mostly fuel, breaking down the food into its basic components: proteins, carbohydrates, and fats. Proteins are coiled chains of amino acids; carbohydrates are long chains of sugar molecules; fats are made of three long molecules called fatty acids held together by a single colorless, odorless molecule known as glycerol (as a result, the chemical term for fats is *triglycerides*). Protein is primarily used for building and maintaining tissues and is broken down for fuel relatively rarely. In contrast, carbohydrates and fats are stored and burned for energy, but in different ways. The key difference to remember is that carbohydrates are much easier and quicker to burn than fats, but they store energy less densely. A gram of sugar contains four calories of energy, but a gram of fat has nine calories. In the same way that you store more money efficiently in large denominations, your body sensibly stores most excess energy as fat, and very little as carbohydrate, which it does in the form of *glycogen*, a large, soggy molecule. Plants store excess carbohydrates much more densely as starch.

The different properties of fat and carbohydrates are reflected in how the body uses and stores them as fuel. Imagine that you just devoured a large slice of chocolate cake, whose primary ingredients were flour, butter, eggs, and sugar. As soon as the cake is inside you, your digestive system starts to break down its constituent fats and carbohydrates, which are transported from the small intestine into the bloodstream, whereupon they undergo different fates. The fat's fate is mostly orchestrated by the liver. Some fat gets stored within the liver, some is immediately burned, and some is stored in muscles, but the remainder is conveyed by blood to specialized fat cells (adipocytes) all over the body. A typical human has tens of billions of these cells, each of which contains a single droplet of fat. As more fat is added to a cell, it swells like a balloon. Fat cells divide if they get too voluminous when you are still growing, but most of us maintain a constant number of fat cells once we become adults.[3] Many of these cells are beneath your skin and thus termed *subcutaneous fat*, some of these cells are in muscles and other organs, and some lie around the organs of your abdomen, where they are known as *visceral fat* (colloquially, belly fat). The contrasts between subcutaneous and visceral fat are really important. As we will discuss below, visceral fat cells behave differently from other fat cells, making excessive belly fat a much more serious risk factor than simply being overweight for many diseases associated with obesity.

The other primary components of the cake are carbohydrates. Enzymes in your saliva start to break down the cake's various carbohydrates into their component sugars, and more enzymes continue the job farther down your gut. There are many different kinds of sugars, but the two most common basic forms are *glucose* and *fructose*.[4] Unfortunately, the nutrition labels on the food you buy do not distinguish between these sugars, but your body does. Let's therefore look at how your body manages them differently.

Glucose, which is not very sweet, is the essential sugar that makes up starch, so all the flour from your cake is quickly broken down to glucose. In addition, table sugar (sucrose) and milk sugar (lactose) are both made up of 50 percent glucose. Your cake, therefore, contains lots and lots of glucose, which your intestine shuttles into the bloodstream as quickly as possible because your body requires a steady, uninterrupted supply of glucose. But there is a catch: you always need sufficient glucose in your blood to prevent cells from dying (especially in your brain), but too much glucose is seriously toxic to tissues throughout your body. Your brain and pancreas therefore constantly monitor and stabilize blood glucose levels by regulating levels of the hormone insulin. Insulin is produced by the pancreas and then pumped into the bloodstream whenever blood sugar levels go up, usually right after you digest food. Insulin has several other jobs, but its most critical function is to keep glucose levels from rising too high, which it does in several ways in different organs. One major site of insulin action is the liver, where about 20 percent of the glucose from your cake ends up. Normally, the liver wants to convert this glucose into glycogen, but it cannot store too much glycogen too quickly, so any excess is converted into fat, which either accumulates within the liver or gets dumped into the blood. The other 80 percent of the glucose from your cake travels around your body and is taken up and then burned as fuel by cells in dozens of organs, such as the brain, muscles, and kidneys. Insulin causes the remaining glucose to be taken up by fat cells and also transformed into fat.[5] The key point to remember is that when glucose levels rise after a meal, your body's immediate objective is to get those levels down as expeditiously as possible, causing most of the excess glucose that you cannot use quickly to be stored as fat.

The other kind of sugar in your cake is fructose, which tastes sweet. Fructose, which is often paired with glucose, is naturally present in fruit and honey, as well as table sugar (sucrose, which is 50 percent fructose). Assuming your baker used plenty of sugar, your cake probably has a fair amount of fructose. Unlike glucose, which can be metabolized (essentially burned) by cells throughout the body, fructose is almost entirely metabolized by the liver. The liver, however, can burn only so much fructose at once, so it converts any excess fructose into fat, which again is either stored in the liver or dumped into the bloodstream. As we will see, both of these fates cause problems.

Now that we have reviewed the basics of how you store fats and carbohydrates as energy, what happens when you need to retrieve that energy a few hours later, perhaps when you go to the gym to burn off that cake? As your muscles and other tissues consume more energy, your blood glucose levels fall, causing the secretion of several hormones whose job is to release stored energy. One of these hormones, glucagon, is also produced by the pancreas, but it has the reverse effects of insulin on the liver, causing it to transform both glycogen and fats into sugar. Another key hormone, cortisol, is produced by the adrenal glands, which sit atop the kidneys. Cortisol has many effects, including to block the action of insulin, to stimulate muscle cells to burn glycogen, and to cause fat and muscle cells to release triglycerides into the bloodstream. If you were to leap up now and run a few miles, your glucagon and cortisol levels would soar, causing your body to release lots of stored energy.[6]

Stepping back from the details, the bottom line is that your body functions like a fuel bank, storing energy after you eat food and withdrawing energy for use during times of need. This exchange, which is mediated by hormones, occurs through an endless flux of fat and carbohydrates to and from the liver, fat cells, muscles, and other organs. Humans, like other animals, are therefore marvelously adapted to remaining active even during long periods of negative energy balance. You can hunt and gather on an empty stomach. Remember, however, that your body stores only a modest supply of glycogen, which you burn primarily when you need energy soon or fast. You therefore store the vast majority of surplus energy as fat, which you burn slowly to get lots of sustained energy. Consequently, when you don't have enough food to keep your weight constant (maintain energy balance), you can survive for weeks or months if you slowly burn your fat reserves and reduce activity levels. In fact, when glycogen levels in the liver fall too much, your body automatically switches to burning mostly fat (and, if needed, some protein) to keep feeding your brain, which has no energy stores of its own.

Until recently, most people regularly endured long periods of negative energy balance. Being hungry was normal. Even though one in eight people today faces a shortage of food, billions of others now face the evolutionarily unusual circumstance of never wanting for food. This embarrassment of riches can be a problem because consuming more calories than you expend over long periods of time causes your body to store additional fat. But it is much more complicated because much of that food (including that piece of cake) is highly processed to contain copious quantities of sugar and fat, and to remove the fiber. Although this processing enhances tastiness, it creates a double whammy for your body. Not only are you getting more calories than you need, but the lack of fiber causes you to absorb the calories faster than your liver and pancreas can handle them. Our digestive systems never evolved to burn that much sugar that fast, and they respond in the only way they can: shuttling much of the excess sugar into visceral fat. A little visceral fat is fine, but unfortunately too much causes a suite of symptoms collectively known as metabolic syndrome. These symptoms include high blood pressure, high levels of triglycerides and glucose in your blood, too little of a protein called HDL (often known as good cholesterol), and too much of another protein called LDL (bad cholesterol). Having three or more of these symptoms strongly increases the risk of many illnesses, the biggest being cardiovascular disease, type 2 diabetes, reproductive tissue cancers, digestive tissue cancers, and diseases of the kidney, gallbladder, and liver.[7] Since obesity is a major risk factor for metabolic syndrome, having a high body-mass index (BMI, weight relative to height) increases the risk of dying from these diseases.[8] If your BMI exceeds 35, you have a 4,000 percent greater chance of developing type 2 diabetes and about a 70 percent greater chance of getting heart disease than if you have a healthy BMI of 22.[9] These probabilities, however, are altered by physical activity and other factors, including your genes and how much of your fat is visceral or subcutaneous.

With this information under our belt, let's now address why humans today are so liable to gain weight when they have extra energy, why losing weight is so difficult, and why different diets have varying effects on the ability to gain or lose weight.

Why Are We So Prone to Plumping?

From a primate's perspective, all humans—even skinny people—are relatively fat. Other primates generally average about 6 percent body fat when they are adults, and their infants are born with about 3 percent body fat, but the percentage of body fat among human hunter-gatherers is typically 15 percent in newborns, rises to about 25 percent during childhood, and then falls to about 10 percent in males and about 20 percent in females.[10] From an evolutionary perspective, having lots of fat makes sense for the reasons discussed in chapter 5. Briefly, humans have voluminous brains that require an unceasing supply of plentiful energy, about 20 percent of resting metabolism. Human babies therefore benefit from ample fat reserves to ensure that they can always feed their big brains. On top of this demand, human mothers wean their children at a relatively young age and thus have to feed not just their own big-brained bodies but also their big-brained infants, as well as other, older, even bigger-brained children. Just producing milk requires a mother to spend 20 to 25 percent more calories a day, and her milk still needs to flow during times when she lacks enough food.[11] A mother's reserve of body fat is thus a critical insurance policy to help her children survive and thrive. Finally, hunting and gathering requires traveling long distances every day, often when hungry. Hunter-gatherers thus profit immensely from having plentiful energy reserves to forage and feed their kids during inevitable times when they don't have enough food to maintain a constant body weight. Having a few pounds of extra body fat can make the difference between life and death, strongly affecting reproductive success.

During the evolution of the human genus, natural selection favored humans with more body fat than other primates, and since fat is so critical for reproduction, natural selection particularly shaped women's reproductive systems to be exquisitely tuned to their energetic status, especially to changes in energy balance.[12] When she is pregnant, a woman must consume enough calories to nourish herself and her fetus, and after she gives birth, she must produce lots of milk, which is energetically expensive. In subsistence economies, where food is limited and people are very physically active, potential mothers are less likely to conceive when they are losing weight. If a normal-weight woman loses even one pound over the course of a month, her ability to get pregnant declines considerably in the subsequent month. Because women who have stored up more energy as fat are more likely to have more surviving offspring, natural selection favored 5 to 10 percent more body fat in women than men.[13]

The bottom line is that fat is vital for all species, but especially for humans. The evolutionary importance of human body fat has given rise to many theories about why humans so easily become obese and get metabolic diseases like diabetes and why some people have a greater susceptibility to these illnesses than others. The first of these theories, still invoked, is the thrifty genotype hypothesis, proposed by James Neel in 1962.[14] This landmark paper reasoned that natural selection during the Stone Age favored thrifty genes that gave their owners a propensity to store as much as fat as possible. Since farmers have more food than hunter-gatherers and might benefit from losing these genes, Neel predicted that populations that started farming more recently are more likely to retain thrifty genes. These individuals are therefore more mismatched with modern environments with plentiful energy-rich food. The thrifty genotype hypothesis is often invoked to explain why populations such as South Asians, Pacific Islanders, and native Americans who recently started to eat Western diets are especially susceptible to obesity and diabetes. One well-studied group is the Pima Indians, who live along the border between Mexico and the United States. Whereas approximately 12 percent of adult Pima living in Mexico have type 2 diabetes, more than 60 percent of Pima living in the United States have the disease.[15]

Neel was right that humans generally have a thrifty genotype that allows us to store fat readily, but decades of intensive research have not supported many of the thrifty genotype hypothesis's predictions. One problem is that a number of thrifty genes have been identified, but none appear to be more prevalent in populations such as the Pima, and these genes do not seem to have strong effects.[16] Genes do matter, but diet and physical activity are far more potent predictors of obesity and illness. A second problem with the thrifty genotype hypothesis is that there is little evidence for regular famines during the Stone Age. Hunter-gatherers rarely have massive food surpluses, but they seldom run out of food either, and their body weights fluctuate only modestly between seasons.[17] As chapter 8 reviewed, famines became much more common and severe *after* farming began. One would thus expect thrifty genes to be more prevalent in populations that started farming earlier rather than later. The evidence also fails to support this prediction. Although some populations with high rates of obesity and metabolic syndrome, such as Pacific Islanders, adopted agriculture fairly recently, others such as South Asians did not. Instead, the most common characteristics of at-risk populations is that they tend to be economically poor and eat cheap, starchy foods, they transitioned to these diets very recently, and they lack genes that protect them from becoming insensitive to insulin (see below).[18]

An important alternative explanation for these and other data is the thrifty phenotype hypothesis, proposed by Nick Hales and David Barker in 1992.[19] The basis for this idea is the observation that babies born with low birth weights are much more likely to become obese and develop symptoms of metabolic syndrome when they are adults. A well-studied example is the Dutch famine, which lasted from November 1944 until May 1945. People who were in utero during this intense famine had significantly higher rates of health problems as adults, including heart disease, type 2 diabetes, and kidney disease.[20] Rodents experimentally subjected to energy deprivation in the womb have similar outcomes. These effects make sense from both developmental and evolutionary perspectives. If a pregnant mother does not have enough energy, her unborn child adjusts by growing smaller with less muscle mass, fewer pancreatic cells that make insulin, and smaller organs, such as kidneys. Such smaller individuals are then adapted to coping with an energy-poor environment not only in the womb but also after they are born. However, these individuals are less well suited to coping with an energy-rich environment as adults because they develop thrifty features, such as a propensity to store abdominal fat.[21] In addition, because they have smaller organs, they have less capacity to deal with the metabolic demands of a surfeit of energy-rich foods.[22] Consequently, when low birth weight babies grow up to be short, thin adults, they tend to be healthy, but if they become big and tall, they are at higher risk of metabolic syndrome.[23] The thrifty phenotype hypothesis therefore explains why adaptations for energy-poor environments make people more susceptible to mismatch illnesses in energy-rich environments.

The thrifty phenotype hypothesis is an important idea because it considers how genes and environments interact during development to mold the body, and it accounts for the prevalence of metabolic syndrome among low birth weight babies and perhaps among small-bodied populations. But the thrifty phenotype hypothesis doesn't explain why so many children born to healthy or overweight mothers also develop diseases of affluence. Most of the people in developed countries who develop metabolic syndrome were not small at birth. Instead these individuals were born with high birth weights (especially from an evolutionary perspective of what is normal), and rather than developing thrifty phenotypes, they develop prodigal phenotypes. By this I mean that children with high birth weights are large primarily because they have lots of body fat, often twice what used to be normal. Long-term studies show that such babies typically have healthy outcomes if they do not stay overweight, but they have much higher chances of developing metabolic syndrome if they continue to gain a disproportionate amount of weight as they mature.[24]

Putting the evidence together, the key point is that excessive weight gain relative to height during childhood is a strong risk factor for future diseases associated with metabolic syndrome. A major reason that overweight children have a propensity to become overweight or obese adults is that they develop and then retain for life more fat cells than average-weight children. Crucially, these extra fat cells are often inside the abdomen, packed around organs such as the liver, kidneys, and intestines. These visceral (belly) fat cells behave differently than fat elsewhere in the body in two important ways.[25] First, they are several times more sensitive to hormones and thus tend to be more metabolically active, which means they are capable of storing and releasing fat more rapidly than fat cells in other parts of the body. Second, when visceral cells release fatty acids (something fat cells do all the time), they dump the molecules almost straight into the liver, where the fat accumulates and eventually impairs the liver's ability to regulate the release of glucose into the blood. An excess of belly fat (a paunch) is therefore a much greater risk factor for metabolic disease than a high BMI.[26]

Although we still don't understand why some people store fat more readily than others, it is uncontroversial to state that all humans are adept at storing extra energy as fat and that all of us inherited trade-offs in the ways we use energy to grow and reproduce that did not adapt us to thrive in conditions of too much energy. However, if you look at any graph of obesity rates over the last few decades, it is evident that the percentage of overweight people has remained constant while the percentage of obese people started rising rapidly in the 1970s and 1980s. What changed?

How and Why Are We Getting Fatter?

The most widespread, partly true, yet overly simplistic explanation for why more people than ever are getting fatter is that more people than ever are eating more and being less active. As chapter 9 described, there is plenty of evidence that food industrialization over the last few decades has increased portion sizes and made food denser in calories. Other industrial "advances," such as the proliferation of cars and labor-saving devices, as well as more sitting, cause people to be less active. If you add up how many extra calories people consume and how many fewer they expend, then you get larger energy surpluses, which translate into more fat.

The "calories in versus calories out" explanation for the obesity epidemic is not entirely wrong, but the situation is more complicated because we have also changed *what* we are eating. Remember that energy balance is regulated by hormones, especially insulin. Insulin's chief function is to shuttle energy from the food you have digested into your body's cells. It bears repeating that insulin rises when blood glucose levels rise, causing muscle and fat cells to take up and store some fraction of that sugar as fat. Insulin also causes fat (triglycerides) in the bloodstream to enter fat cells and simultaneously inhibits fat cells from releasing triglycerides back into the bloodstream.[27] Insulin thus makes you fatter, regardless of whether the fat comes from eating carbohydrates or fat. According to some estimates, twenty-first-century adolescents in the United States secrete far more insulin than their parents produced when they were the same age in 1975.[28] It's no wonder more of them are overweight. Since insulin rises only after you eat foods that contain glucose, one obvious culprit for higher levels of insulin and more fat must be eating more glucose-rich foods, such as soda and cake. There are, however, many other factors that promote obesity, including two additional factors related to sugar. One is the rate at which you break foods down into glucose, which determines how quickly your body produces insulin. The other factor, which is more indirect, is how much fructose you eat, and how fast it hits your liver.

To explore these effects of sugar on obesity let's compare how your body responds to eating a raw apple that weighs 100 grams (3.5 ounces) and a 56 gram (2 ounce) pack of fruit rolls that once upon a time were apples but then were industrially processed with sugar added for sweetness and any fiber removed (along with the apple's nutrients) to improve the product's shelf life. If we focus only on the sugar, one major difference evident between these two foods is that the apple has about 13 grams (a bit less than half an ounce) of sugar, whereas the fruit rolls have been packed with 21 grams (three quarters of an ounce) of sugar, hence nearly twice the calories. A second difference is the percentage of sugar types. The apple is about 30 percent glucose: the fruit roll is about 50 percent glucose. So eating the fruit rolls yields about the same amount of fructose and more than twice the glucose. Finally, the apple comes with a skin, and the apple's sugar resides within cells, both of which contain fiber. Fiber, also known as roughage, is the portion of the apple you cannot digest, but it plays a crucial role in how you digest the apple's sugars. Fiber makes up the walls of the cells that encase the sugars in the apple, slowing the rate at which you break down carbohydrates into sugars. Fiber also coats the food and the walls inside your gut, functioning as a barrier to slow the rate at which your intestine transports all those calories, especially the sugar, from your gut to your bloodstream and organs. Finally, fiber speeds the rate at which food passes through your gut, and it makes you feel full. As a result, when we compare the two apple products, the real apple not only supplies less sugar, but it makes you feel more sated and causes you to digest those sugars at a much more gradual rate. In contrast, the fruit rolls are termed *high glycemic* because they rapidly and markedly elevate blood sugar levels (a condition known as hyperglycemia).[29]

It is possible to get fat by eating too many apples, but you now have enough information to appreciate why the fruit roll is so much more likely to cause weight gain. Most obviously, the fruit roll has more calories. A second problem is the rate at which you get those calories. When you eat the apple, your insulin levels rise, but they rise gradually because the apple's fiber slows the rate at which you extract the glucose. As a result, your body has plenty of time to figure out how much insulin to make to keep your blood glucose levels steady. In contrast, the fruit roll's double load of glucose passes rapidly into your bloodstream, causing your blood sugar levels to skyrocket, in turn causing your pancreas to frantically pump out lots of insulin, often too much. This overshoot commonly causes your blood sugar levels to subsequently plummet, and you then become ravenous, causing you to crave more fruit rolls or other calorie-dense foods to raise your blood sugar quickly back to normal again. Put simply, foods rich in rapidly digested glucose supply lots of calories and make you hungrier sooner. People who eat meals with a higher percentage of calories from protein and fat are less hungry for longer and thus eat less food overall than people whose calories come mostly from sugary and starchy foods.[30] Less processed food with more fiber also induces hunger less quickly because the food remains longer in the stomach, which releases appetite-suppressing hormones.[31]

Glucose, however, is not the whole story, and the other sweet elephant in the room (or apple) is fructose. It has become common (sometimes justifiably) to demonize fructose, in large part because the invention of high fructose corn syrup has made sugar ridiculously cheap and abundant. But I hope you noted that the apple and the fruit roll contain about the same dose of fructose. In fact, chimps eat a diet of almost entirely fruit, so they must digest lots of fructose. Yet they and other fruit lovers don't get fat. Why is the fructose in raw fruit less likely to promote obesity than the fructose in the processed fruit or other fructose-laden foods like soda and fruit juice?

The answer again has to do with the combination of the quantity and the rate at which the fructose is handled by the liver. In terms of quantity, one factor is domestication. Most of the fruits we eat today have been heavily domesticated to be much sweeter than their wild progenitors. Until recently, most apples were like crab apples and had considerably less fructose. In fact, almost all the fruits our ancestors ate were about as sweet as carrots—hardly a food that promotes obesity. Even so, domesticated fruits are not pumped up on fructose compared to processed foods like fruit rolls and apple juice, and they also contain lots of fiber, which as we have discussed is removed from many industrial foods. Because of fiber, a raw apple's fructose is digested gradually and thus arrives more slowly at the liver. As a result, the liver has plenty of time to cope with the apple's fructose and can readily burn it at a leisurely pace. However, when processed foods deluge the liver with too much fructose too quickly, the liver is overwhelmed and converts most of the fructose into fat (triglycerides). Some of this fat fills up the liver, causing inflammation, which then blocks the action of insulin in the liver.

This sets off a harmful chain reaction: the liver releases its stores of glucose into the bloodstream, which in turn drives the pancreas to release more insulin, which then shuttles the extra glucose and fat into cells.[32] The rest of the fat the liver produces from rapid doses of fructose gets dumped into the bloodstream, where it, too, ends up in fat cells, your arteries, and other potentially bad places.

If fructose sounds dangerous, it can be, but only in fast and large doses. For most of human evolution the only big, rapidly digestible source of fructose that our ancestors could acquire was honey. As chapter 9 described, gargantuan, cheap quantities of fructose first became available in the 1970s because of high fructose corn syrup. Before World War I, the average American consumed about 15 grams of fructose (half an ounce) a day, mostly from eating fruits and vegetables that surrender the fructose slowly; the average American today consumes 55 grams (almost 2 ounces) per day, much of it from soda and processed foods made with table sugar.[33] All in all, the chief reason why more people are getting fatter, especially in our bellies, is that processed foods are supplying them with too many calories, many from sugar—both glucose and fructose—in doses that are both too high and too rapid for the digestive systems we inherited. Although we evolved to eat plenty of carbohydrates and to store them efficiently, we are not well adapted to consume them so plentifully in the raw form that is found in sweet beverages like soda and juice (yes, fruit juice is a junk food), as well as cake, fruit rolls, candy bars, and countless other industrial foods. The problems caused by industrial diets explain why many traditional diets that evolved independently in different farming societies around the globe all seem to do a good job of preventing weight gain. Classic Asian and Mediterranean diets, for example, seem to have little in common and both include plenty of starch (rice, or bread and pasta), yet both cuisines incorporate lots of fresh vegetables that contain fiber, and both are rich in protein as well as healthy fats, such as fish and olive oil (more on fats later). These diets also tend to be rich in other health-providing nutrients (another important topic). In short, it is harder to become overweight and easier to keep weight off if you get your carbohydrates from an old-fashioned, commonsensical diet with lots of unprocessed fruits and vegetables.[34]

Diet plays a dominant role in explaining why more people around the globe are getting fatter, but there are several additional factors that are also important: genes, sleep, stress, the bacteria in your gut, and exercise.

First: genes. Wouldn't it be nice if we found a gene that causes obesity? If so, we could devise a drug to turn the gene off and solve the problem. Unfortunately, no such gene exists, but since every aspect of the body derives from interactions between genes and the environment, it should be no surprise that dozens of genes have been identified that do increase people's susceptibility to gaining weight, mostly by affecting the brain.[35] The most potent gene so far discovered, *FTO,* affects how the brain regulates appetite. If you have one copy of this common gene, the chances are you weigh an average of 1.2 kilograms (2.6 pounds) more than someone without the gene; if you are unlucky enough to have two copies you are probably 3 kilograms (6.6 pounds) heavier.[36] Carriers of the *FTO* gene struggle a little more to control their appetite, but they are otherwise no different from noncarriers when they try to lose weight by exercising or dieting.[37] Moreover, *FTO* and other genes associated with being overweight long precede the recent rise in human obesity. Weight-gaining genes did not sweep through the human species in the last few decades. Instead, for thousands of generations, almost all the people who carried these genes had normal body weights, emphasizing that what has most changed are environments, not genes. It follows that if we are to quell this epidemic, we need to focus not on genes but on environmental factors.

And boy have our environments changed in more ways than diet. As noted in chapter 9, one major realm of change is that we are more stressed and we sleep less—two related factors that contribute to weight gain in pernicious ways. The word "stress" has negative connotations, but stress is an ancient adaptation to save you from dangerous situations and to activate energy reserves when you need them. If a lion roars nearby, a car nearly runs you over, or you go for a run, your brain signals your adrenal glands (which are on top of your kidneys) to secrete a small dose of the hormone cortisol. Cortisol doesn't *make* you stressed; it is released *when* you are stressed. Among its many functions, cortisol gives you needed, instant energy: it causes your liver and fat cells, especially visceral fat cells, to release glucose into the bloodstream, it increases your heart rate and increases your blood pressure, and it makes you more alert and inhibits sleep. Cortisol also gets you ready to recover from stress by making you crave energy-rich foods. All in all, cortisol is a necessary hormone that helps keep you alive.

Stress, however, has a dark and fattening side when it doesn't abate. One of the problems of chronic, long-term stress is that it elevates cortisol levels for extended periods of time. Many hours, weeks, and even months of too much cortisol is harmful for many reasons, not the least of which is by promoting obesity through a vicious circle that works as follows: First, cortisol causes you not only to release glucose but also to crave calorie-rich food (this is why stress makes you yearn for comfort food).[38] As you now know, both responses elevate your insulin levels, which then promote fat storage, especially in visceral fat, which is about four times more sensitive to cortisol than subcutaneous fat.[39] To make matters worse, constantly high levels of insulin also affect the brain by inhibiting its response to another important hormone, leptin, which fat cells secrete to signal satiety. As a result, the stressed brain thinks you are starving, so it activates reflexes to make you hungry while simultaneously activating other reflexes to make you less active.[40] Finally, as long as the environmental causes of stress remain (your job, poverty, commuting, and so on), you keep on secreting too much cortisol, which then leads to too much insulin, which then increases appetite and decreases activity. Another vicious circle is sleep deprivation, which is sometimes caused by elevated levels of stress, hence high levels of cortisol, but which then increases cortisol. Insufficient sleep also elevates levels of yet another hormone, ghrelin. This "hunger hormone" is produced by your stomach and pancreas and stimulates appetite. Numerous studies find that people who sleep less have higher ghrelin levels and are more likely to be overweight.[41] Apparently, our evolutionary history did not adapt us well to cope with relentless, endless stress and sleep deprivation.

We also were never adapted to be physically inactive, but the relationship between exercise and obesity is often misunderstood, sometimes grievously so. If you were to leap up right now and jog three miles, you'd burn about 300 calories (depending on your weight). You might think that these extra spent calories will help you lose weight, but numerous studies have shown that regular moderate to vigorous exercise leads only to modest reductions in weight (typically 2 to 4 pounds).[42] One explanation for this phenomenon is that burning an additional 300 calories a few times a week amounts to a relatively small number of calories compared to your body's overall metabolic budget, especially if you are already overweight. What's more, exercise stimulates hormones that temporarily suppress appetite but also stimulate other hormones (like cortisol) that make you hungry.[43] So if you run 16 kilometers (10 miles) a week, you'll lose weight only if you can manage to override the natural urge to eat or drink an additional 1,000 calories (about two or three muffins) that keep you in energy balance.[44] In addition, some forms of exercise replace fat with muscle, leading to no net loss of weight (albeit in a healthy manner). Being physically active may not help you shed pounds easily, but it does help you avoid gaining weight. One of the most important mechanisms of physical activity is to increase the sensitivity of muscle but not fat cells to insulin, causing fat uptake in your muscles rather than your belly.[45] Physical activity also increases the number of mitochondria that burn fat and sugar. These and other metabolic shifts help explain why very active people can eat so much with no seeming ill effects.

A final environmental factor, barely explored, is that we are not the only organisms dining on the food we eat. Your intestine is filled with billions of microbes (your microbiome) that digest proteins, fats, and carbohydrates, provide enzymes that help you absorb calories and certain nutrients, and even synthesize vitamins. They are as natural and critical a part of your environment as the plants and animals you observe every day. There is good evidence that dietary shifts as well as the use of broad-spectrum antibiotics may contribute to obesity by abnormally altering people's microbiomes.[46] In fact, one of the reasons industrially raised animals are given antibiotics is to promote weight gain.

No matter how you look at it, humans are adapted to store fat, lots of it, but mostly subcutaneous fat. An evolutionary perspective on human metabolism also helps explain why it is so difficult for overweight people to lose weight. Consider that people who are overweight or obese are not in positive energy balance if they are not still gaining weight. They are in neutral energy balance just as much as a skinny person. If they go on a diet or start to exercise more, which means eating fewer calories than they expend, they inevitably become hungry and tired, activating primal urges to eat more and exercise less. Hunger and lethargy are ancient adaptations. There was probably no time in our evolutionary history when it was adaptive to ignore or override hunger. But that doesn't mean we are adapted to being too fat. As we will see below, some people manage to be overweight and fit, but obesity, especially from too much visceral fat, is associated with metabolic diseases such as type 2 diabetes, cardiovascular disease, and reproductive cancers. Why? And how do the ways we treat the symptoms of those diseases sometimes contribute to dysevolution?

Type 2 Diabetes: A Preventable Disease

One of my grandmothers suffered from type 2 diabetes for decades and considered sugar to be a form of poison on a par with deadly nightshade. To teach my brother and me about its dangers, she kept a sugar bowl on the kitchen table as a lure and then scolded us whenever we dared to sweeten our tea or breakfast cereal. My grandmother's attitude made a certain degree of sense given that too much sugar in the blood—the diagnostic hallmark of diabetes—is toxic to tissues throughout the body. But as a child I paid no heed to my grandmother's warnings. Everyone else I knew, including my other grandparents, consumed plenty of sugar and none of them had diabetes.

Diabetes is actually a group of diseases, all of which are characterized by the inability to produce enough insulin. Type 1 diabetes, which mostly develops in children, occurs when the immune system destroys cells in the pancreas that make insulin. Gestational diabetes arises occasionally during pregnancy when a mother's pancreas produces too little insulin, giving both her and the fetus a dangerous, prolonged sugar rush. My grandmother had the third and most common form of the disease, type 2 diabetes (also called adult onset diabetes or diabetes mellitus type 2), which is the focus of this discussion because it is a formerly rare mismatch disease associated with metabolic syndrome that is now one of the fastest growing diseases in the world. Between 1975 and 2005, the worldwide incidence of type 2 diabetes increased by more than sevenfold, and the rate keeps accelerating, not just in developed countries but also in developing nations.[47] My grandmother was partly right that type 2 diabetes is caused by too much sugar, but it is also caused by too much visceral fat and by too little physical activity.

At a fundamental level, type 2 diabetes begins when fat, muscle, and liver cells become less sensitive to the effects of insulin. This loss of sensitivity, known as insulin resistance, triggers a perilous feedback loop. Normally, after you eat a meal, your blood glucose levels rise, causing the pancreas to produce insulin, which then directs liver, fat, and muscle cells to pull the glucose out of the bloodstream. But if these cells fail to respond adequately to the insulin, then blood glucose levels will stay high (or keep rising if you eat more), stimulating the pancreas to respond by producing even more insulin to compensate. A type 2 diabetic thus suffers from high levels of blood sugar, which cause a frequent need to urinate, excessive thirst, blurry vision, palpitations, and other problems. During the disease's early stages, diet and exercise can reverse or halt its progression, but if the feedback loop continues for a long time, insulin resistance gradually intensifies throughout the body, and the pancreas cells that synthesize insulin become exhausted from overwork. Eventually, these cells cease to function, requiring patients with advanced type 2 diabetes to get regular insulin injections to keep blood sugar levels in check and to avoid heart disease, kidney failure, blindness, loss of sensation in the limbs, dementia, and other horrid complications. Diabetes is a leading and costly cause of death and disability in many countries.

Type 2 diabetes is a distressing disease because of the suffering it causes and a frustrating disease because it is mostly avoidable, used to be rare, and is now considered a nearly inevitable consequence of affluence—a by-product of the epidemiological transition discussed in chapter 9. In fact, we already know how to prevent the majority of cases and even how to cure the disease in its early stages. In the search for treatments, many medical scientists focus on ways to help diabetics cope with their illness and on why some but not others get the disease. These are key questions, but there has been less serious consideration of how to prevent the disease from occurring in the first place. How can an evolutionary perspective inform this issue?

To evaluate these questions, let's look at interactions between genes and environmental factors that cause insulin resistance, the fundamental cause of type 2 diabetes. As we have repeatedly discussed, blood glucose levels rise as you digest a meal, providing fuel for your cells to burn. In order to get the glucose from the blood into each cell, the glucose needs to be conveyed across the cell's outer memberane by special proteins called glucose transporters, which are present in nearly every cell in the body. Glucose transporters on liver and pancreas cells are passive and simply let glucose flow in freely, the way small particles pass through a sieve. However, glucose transporters on fat and muscle cells won't let any glucose molecules into a cell unless insulin binds to nearby receptors. As figure 23 shows, when an insulin molecule binds to one of these receptors, a cascade of reactions occurs inside the cell that causes the glucose transporter to allow blood sugar into the cell. Once inside the cell, glucose molecules are either burned quickly or converted into either glycogen or fat (also guided by insulin). In summary, under normal conditions, fat, liver, and muscle cells take up sugar whenever insulin is present, especially after a meal.

Insulin resistance can happen in many different kinds of cells, including in muscles, fat, the liver, and even the brain. Although the precise causes of insulin resistance are incompletely understood, insulin resistance in muscle, fat, and liver cells is strongly associated with high levels of triglycerides from excess visceral fat. Most notably, people with abundant visceral fat, especially fatty livers, and whose diets lead to high triglyceride levels in the blood have a significantly greater risk of developing insulin resistance.[48] In practical terms, apple-shaped people, who mostly store belly fat around the abdomen, tend to be at a higher risk for diabetes than pear-shaped people, who mostly store fat on the bottom or thighs. In fact, some people who get insulin resistance are not overtly obese (they have normal BMIs), but they do have fatty livers (and are described as TOFI, thin outside, fat inside).[49] As we have already

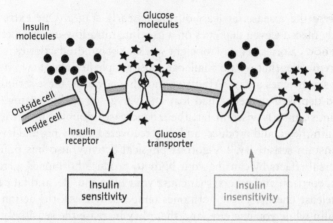

FIGURE 23. How insulin affects glucose uptake in cells. Muscle, fat, and other cell types have insulin receptors that are located near glucose transporters on the cell surface. Normally, insulin in the bloodstream binds to the insulin receptor, which then signals the glucose transporter to admit glucose. During insulin resistance (as shown on the right), the insulin receptor becomes insensitive, preventing the glucose transporter from taking in glucose, leading to elevated levels of blood sugar.

seen, the biggest contributors to fatty livers and other forms of visceral fat are foods with large quantities of rapidly digestible glucose and fructose, often from lots of high fructose corn syrup or table sugar (sucrose). In this regard, soda, juice, and other sugary foods with lots of fructose and no fiber are especially dangerous because the liver readily converts most of the fructose into triglycerides, which build up in the liver and are also dumped straight into the bloodstream.[50] Lack of physical activity and diets that are low in unsaturated fats also contribute to visceral fat, hence insulin resistance (more on these factors below).

Recognizing that excess visceral fat provokes insulin resistance, which in turn underlies type 2 diabetes, explains why this mismatch disease is almost entirely preventable and why several interrelated factors cause some people to get the disease and others to avoid it. You cannot control two of these factors: your genes and your prenatal environment. But you have some degree of control over the other two, more important factors that determine your energy balance: diet and activity. In fact, several studies have shown that losing weight and exercising vigorously can sometimes actually *reverse* the disease, at least during its early stages. One extreme study placed eleven diabetics on a grueling ultra-low-calorie diet of just 600 calories per day for eight weeks. Six hundred calories is an extreme diet that would challenge most people (it's about two tuna fish sandwiches a day). After two months, however, these seriously food-deprived diabetics had lost an average of 13 kilograms (27 pounds), mostly visceral fat, their pancreases doubled how much insulin they could produce, and they recovered nearly normal levels of insulin sensitivity.[51] Vigorous physical activity also has potent reversal effects by causing your body to produce hormones (glucagon, cortisol, and others) that cause your liver, muscle, and fat cells to release energy. These hormones temporarily block the action of insulin while you exercise, and then they increase the sensitivity of these cells to insulin for up to sixteen hours following each bout of exercise.[52] When obese adolescents with high levels of insulin resistance are enticed to exercise moderately (thirty minutes a day, four times a week, for twelve weeks), their insulin resistance decreases to nearly normal levels.[53] Stated simply, increasing levels of physical activity and decreasing levels of visceral fat can reverse early type 2 diabetes. In one remarkable study, ten middle-aged, overweight Australian Aborigines with type 2 diabetes were asked to return to a hunting and gathering lifestyle. After seven weeks, the combination of diet and exercise had reversed the disease almost completely.[54]

More research is needed on the long-term effects of diet and exercise interventions on type 2 diabetics, but these and other studies beg the question, Why aren't we more successful at following prescriptions for vigorous physical activity and better diets to prevent the disease from arising or progressing? The biggest problem, of course, is the environment we have created. Because of industrialization, the cheapest and most abundant foods are low in fiber and rich in simple carbohydrates and sugar, especially high fructose corn syrup—all of which promote obesity, especially visceral obesity, hence insulin resistance. Robert Lustig and colleagues have found that for every 150 calories of increased sugar consumption per day, the prevalence of type 2 diabetes increases 1.1 percent after correcting for factors such as obesity, physical activity, and alcohol use.[55] Cars, elevators, and other machines have reduced physical activity levels, compounding the problem. By the time people become overweight or obese, let alone contract type 2 diabetes, it is difficult, expensive, and time-consuming to change their diet and exercise habits.

A secondary problem may be the way we treat the disease. Many doctors never see patients until they are sick, at which point they have little choice but to take what is widely considered a sensible two-pronged approach to treating the illness. First, they encourage patients to increase physical activity and to reduce their calorie intake, especially by avoiding too much sugar, starch, and fat. At the same time, most physicians also prescribe medications, allowing patients to counter the disease's symptoms. Some popular anti-diabetes drugs improve insulin sensitivity in fat and liver cells, some drugs increase pancreatic cells' abilities to synthesize insulin, and others block the gut from absorbing glucose. Although these medicines can keep the symptoms of type 2 diabetes at bay for years, many of them have nasty side effects, and they are only partially effective. A large study that compared the effectiveness of the most popular drug, metformin, with a lifestyle intervention in more than three thousand people found that changing diet and exercise was almost twice as effective and had more long lasting effects.[56]

Viewed in this light, type 2 diabetes is a case of dysevolution, in which the disease is increasing in prevalence from one generation to the next because we are not preventing its causes. The illness is first and foremost a mismatch disease that is rapidly becoming more common as chronic energy surpluses contribute over many years to obesity, especially visceral obesity, and insulin resistance. Although good old-fashioned diet and physical activity are by far the best ways to prevent and treat type 2 diabetes, too many people wait until they experience the disease's symptoms before acting. Some diabetics cure themselves through drastic changes in diet and exercise; others are too infirm to exercise vigorously or alter their diet very much; most diabetics combine medications with moderate shifts in diet and exercise, thus managing the disease for decades. To some extent, this approach makes sense for many people because it is pragmatic, framed by the immediate needs and capabilities of those who may not be able to embark on a course of drastic exercise and dieting. Further, after years of futilely trying to help patients to lose weight and exercise more, many doctors have become pessimistic (or realistic) and suggest only modest goals for weight loss and exercise because more extreme prescriptions tend to fail and then backfire. Unfortunately, resigning ourselves to primarily managing the disease's symptoms in a growing number of people perpetuates this unfortunate cycle. To make matters worse, many people who cope with diabetes also suffer from other related diseases, the most prevalent of which is heart disease.

The Silent, Inflammatory Killer

Most of the time, even when exercising, we pay little attention to our hearts. They just pump away, forcing blood to and from the lungs and through every artery and vein. About one-third of us, however, will die because our circulatory systems deteriorate gradually and silently over decades. Some forms of heart disease, like congestive heart failure, can kill you very slowly, but the most common cause of death from cardiovascular disease is a heart attack. Often, this crisis starts with tightness in the chest, pain in the shoulders and arm, nausea, and shortness of breath. Without prompt treatment these symptoms intensify into searing pain before a loss of consciousness and then death. A related kind of killer is a stroke. You can't feel when a vessel bursts in your brain, but all of a sudden, you may experience a headache, a part of your body becomes weak or numb, and you become confused and unable to speak, think, or function.

At a proximate level, heart attacks and strokes occur from what seems to be an obvious design flaw in the circulatory system. The heart and brain, like other tissues, are supplied with extremely narrow vessels that deliver oxygen, sugar, hormones, and other needed molecules. As we age, their walls harden and thicken. If a clog occurs in one of the slender coronary arteries that supply the muscles of the heart, then that region dies, and the heart stops. Similarly, if one of the thousands of tiny vessels that supply the brain becomes obstructed, it bursts, killing vast numbers of brain cells. Why are these and other critical vessels so small and thus liable to get blocked? Why do strokes and heart attacks occur so often in humans? And to what extent is cardiovascular disease an example of dysevolution: a mismatch condition we allow to persist and proliferate because we too often fail to treat its causes? To answer these and other related questions, let's first consider the basic mechanisms that cause cardiovascular disease and how these diseases are mismatch conditions caused by too much energy.

A stroke or a heart attack may seem like a sudden event, but in most cases these crises are partly the end of a lengthy, gradual process of hardening of the arteries called *atherosclerosis*. Atherosclerosis is a chronic inflammation of the arterial walls that results from how you shuttle cholesterol and triglycerides (fats) around your body. Cholesterol—a much-maligned molecule—is a small, waxy, fatlike substance. All your cells use cholesterol for many vital functions, so if you don't eat enough cholesterol, your liver and intestine synthesize it readily from fat. Since neither cholesterol nor triglycerides are soluble in water, they need to be transported within the bloodstream by special proteins known as lipoproteins. This transport system is complex, but a few facts are worth knowing. First, low-density lipoproteins (LDLs, often called "bad" cholesterol) carry cholesterol and triglycerides from the liver to other organs, but they vary importantly in size and density: LDLs that carry mostly triglycerides are denser and smaller than those that carry mostly cholesterol, which are larger and more buoyant. High-density lipoproteins (HDLs, or "good" cholesterol) primarily carry cholesterol back to the liver.[57] Figure 24 diagrams how atherosclerosis starts when LDLs (especially the smaller, denser ones) get stuck on an artery wall and then react with passing oxygen molecules. They burn slowly like apple flesh turning brown.

If slowly burning the walls of your arteries sounds bad, you are right. This oxidation is one of a number of processes that cause chronic inflammation in various tissues of body that contribute to aging and to a wide range of diseases. In the case of arteries, oxidization of LDLs causes an inflammation in the cells that make up the arterial wall, which then triggers white blood cells to come and clean up the mess. Unfortunately, the white blood cells trigger a positive feedback loop because part of their response is to create a foam that traps more small LDLs, which then also get oxidized. Eventually, this foamy mixture coagulates into a stiffened accumulation of crud on the artery wall, known as a plaque. Your body fights plaques primarily with HDLs, which scavenge cholesterol from the plaque and return it to the liver. Plaques thus develop not just when LDL levels (again, mostly the small ones) are high but also when HDL levels are low. If the plaque expands, the artery's wall sometimes grows over the plaque, permanently narrowing and hardening the artery. Plaques also increase the chances of blockage or that a chunky clot of plaque will be released into the bloodstream. Floating clots are dangerous because they can get lodged in a smaller artery, often in the heart or brain, causing a blockage that leads to a heart attack or stroke. To make matters worse, when pipes narrow, higher pressures are needed to deliver the same volume of flow. A vicious circle ensues as stiffer, narrower arteries increase blood pressure (hypertension), requiring the heart to work harder and raising the chances of a clot or rupture.

The way plaques form and cause cardiovascular disease is unquestionably an example of unintelligent design. How and why did natural selection flub up so much? As you might expect for a complex disease, certain gene variants can modestly increase your risk, but the disease is mostly caused by other factors, including that inevitable enemy: age. As the years advance, damage to your arteries accumulates relentlessly, causing them to harden throughout your body. Studies of ancient mummies whose hearts and vessels were imaged with CT scans (a form of three-dimensional X-ray) confirm that this form of aging also occurred in ancient populations, including Arctic hunter-gatherers.[58] Although some degree of atherosclerosis is inevitable and certainly not novel, there is nonetheless good evidence that most forms of cardiovascular disease are partly if not largely mismatch diseases. For one, diagnoses of atherosclerosis among ancient mummies are not evidence that these individuals actually died from heart attacks, and every study (including autopsies) so far conducted on hunter-gatherers and other traditional populations confirms that, despite having some degree of atherosclerosis, they apparently do not suffer from heart attacks or other markers of heart disease, such as high blood pressure.[59] In addition, heart attacks are caused specifically by atherosclerosis in the tiny coronary arteries that feed the heart, and the incidence of

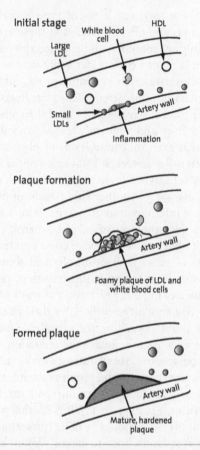

FIGURE 24. Plaque formation in an artery. First, the oxidation of low-density lipo-proteins (LDLs, usually the smaller ones, which transport mostly triglycerides) triggers an inflammation in the wall of the artery. The inflammation attracts white blood cells and a foamy plaque develops, which then narrows and hardens the artery.

coronary atherosclerosis among the scanned mummies was at least 50 percent lower than in Western populations. The most reasonable hypothesis is that, until recently, humans rarely developed enough atherosclerosis to cause heart attacks. Today, however, heart disease is rampant because of the same novel environmental conditions that contribute to rising levels of type 2 diabetes: physical inactivity, poor diet, and obesity. Added to these are new risk factors, notably drinking, smoking, and emotional stress.

The first of these factors to consider is physical activity, which is required for the cardiovascular system to grow and function properly. Aerobic activity not only strengthens your heart but also regulates how fats are stored, released, and used throughout your body, including in your liver and your muscles. Many studies have consistently found that even moderate levels of physical activity such as walking fifteen miles a week substantially raises levels of HDLs and lowers levels of triglycerides in the blood—both of which lower the risk of heart disease.[60] Another vital benefit of physical activity is to lower levels of inflammation in arteries, which as we have seen is the real culprit for atherosclerosis.[61] In general, the duration of activity appears to have more beneficial effects on these risk factors than the intensity of activity. Vigorous physical activity also lowers your blood pressure by stimulating the growth of new vessels, and it strengthens muscles in your heart and the walls of your arteries. Adults who exercise regularly nearly halve their chances of a heart attack or stroke (after correcting for other risk factors), and the more intense the exercise the greater the reduction in risk.[62] From an evolutionary perspective, these statistics make sense because the cardiovascular system expects and requires stimuli that come from physical activity to stimulate its normal repair mechanisms (more on how and why this is in chapter 11). It is normal to be vigorously active throughout life, so it should be no surprise that an absence of physical activity permits the body to accrue various kinds of pathology, including atherosclerosis.

Diet, the other major determinant of energy balance, also has potent effects on atherosclerosis and heart disease. A common opinion is that high levels of dietary fat contribute to high levels of LDLs (a.k.a. "bad" cholesterol), low levels of HDLs (a.k.a. "good" cholesterol), and high levels of triglycerides—a trio of symptoms collectively termed dyslipidemia, which means "bad fat." Consequently, most people believe that a high percentage of dietary fat is unhealthy. In reality, the extent to which fat contributes to atherosclerosis is much more complicated for a number of reasons, not the least of which is that not all fats are the same. Recall that fats contain molecules known as fatty acids that have long chains of carbon and hydrogen atoms. Differences in the structure of these chains yield alternative types of fatty acids with critically different properties. Fatty acids with fewer hydrogen atoms are unsaturated oils that are liquid at room temperature; fatty acids with a full set of hydrogen atoms are saturated fats that are solid at room temperature. After digestion, these seemingly unimportant differences matter because saturated fatty acids stimulate the liver to produce more supposedly unhealthy LDLs, whereas unsaturated fatty acids cause the liver to produce more healthy HDLs.[63] This difference underlies the general consensus that consuming diets higher in saturated fats elevates the risk of atherosclerosis, hence heart disease.[64] It also explains the clear benefits of eating unsaturated fats, especially those comprised of omega-3 fatty acids, which are common in fish oil, flaxseeds, and nuts. Diets rich in these and other foods with abundant unsaturated fatty acids have been shown to elevate HDLs and lower LDLs and triglycerides, reducing the risk factors associated with cardiovascular disease.[65] The worst of all possible fats are unsaturated fats that have been industrially converted into saturated fats under high heat and pressure. These unnatural trans fats don't go rancid (hence their use in many packaged foods), but they wreak havoc on the liver: they raise LDLs, lower HDLs, and interfere with how the body uses omega-3 fats.[66] Trans fats are essentially a form of slow-acting poison.

If you are reading this skeptically (as anyone should), you may think, Aha, but how did hunter-gatherers in Africa and elsewhere get foods containing heart-healthy fats like olive oil, sardines, and flaxseed? Weren't they eating lots of red meat? There are two answers to this question. The first is that studies of hunter-gatherer foods reveal that hunter-gatherer diets are actually dominated by unsaturated fats, including omega-3 fatty acids. These fatty acids are plentiful in seeds and nuts and also come from the meat they consume, because wild animals that eat grass and shrubs instead of corn store more unsaturated fatty acids in their muscle. The meat of grass-fed animals is leaner and five to ten times lower in saturated fats than that of corn-fed animals.[67] In addition, even though Arctic hunter-gatherers, such as the Inuit, eat large quantities of animal fat, they also eat plenty of healthy fish oils, which help to keep their cholesterol ratios in a healthy range.[68]

Another, and frankly controversial, answer is that we may have overdemonized saturated fats, which are possibly not as detrimental as the consensus view would have it. Eating saturated fats elevates LDL levels, but it has long been known and repeatedly shown that low levels of HDL are much more strongly associated with heart disease than high levels of LDL.[69] Remember also that atherosclerosis is caused by a combination of high LDL levels along with low HDL levels plus high triglyceride levels. People who eat diets high in fat but low in carbohydrates (such as the Atkins Diet) tend to have higher HDL levels and lower triglyceride levels than people who eat low fat, high carbohydrate diets.[70] As a result, people on low carbohydrate diets may be better protected from atherosclerosis than people who eat diets low in fat but high in simple carbohydrates (such diets lower LDL levels but also lower HDL and raise triglyceride levels). Another very important factor is that smaller, denser LDLs cause much more inflammation in arterial walls than larger, less dense LDLs, but diets that are high in saturated fat tend to increase the size of the less unhealthy larger LDLs.[71] Although unsaturated fats are generally healthier than saturated ones, saturated fat may not be as evil as some think.[72]

Finally, remember that not all carbohydrates in your diet are the same, and that many carbohydrates are converted into fats that, in turn, can increase your risk of atherosclerosis. As we have already discussed, foods that rapidly deliver big quantities of glucose into the bloodstream and fructose into the liver are especially lethal because they impair liver function and increase triglyceride levels in the bood. These junk foods are the ones that contribute the most to excess visceral fat, the real archenemy, because it is visceral fat that chiefly dumps into your bloodstream the triglycerides that eventually end up causing inflammation, hence atherosclerosis. For this reason, a diet that is rich in fresh vegetables and fruits, which are mostly complex carbohydrates and contain few simple carbohydrates, is unquestionably healthy. Such foods not only prevent the buildup of visceral fat but also provide antioxidants that help reduce inflammation.[73]

Leaving aside the fight over fats, other characteristics of modern lifestyles also differ from those of our ancestors in ways that contribute to atherosclerosis and heart disease. One of these is overconsumption of salt—the only rock we eat. Most hunter-gatherers get sufficient salt, about 1 to 2 grams a day, from meat, and they have few other natural sources of this mineral unless they live near the ocean.[74] Today we have salt in overabundance; we use it to preserve food, and it tastes so good that many people consume more than 3 to 5 grams a day. Excess salt, however, ends up in the blood, where it draws in water from the rest of the body. Just as more air in a balloon increases pressure, more water in the circulatory system elevates blood pressure in your arteries. Chronic high blood pressure, in turn, stresses the heart and arterial walls, which leads to damage and then to the inflammation that causes plaque formation as described earlier.[75] Chronic emotional stress has similar effects by elevating blood pressure. Another problem is too little fiber from overly processed foods. Ample digested fiber keeps LDL levels low by speeding the passage of food through the lower intestine and soaking up saturated fats.[76] And, finally, let's not forget alcohol and other drugs. Moderate alcohol consumption lowers blood pressure and improves cholesterol ratios, but overconsumption damages the liver, which then ceases to function properly to regulate fat and glucose levels. Tobacco smokers also damage their livers, elevating LDL levels, and the toxins they inhale inflame artery walls, stimulating plaque formation.

Putting the evidence together, it should be no surprise that surveys of hunter-gatherer health indicate that they are much less likely to get heart disease as they age because they are physically active and eat a naturally healthy diet. Our Paleolithic ancestors had no access to cigarettes, either. Despite a diet of plentiful meat, cholesterol levels measured in hunter-gatherers are far healthier than those of industrialized Westerners.[77] Moreover, as noted above, assessments of hunter-gatherer health both in clinical settings and from autopsies have yielded little evidence for heart disease, even in elderly individuals. These data are necessarily limited, and they don't come from randomized controlled studies, but one can only conclude that heart attacks and strokes are primarily evolutionary mismatches, largely caused by the combination of agricultural (especially industrial) diets plus sedentary lifestyles. Subsistence farmers who are very physically active also don't have much risk of these diseases, and the proclivity to fall victim to heart disease probably didn't pick up until civilization permitted the emergence of upper classes. One of the oldest known cases of atherosclerosis (revealed by a CT scan) is an Egyptian mummy, Princess Ahmose-Meryet-Amon, who died in 1550 BC.[78] This wealthy princess, the pharaoh's daughter, presumably led a coddled, sedentary life and consumed an energy-rich diet.

The Nuns' Disease

If there is any disease everyone should worry about, it is cancer. Approximately 40 percent of Americans will receive a diagnosis of cancer at some point in their lives, and about one-third of them will die from the illness, making cancers the second leading cause of death, behind heart disease, in the United States and other Western countries.[79] Cancers are an ancient problem hardly unique to humans. They can develop in other mammals, such as apes and dogs (although less frequently)[80], and some cancers have been afflicting humans for millennia. In fact, cancer was first named and described by the ancient Greek physician Hippocrates (460–370 BC). Despite its antiquity, there is little doubt that cancer is more prevalent today than in the past. The first analysis of cancer rates was published in the mid-nineteenth century by Domenico Rigoni-Stern, chief physician of the Verona Hospital.[81] Of the 150,673 Veronese deaths that Rigoni-Stern documented between 1760 and 1839, less than 1 percent (1,136) were from cancer, and of these, 88 percent were in women. Even if one assumes that Rigoni-Stern and his colleagues missed many cancer diagnoses, and that the disease's prevalence would have been higher had more Veronese lived to be older, these rates are at least ten times lower than contemporary cancer rates.

Cancer is a tricky class of diseases to understand and treat because there are many types, each with different causes. All cancers, however, start from chance mutations in some errant cell. You probably have several of these potentially lethal cells already. Fortunately, most of them will remain dormant, doing nothing, but sometimes one of them undergoes additional mutations that cause it to function abnormally, cloning without restraint to form a tumor. Even more mutations enable these cells to spread like wildfire from tissue to tissue, consuming resources meant for other cells, eventually causing organs to fail. As Mel Greaves has pointed out, cancer is actually a type of unrestrained natural selection gone awry within the body because cancers are selfish cells whose mutations give them a reproductive advantage over other, normal cells.[82] In addition, just as environmental stresses promote evolution within a population, toxins, hormones, and other factors that stress the body set up conditions that favor cancerous cells to reproduce more effectively than normal cells and to invade tissues and organs where they don't belong. Here, however, the comparison to natural selection ends because the comparative advantages of cancer cells are short-lived and ultimately counterproductive. The factors that cause mutant cells to thrive within an organism also cause their host to die, and they are rarely passed from one generation to the next. With the exception of the few cancers borne by viruses, cancer is thus a disease that recapitulates itself independently and slightly differently in almost every individual in which it occurs.

Cancers have many causes. One is simply the process of aging, which gives more time for mutations to occur, which explains why cancer risk increases with age. In addition, some cancers occur from inheriting unlucky genes that interfere with your cells' ability to repair mutations or to stop replicating.[83] Another common and widespread set of causes for cancer includes toxins, radiation, and other environmental agents that provoke potentially carcinogenic mutations. A few cancers are caused by viruses. Here, however, our focus is on cancers caused by long-term positive energy balance and obesity. These cancers of affluence are most common in reproductive organs—especially breasts, the uterus, and ovaries in women, and the prostate gland in men—but cancers of other organs such as the colon are sometimes also affected by a chronic surfeit of energy.

How and why energy balance contributes to reproductive cancers has been difficult to fathom because the causal relationship is indirect and complex. The first clues to an energy-related pathway for cancer appeared in the form of puzzling correlations between babies and breast cancer. Early physicians such as Rigoni-Stern noticed and wondered why nuns were far more likely to get breast cancer than married women (for years, breast cancer was known as the "nuns' disease"). These observations were later bolstered by large-scale studies that showed that a woman's chances of developing breast, ovarian, or uterine cancer increase significantly with the number of menstrual cycles she experiences and decrease with the number of children she bears.[84] Decades of research now indicate that cumulative exposure to high levels of reproductive hormones, especially estrogen, is a major cause of these associations. Estrogen acts widely throughout the body but is a particularly potent stimulator of cell division in a woman's breasts, ovaries, and uterus. During each menstrual cycle, levels of estrogen rise (as do other related hormones, such as progesterone), causing cells that line the wall of the uterus to multiply and enlarge in preparation for a fertilized embryo to implant. These surges also stimulate breast cells to divide. Thus, when women cycle they repeatedly experience high doses of estrogen, which cause reproductive cells to proliferate, each time increasing the chances for cancerous mutations to occur and increasing the number of copies of any mutant cells. However, when a woman becomes a mother, by getting pregnant and then nursing, she lowers her risk of breast and other reproductive tissue cancers by reducing her exposure to reproductive hormones.[85] Breast-feeding may also help flush out the lining of the mammary ducts, removing potentially mutant cells.[86]

The association between estrogen and some other estrogen-related homones with reproductive cancers highlights why these diseases are evolutionary mismatches influenced by a chronic state of positive energy balance. Remember that for millions of years natural selection favored women who devoted whatever extra energy they had toward reproduction, partly through the action of reproductive hormones such as estrogen. Natural selection, however, never geared women's bodies for coping with long-term surfeits of energy, estrogen, and other related hormones. As a result, women today are very different and vastly more at risk of developing cancer than mothers from long ago because their bodies are still functioning as they evolved to have as many surviving children as possible. The result is that women who have more energy also have a greater cumulative exposure to reproductive hormones that, in abundance, elevate the risk of cancer.[87]

Looking more closely, there are two pathways that link energy and estrogen to higher rates of reproductive cancers among women in developed countries. The first is how many menstrual cycles women experience. The average woman in countries such as the United States, England, and Japan starts menstruating when she is twelve or thirteen years old, and she continues to menstruate until her early fifties. Because she has access to birth control, she gets pregnant only once or twice over her lifetime. Further, after she gives birth, she probably breast-feeds her babies for less than a year. All told, she can expect to experience approximately 350 to 400 menstrual cycles during her life. In contrast, a typical hunter-gatherer woman starts menstruating when she is sixteen, and she spends the majority of her adult life either pregnant or nursing, often struggling to get enough energy to do so. She thus experiences a total of only about 150 menstrual cycles. Since each cycle floods a woman's body with powerful hormones, it is not surprising that reproductive cancer rates have multiplied in recent generations as birth control and affluence has spread.

The other key pathway that links chronic positive energy balances with reproductive cancers among women is through fat. Earlier, I discussed how human females are especially well adapted to store extra energy in fat cells, which collectively act as a sort of endocrine organ to synthesize estrogen that is released into the bloodstream. Obese women can have 40 percent higher estrogen levels than nonoverweight women.[88] As a result, rates of reproductive tissue cancers among women are strongly correlated with obesity after menopause. In a study of more than 85,000 American women who were postmenopausal, those who were obese had a 2.5 times greater risk of developing breast cancer than those who were not overweight.[89] These relationships explain why rising rates of many reproductive cancers closely mirror rising rates of obesity.

A relationship between energy surpluses and reproductive cancer may also apply to men, although less strongly. One of the many functions of the major male reproductive hormone, testosterone, is to stimulate the prostate gland to produce a milky fluid that helps protect sperm. Prostate glands are constantly producing this fluid.

Several studies show that lifetime exposure to high levels of testosterone increases the risk of prostate cancer, especially in men who live in developed countries and in frequent positive energy balance.[90]

Because reproductive cancers are mismatch diseases that are linked via reproductive hormones to a surfeit of energy, physical activity has potent effects on the rates of some cancers. This makes sense: the more energy your body spends on physical activity the less it can spend on pumping out reproductive hormones. Women who are physically active have estrogen rates about 25 percent lower than those who are sedentary.[91] These differences may partially account for why several studies have documented that just a few hours a week of moderate exercise substantially lowers the rates of many cancers, including those of the breast, uterus, and prostate.[92] Several of these studies have found that the more intensive the exercise, the lower the cancer risk. In one study of more than 14,000 women divided into low, moderate, and high fitness groups, those who were moderately fit had 35 percent lower rates of breast cancer, and those who were very fit had more than 50 percent lower rates of breast cancer (after controlling for age, weight, smoking, and other factors).[93]

In short, an evolutionary perspective explains why the embarrassment of riches many people now enjoy elevates their levels of reproductive hormones, which, along with birth control, then increases the likelihood that cancers will evolve in their breasts, ovaries, uteruses, and prostates. Many reproductive cancers are thus mismatch diseases that are ultimately linked to having lots of energy to spare. As economic development and processed food diets sweep the globe, more people are shifting into positive energy balance, often extremely so, driving up the percentage of women and men with reproductive tissue cancers.[94] But are these cancers examples of dysevolution? Are we making reproductive cancers worse or more prevalent by the way we treat them?

In most respects, the answer appears to be no. Although some people can lower their chances of getting reproductive tissue cancers by exercising more and eating less, the way we treat cancers seems sensible. Should I ever get a diagnosis of cancer, I suspect I will want to employ every available weapon—drugs, surgery, and radiation—to kill those mutant cells as early as possible and to prevent them from spreading throughout my body. These approaches have increased the rate of survivorship for a few types of cancer, including breast cancer. In two important respects, however, our approach to treating cancer may sometimes be dysevolutionary. The first is that cancer is more preventable than we often suppose. Reproductive cancer rates could be significantly reduced through more physical activity and changes in diet, and other types of cancers caused by the carcinogenic substances that we breathe and ingest could be reduced dramatically if we did more to regulate pollution and halt smoking. In addition, remember that cancer is basically a kind of evolution gone wild, in which mutant cells reproduce without restraint in a body. Just as treating bacteria with antibiotics sometimes creates conditions that encourages resistant strains of bacteria to evolve, treating cancers with poisonous chemicals may sometimes favor new drug-resistant cancer cells.[95] It follows that thinking about cancers from an evolutionary perspective may help us devise more effective strategies to fight the disease. One idea is to promote benign cells to outcompete harmful cancerous ones; another is to trap cancer cells by first promoting those that are sensitive to a particular chemical and then attacking them when they are in a vulnerable state. Since cancers are a kind of evolution within the body, perhaps evolutionary logic may help us find a way to better combat this scary disease.

Are Too Many Riches an Embarrassment?

Type 2 diabetes, heart disease, and reproductive tissue cancers are not the only diseases of affluence. Others include gout and fatty liver syndrome (whose name says it all). Being overweight also contributes to a host of other ailments, such as interrupted breathing during sleep (apnea), kidney and gallbladder disease, and increased chances of getting back, hip, knee, and foot injuries. As people across the globe exercise less and eat more calories, especially sugars and simple carbohydrates, these and other diseases of affluence—all mismatch diseases previously rare during human evolution—will continue to rise as they have done in recent years.[96]

How much are diseases of affluence examples of dysevolution in which we get sick from evolutionary mismatches and allow the diseases to remain prevalent or worsen by failing to treat their causes? Chapter 7 concluded with three characteristics of such mismatch diseases. First, they tend to be chronic, noninfectious diseases with multiple interacting causes that are difficult to treat or prevent. Second, these diseases tend to have a low or negligible effect on reproductive fitness. Third, the factors that contribute to these diseases have other cultural values, leading to trade-offs between their costs and benefits.

Type 2 diabetes, heart disease, and breast cancer have all these qualities. All of them are promoted by numerous complex environmental stimuli, most especially by novel diets and physical inactivity, but also by living longer, maturing younger, using more birth control, and other factors. In addition, these diseases usually don't occur until middle age, causing them to have negligible effects on how many children people have (most women diagnosed with breast cancer are in their sixties).[97] Finally, it is hard to tally up the costs and benefits of farming, industrialization, and other cultural developments that have played a major role in fostering diseases of affluence. For example, farming and industrialization have made food less expensive and more plentiful, enabling us to feed billions more people. At the same time, many of these inexpensive calories come from sugars, starch, and unhealthy fats. Can we afford to feed the world with healthy fruits and vegetables, not to mention meat from grass-fed animals? Economic forces are also factors. On the one hand, market systems have made possible many forms of progress that allow more people in the developed world to live longer, healthier lives than their grandparents did. However, not all capitalism has been beneficial for the human body, because marketers and manufacturers prey on people's urges and ignorance. For example, deceptive advertising of "fat-free" food entices people to buy calorically dense products rich in sugar and simple carbohydrates that actually make the consumer fatter. Paradoxically, it now requires more effort and money to consume food that has fewer calories. A quick glance tells me that the seemingly healthy and modest 15-ounce bottle of cranberry juice in my refrigerator contains 120 calories, but closer perusal reveals that the bottle is improbably considered to have two servings. So you actually ingest 240 calories when you drink it, as much as a 20-ounce bottle of Coca-Cola. We have also filled our environments willingly with cars, chairs, escalators, remote controls, and other devices that decrease our physical activity levels, calorie by calorie. Our environment is needlessly *obesogenic*. And at the same time, the pharmacological industry has developed a stunning array of drugs, some extremely effective, to treat the symptoms of these diseases. These drugs and other products save lives and reduce disability, but they can also be permissive and enabling. All in all, we have created an environment that makes people sick through a surfeit of energy and then keeps them alive without having to turn down the energy flow.

What do we do? The obvious, fundamental solution is to help more people eat a healthier diet and to exercise more, but this is one of the greatest challenges our species faces (and the subject of chapter 13). The other key solution is to focus more intelligently and rationally on the causes rather than the symptoms of these diseases. Having too much fat, especially visceral fat, is a health risk for many diseases and a symptom of energy imbalance, but being overweight or obese are not diseases. Most people who are overweight or obese are justifiably fed up with those who focus on weight rather than health and who stigmatize or blame obese people for being obese. The same despicable logic leads to blaming poor people for being poor. In fact, such condemnations are often linked because obesity is strongly correlated with poverty.[98]

Widespread obsession with the obesity "epidemic" has led to an understandable backlash. Some wonder if alarmists have exaggerated the problem.[99] According to this view, we not only stigmatize people unnecessarily, we also waste billions of dollars to fight an invented crisis. To some extent, the anti-alarmists have a point. Exceeding a recommended body weight is not necessarily unhealthy, as is evident from the many overweight individuals who live long and reasonably healthy lives. About one-third of people who are overweight show no sign of metabolic disturbance, perhaps because they have genes that adapt them to being heavy.[100] But as this chapter has repeatedly stressed, what matters most for health is not fat per se. Even more important predictors of health and longevity are where you store your body fat, what you eat, and how physically active you are.[101] One landmark study, which followed nearly 22,000 men of all weights, sizes, and ages for eight years, found that lean men who did not exercise had twice the risk of dying as obese men who engaged in regular physical activity (after adjusting for other factors, such as smoking, alcohol, and age).[102] Being fit can mitigate the negative effects of being fat. Therefore, a sizeable percentage of fit but overweight and even mildly obese individuals do not have a greater risk of premature death.

To understand better how and why adequate physical activity is so important for health, it's time to consider another class of mismatch conditions that are subject to dysevolution: diseases of disuse. These illnesses are caused by too little, rather than too much, of a good thing.

11

Disuse

Why We Are Losing It by Not Using It

> For unto every one that hath shall be given, and he
> shall have abundance: but from him that hath not
> shall be taken away even that which he hath.
>
> —MATTHEW 25:29

Have you ever been caught in a traffic jam on a bridge and won-dered if it is strong enough to hold the weight of all those cars and people? Imagine the chaos and horror of the bridge collapsing, plunging everyone into the river below in a deadly shower of metal, bricks, and concrete. Fortunately, this sort of accident is extraordi-narily unlikely because most bridges are built to withstand many more cars and people than they actually carry. For instance, John Roebling intentionally designed the Brooklyn Bridge to support six times more weight than he expected it would ever hold. In the par-lance of engineering, the Brooklyn Bridge has a safety factor of six.[1] We can take comfort in knowing that engineers are usually required to use similarly high safety factors when they design all sorts of important structures, like bridges, elevator cables, and air-plane wings. Although safety factors increase construction costs, they are sensible and necessary because we never really know how strong to make things.

What about your body? As anyone who has broken a bone or snapped a ligament or tendon can attest, natural selection apparently failed to give these structures a large enough safety factor to cope with some of your activities. Obviously, evolution did not adapt human bones and ligaments to resist the forces caused by high-speed car crashes and bicycle accidents, but why do so many people fracture their wrists, shins, and toes from a simple fall when walking or running? Even more concerning is the prevalence of osteoporosis, a disease in which bones gradually waste away, becoming so brittle and fragile that they crack and then collapse. Osteoporosis causes more than one-third of elderly women in the United States to fracture bones, but the disease was rare among the elderly until recently. As chapter 4 described, human grandmothers didn't evolve to hobble around with a cane or take bed rest during old age, but instead to actively help provision their children and grandchildren.

Sadly, mismatches of inadequate capacity relative to demand show up in more than just the skeleton. Why do some people constantly get colds but others have immune systems better able to ward off infections? Why are some people less able to adapt to extreme temperatures? Why can some people breathe in oxygen fast enough to win the Tour de France, but others can barely draw enough air to climb a flight of stairs? Why are these and other such mismatches so widespread despite their important consequences for survival and reproduction?

Having insufficient capacity to cope with the demands we place on our bodies, like all mismatches, is often a consequence of altered gene-environment interactions, in which environments have recently changed in ways for which our bodies are inadequately adapted. As we age, the genes we inherited interact intensely and constantly with the environment to affect how our bodies grow and develop. However, in contrast to the diseases of affluence discussed in chapter 10, which result from too much of a formerly rare stimulus (like sugar), these diseases result from too little of a formerly common stimulus. If you don't load your skeleton when you are young, it will never grow to be strong, and if you don't stimulate your brain sufficiently as you age, you are at risk of losing cognitive function more rapidly, potentially leading to diseases like dementia.[2] When we then fail to prevent the causes of these diseases we allow the pernicious feedback loop of dysevolution to occur in which we pass on the same environments to our children, enabling the disease to remain common or grow in prevalence. Diseases of disuse account for considerable disability and illness in developed nations. Once they arise, these diseases tend to be difficult to treat, but they are largely preventable if we pay attention to how our bodies evolved to grow and function.

Why Growing Up Needs to Be Stressful

As a thought experiment, imagine you are a robot engineer in the distant future, able to build technologically wondrous robots that can talk, walk, and do other sophisticated tasks. You would probably build each robot for specific purposes, tailoring its capacities to its intended functions (a police robot would have weapons, a waiter robot would have a tray). You would also design each robot for particular environmental conditions, such as extreme heat, freezing cold, or being under water. Now imagine being tasked to design robots without knowing what functions they would serve or which environmental conditions they would experience. How would you create a superadaptable robot?

The answer is you would design each robot to develop dynamically so it adjusts its capacity and function to its conditions. If the robot encountered water, it would develop waterproofing, and if it needed to rescue people from fires, it would develop the ability to resist burning. Since robots are made of a multitude of integrated parts, you would also need to make the robot's components interact with one another as they developed, permitting everything to fit and work together. That way, for example, its waterproofing would not interfere with the movements of its arms or legs.

Perhaps engineers of the future may acquire such abilities, but thanks to evolution, plants and animals already do. By developing through myriad interactions between genes and environments, organisms are able to build extremely complex, highly integrated bodies that not only work well, but also can adapt to a wide range of circumstances. To be sure, we can't just grow new organs at will, but many organs do adapt their capacities to demands by responding to stresses as they grow. For example, if you run around more as a child, you load your leg bones and they grow thicker. Another less appreciated example is the capacity to sweat. Humans are born with millions of sweat glands, but the percentage of glands that actually secrete sweat when you get hot is influenced by how much heat stress you experienced in the first few years of life.[3] Other adjustments respond dynamically throughout life to environmental stresses, even in adults. If you were to lift weights regularly over the next few weeks, your arm muscles would get tired and then get bigger and stronger. Conversely, if you were confined to bed for months or years, your muscles and bones would waste away.

The capacity for bodies to adjust their observable characteristics (their phenotype) in response to environmental stresses is formally known as *phenotypic plasticity*. All organisms require phenotypic plasticity to grow and function, and the more biologists look, the more examples they discover.[4] It makes sense for my body to develop more sweat glands if I am going to live in a really hot environment, to have thicker bones if I am more likely to break my legs or arms, and to have darker skin during the summer when my skin is more likely to burn. However, relying on these interactions has drawbacks that potentially lead to mismatches when critical environmental cues are absent, reduced, or abnormal. As winter turns to spring, I normally develop a tan, which prevents my skin from burning, but if I get on an airplane during the winter and fly to the equator, my usually pale skin will burn in a trice unless I protect it with clothes or sunblock. An evolutionary perspective on the body suggests that such mismatches are more common now than ever because in the last few generations we have changed the conditions in which we develop, sometimes in ways for which natural selection never prepared us (like jet travel). These mismatches can be pernicious because they sometimes arise early in life and then cause problems many years later, when it is too late to correct the problem.

Which brings us back to safety factors. Why doesn't nature build bodies the way engineers build bridges—with generous safety factors so we can adapt to a wide range of conditions? The primary explanation is trade-offs. Everything involves compromises: more of one thing means less of something else. Thicker leg bones, for example, are less likely to break, but they cost more energy to move. Dark skin prevents your skin from burning, but limits how much vitamin D you synthesize.[5] By favoring mechanisms that adjust phenotypes to particular environments, natural selection helps bodies find the right balance between diverse tasks and attain the right level of function: enough but not too much.[6] Some features, like skin color and muscle size, can thus adapt throughout life. Muscle, for example, is an expensive tissue to maintain, consuming about 40 percent of your body's resting metabolism. So it makes sense to let your muscles atrophy when you don't need them and to build them up when you do. However, most features such as leg length or brain size cannot adapt continuously to changes in the environment because they cannot be restructured after they have grown. For these features, the body has to use environmental cues—stresses—to predict the structure's optimal adult configuration during early development, often in utero or during the first few years of life. Although these predictions help you adjust appropriately to your particular environment, structures that didn't experience the right stimuli during early life might end up being poorly suited for conditions you experience later.

To sum up, we really did evolve to "use it or lose it." Because bodies are not engineered but instead grow and evolve, your body expects and indeed requires certain stresses when you are maturing in order to develop appropriately. Such interactions are widely appreciated in the brain: if you deprive a child of language or social interactions his or her brain will never develop properly, and the best time to learn a new language or the violin is when you are young. Similarly important interactions also characterize other systems that interact intensively with the outside world, such as your immune system and the organs that help you digest food, maintain a stable body temperature, and more.

Viewed in this light, one predicts many mismatch diseases to occur when growing bodies fail to experience as much stress as natural selection geared them to expect. Some of these mismatches manifest themselves early in development, but others, such as osteoporosis, do not begin to cause troubles until old age. To be sure, osteoporosis and other age-related illnesses are more common because humans are now living to be older, but the evidence suggests that such diseases are preventable and hardly inevitable. Brittle bones in a sixty-year-old body is an evolutionary mismatch. These mismatches, moreover, are susceptible to dysevolution when we fail to prevent their causes. There are many diseases of disuse, but this chapter focuses on a few prevalent and illustrative examples. Let's begin with two examples in the skeleton: why people get osteoporosis and why we get impacted wisdom teeth. Both follow from how bones grow in response to stress.

Why Bones Need to Be Stressed Enough (but Not Too Much)

Your bones, like the beams in a house, have to bear a lot of weight. But unlike a building's beams, your bones also have to be moved, store calcium, house bone marrow, and provide sites for muscles, ligaments, and tendons to attach. In addition, your bones have to grow and thus change size and shape throughout life without compromising your ability to function. When damaged, they also need to repair themselves. No engineer has ever managed to create a material as versatile and functional as bone.

Bones accomplish so much and do it so well because of natural selection. Over hundreds of millions of years, bone evolved to be a single tissue with multiple components that work together like reinforced concrete to create a material that is both stiff and strong and that also grows dynamically in response to a combination of genetic and environmental cues. The initial shape of a bone is highly controlled by genes, but for the bone to develop properly it needs appropriate nutrients and hormones to grow in concert with the rest of the body. In addition, for an adult bone to achieve the right shape it must experience certain mechanical stresses while it grows. Every time you move, your body's weight and muscles apply forces to your bones, which in turn generate very small deformations. These deformations are so slight that you don't notice them, but they are large enough that cells in your bones constantly measure and react to them. In fact, these deformations are necessary for a bone to develop its appropriate size, shape, and strength. A growing bone that doesn't experience enough load will remain weak and fragile, like the leg bones of a child confined to a wheelchair. In contrast, if you load a bone a lot during development, it will grow thicker, hence stronger. Tennis players' arms illustrate this principle nicely. People who played lots of tennis as youngsters have bones in their dominant, racket-swinging arm that are up to 40 percent thicker and stronger than in their other arm.[7] Other studies show that children who run and walk more develop thicker leg bones, and children who chew harder, tougher food develop thicker jaw bones.[8] No strain, no gain.

Factors such as genes and nutrition also have important effects on how bones grow, but your skeleton's ability to respond to mechanical loads during development is especially adaptive. Without such plasticity, your bones would need to be like the Brooklyn Bridge and be heavily overbuilt to avoid failure, making them bulkier and costlier to move. However, the way the skeleton adapts to its mechanical environment has one unfortunate constraint: once the skeleton stops growing up, bones can no longer grow much thicker. If you start whacking lots of tennis balls as an adult, your arm bones might get a little thicker but not as much as a teenage tennis player's would. In fact, your skeleton attains its peak size soon after you become an adult, between eighteen and twenty years in girls and between twenty and twenty-five years in boys.[9] After then, there is little you can do to make your bones bigger, and soon thereafter your skeleton starts to lose bone for the rest of your life.

Your bones might not be able to grow much thicker, but they are hardly inert and you can take comfort in knowing that they retain the ability to repair themselves. As noted above, every time you move, the forces you apply to each bone cause very slight deformations (strains). These deformations are normal and healthy, but if they are too numerous, rapid, and forceful, they can cause damaging cracks to form. Should these cracks accumulate, grow, and start to join into bigger cracks, the bone would snap like a collapsing bridge that has been weighed down by too many cars. Under ordinary circumstances, however, such disasters don't happen because your bones repair themselves. During this repair process, old and damaged bone is tunneled out and replaced by new, healthy bone. In fact, the repair process is often initiated by stressing the bone. Whenever you run, jump, or climb a tree, the resulting deformations generate signals that stimulate repair in exactly the locations where repairs are most needed.[10] The more you use your skeleton, the more it keeps itself in good condition. Unfortunately, the converse is also true: not using your bones enough leads to bone loss. Astronauts who live in the nearly gravity-free environment of space, which places little stress on the skeleton, lose bone at a rapid rate and return from lengthy tours of duty with dangerously weak bones. When they get back, they often need to be carried to prevent their leg bones from snapping when they walk. Obviously, natural selection did not adapt humans to live in space, but not using your bones here on earth as much as evolution geared your body to expect leads to common mismatch diseases of the skeleton, including osteoporosis and impacted wisdom teeth.

Osteoporosis

Osteoporosis is a debilitating disease that often sneaks up with little warning on older people, most often women. An all-too-common scenario is when an elderly woman falls and breaks her hip or wrist. Under ordinary circumstances, her skeleton should be able to withstand a tumble, but her bones have become so thin they lack the strength to withstand the force of the fall. Another common kind of fracture is when a weakened bone in the spine can no longer bear the upper body's weight and suddenly collapses like a pancake. Such compression fractures cause chronic pain, loss of height, and stooped posture. Overall, osteoporosis afflicts at least a third of all women over the age of fifty and at least 10 percent of similarly aged men, and its prevalence is soaring in developing nations.[11] This growing epidemic is a serious social and economic concern, causing much misery and billions of dollars in health-care costs.

On the face of it, osteoporosis is a disease of old age, so its rising prevalence should hardly be surprising as more people live longer. Osteoporosis-related fractures, however, are exceedingly uncommon in the archaeological record, even after farming began.[12] Instead, the evidence suggests that osteoporosis is a mostly modern mismatch disease caused by interactions between the genes you inherited and several risk factors: physical activity, age, sex, hormones, and diet. The worst-case scenario is to be a sedentary postmenopausal woman who didn't exercise much when she was younger, doesn't eat enough calcium, and gets insufficient vitamin D. Smoking also exacerbates the disease.

To understand how age, sex, exercise, hormones, and diet interact to cause osteoporosis, let's explore how these risk factors influence the two major kinds of cells that create your bones: osteoblasts and osteoclasts. Osteoblasts are the cells that make new bone, and osteoclasts are the ones that dissolve and remove old bone. Both cell types are needed, because just as you often must knock down old walls in order to build new ones when you expand or restore a house, both kinds of cells must work in concert to grow and repair bones. When a bone is growing normally, osteoblasts are more active than osteoclasts (otherwise bones wouldn't get thicker). But as you age and your skeleton's growth slows or ceases, osteoblasts produce less bone and they increasingly spend more time regulating bone repair, as shown in figure 25. During this process, osteoblasts first signal osteoclasts to burrow out bone in a particular location, and then osteoblasts fill in the hole with new, healthy bone.[13] Under normal conditions, the osteoblasts replace about as much bone as the osteoclasts remove. However, osteoporosis develops when osteoclast activity outpaces osteoblast activity. Such imbalances make bones thinner and more porous, a serious problem in spongy bone, which fills certain bones such as vertebrae, as well as joints (figure 25). This type of bone consists of a multitude of tiny, lightweight rods and plates. A growing skeleton creates millions of these vital struts, but unfortunately loses the ability to make new struts after the skeleton stops growing. Thereafter, when an overzealous osteoclast removes or severs a strut, it can never be regrown or repaired. Strut by strut, the bone weakens permanently, until one day its safety factor is too low and it fractures.

Seen in this light, osteoporosis is basically a disease caused by too much bone resorption by osteoclasts relative to too little bone deposition by osteoblasts. As you age, the effects of this imbalance

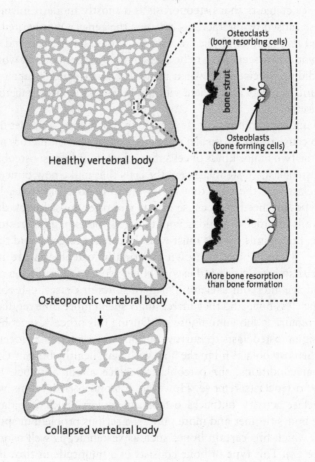

FIGURE 25. Osteoporosis. Schematic illustration of a midsection through the body of a normal vertebra (top), which is full of spongy bone. Detail on the right shows how bone resorbing cells (osteoclasts) remove bone, which is then replaced by bone forming cells (osteoblasts). Osteoporosis occurs when bone resorption outpaces bone replacement, leading to a loss of bone mass and density (middle). Eventually, the vertebra is too weak to support the weight of the body and it collapses (bottom).

cause bones to become fragile and then break. And of all the age-related factors that trigger osteoclasts to outpace osteoblasts, insufficient estrogen is the biggest. Among its many roles, estrogen turns on osteoblasts to manufacture bone and it turns off osteoclasts, keeping them from removing bone. This dual function becomes a liability when women go through menopause and their estrogen levels plummet. All of a sudden, osteoblasts slow down while osteoclasts become more active, causing a rapid rate of bone loss. Giving estrogen to postmenopausal women (estrogen replacement therapy) therefore slows or even halts their rate of bone loss. Men are also at risk, but less so than women because men convert testosterone into estrogen within their bones. Older men don't go through menopause, but as their testosterone levels drop, they create less estrogen and also face rising rates of bone fractures.

Of the various factors that make osteoporosis a modern mismatch disease, one of the biggest is physical activity, whose beneficial effects on bone health are difficult to exaggerate. First, because the skeleton mostly forms before one's early twenties, lots of weight-bearing activity during youth—especially during puberty—leads to greater peak bone mass. As figure 26 graphs, people who are sedentary when they are young commence middle age with considerably less bone than those who were more active. Physical activity also continues to affect bone health as people age. Dozens of studies prove that high levels of weight-bearing activity considerably slow and sometimes even halt or modestly reverse the rate of bone loss in older individuals.[14] Changes in how we mature and age exacerbate this problem, especially in women. Hunter-gatherer girls generally begin puberty about three years later than girls in developed nations, giving them several more years to grow a strong, healthy skeleton primed to withstand years of aging.[15] And of course the longer one lives, the more one's bones become frail and liable to break.

Beyond physical activity and estrogen, the other major factor that increases the risk of osteoporosis is diet, especially calcium. A body needs abundant calcium to function properly, and one of bone's many jobs is to serve as a reservoir of this vital mineral. If calcium levels in the blood drop too much because of insufficient calcium from food, hormones stimulate osteoclasts to resorb

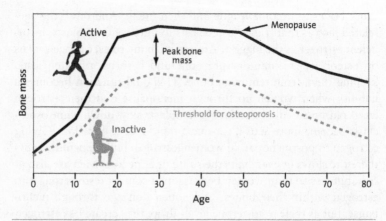

FIGURE 26. General model of osteoporosis. People who are physically inactive develop less bone mass as they mature. After peak bone mass occurs, everyone loses bone, especially women following menopause. Inactive individuals lose bone at a faster rate and cross the threshold for osteoporosis earlier because they started with less peak bone mass.

bone, restoring calcium balance. This response, however, weakens bones if the tissue is not replaced. Consequently, both animals and people whose diets are permanently deficient in calcium develop flimsy bones, and they lose bone more rapidly as they age. Modern grain-based diets, moreover, tend to be woefully deficient in calcium—between two and five times lower than typical hunter-gatherer diets, and only a minority of adult Americans eat sufficient calcium.[16] This problem, moreover, is often exacerbated by low levels of vitamin D, which helps the gut absorb calcium, and by low levels of dietary protein, which is also necessary to synthesize bone.[17] If you are worried about osteoporosis, keep in mind that merely getting enough calcium and vitamin D is not enough to prevent or reverse the disease. You still need to load your skeleton to stimulate your osteoblasts to use that calcium.

All in all, millions of years of natural selection did not gear our skeletons to mature in the absence of plentiful physical activity along with lots of calcium, vitamin D, and protein. Also, until recently, women did not go through puberty until they were sixteen, giving them several extra years to build a larger, stronger skeleton. Genetic variations also play a key role, giving some people a greater predisposition for getting osteoporosis. But, as with many other mismatch diseases, individuals with these genes would be less at risk if our environment hadn't changed so much. One of the biggest problems with this epidemic is that by the time the disease is diagnosed—often because of a broken bone—it is much too late to prevent. At this point, the best strategy is to halt the disease's progress and prevent any more fractures. Doctors usually prescribe a combination of dietary supplements, moderate exercise (vigorous exercise can be dangerous if one's bones are frail), and drugs. Giving postmenopausal women estrogen supplements is highly effective, but these supplements elevate the risks of heart disease and cancer, forcing doctors and patients to balance the risk of osteoporosis against other perils. Several drugs have been developed that slow osteoclast activity, but they often have unpleasant side effects.

Osteoporosis is thus a mismatch disease that is partly a byproduct of people going through puberty younger and living to be older, but people who eat enough calcium and are more physically active when they are young build a more robust, hence more osteoporosis-proof, skeleton. Further, if they continue to be physically active as they age (again, while getting enough calcium), they will lose bone at a much slower rate. Postmenopausal women will always be at higher risk, but evolutionarily normal stresses from youth to old age help their skeletons develop an adequate safety factor. In this respect, osteoporosis is a widespread example of dysevolution, because until we do a better job of getting people, especially young girls, to be more physically active and to eat more calcium-rich foods we will inevitably face rising rates of this unnecessary, debilitating, and costly disease.

Unwise Wisdom Teeth

During my senior year in college, my jaw ached for months. I tried to ignore the discomfort and coped by taking pain relievers until, during a routine tooth cleaning, my dentist ordered me to see an orthodontic surgeon without delay. An X-ray showed that my wisdom teeth (third molars) were unwisely trying to erupt but did not have enough space. They had rotated in the bone and were jamming into the roots of my other teeth. So, like most Americans, I had oral surgery to remove these unwelcome teeth. In addition to being painful, impacted wisdom teeth push other teeth out of proper position, they can cause nerve damage, and they sometimes lead to serious oral infections. Before the invention of antibiotics, such infections could be life-threatening. How and why did evolution design our heads so poorly with insufficient room for all our teeth, putting you and me at risk of severe suffering and sometimes death? What did people do about impacted wisdom teeth before penicillin and modern dentistry?

Evolution turns out not to have been such a bad designer. If you look at lots of recent and modern skulls, you will quickly appreciate that impacted wisdom teeth are another example of an evolutionary mismatch. The museum I work in has thousands of ancient skulls from all over the world. Most of the skulls from the last few hundred years are a dentist's nightmare: they are filled with cavities and infections, the teeth are crowded into the jaw, and about one-quarter of them have impacted teeth. The skulls of preindustrial farmers are also riddled with cavities and painful-looking abscesses, but less than 5 percent of them have impacted wisdom teeth.[18] In contrast, most of the hunter-gatherers had nearly perfect dental health. Apparently, orthodontists and dentists were rarely necessary in the Stone Age. For millions of years, humans had no problem erupting their wisdom teeth, but innovations in food preparation techniques have messed up the age-old system in which genes and mechanical loads from chewing interact to enable teeth and jaws to grow together properly. In fact, the prevalence of impacted wisdom teeth has many parallels to osteoporosis. Just as your limbs and spine will not grow strong enough if you don't sufficiently stress your bones by walking, running, and doing other activities, your jaws won't grow large enough for your teeth and your teeth won't fit properly if you don't stress your face sufficiently from chewing food.

Here's how it works. With every chew, muscles move your lower teeth forcefully against your upper teeth to break down food. Anyone who has stuck a finger accidentally in another person's mouth knows that humans can generate bone-crunchingly high bite forces.[19] These forces not only break down the food, they also stress your face. In fact, such chews cause bones in your jaws to deform as much as your leg bones deform when you walk and run.[20] Chewing also requires that you apply those forces repeatedly. A typical Stone Age meal—especially something tough like a gristly steak—might require thousands of chews. Repeated high forces cause your jaws to adapt over time by growing thicker in the same way that running and playing tennis cause your arm and leg bones to grow thicker. In other words, a childhood spent chewing on hard, tough food helps your jaws grow big and strong. As a test of this hypothesis, my colleagues and I raised hyraxes (small but adorable relatives of elephants that chew like humans) on nutritionally identical hard and soft diets. The hyraxes that chewed harder food developed jaws that were significantly longer, thicker, and wider than the ones who chewed softer food.[21]

The mechanical forces generated by chewing food not only help your jaws grow to the right size and shape, they also help your teeth fit properly within the jaw. Your cheek teeth have cusps and basins that act like little mortars and pestles. During each chew, you pull the lower teeth against the upper teeth with near pinpoint precision so that the cusps of the lower teeth fit perfectly into the basins of the upper teeth and vice versa. Therefore, to chew effectively, your lower and upper teeth need to be just the right shape and in exactly the right place. Tooth shape is mostly controlled by genes, but proper tooth position in the jaw is heavily influenced by chewing forces. As you chew, the forces you apply to your teeth, gums, and jaws activate bone cells in the tooth socket, which then shuttle the teeth into just the right position. If you don't chew enough, your teeth are more likely to be misaligned. Experimental pigs and monkeys raised on ground, softened food that never required them to chew forcefully develop abnormally shaped jaws in which the teeth are improperly aligned and don't fit together.[22] Orthodontists take advantage of the same mechanisms—in which forces push, pull, and rotate teeth—to straighten and align people's teeth using braces. Braces are basically metal bands that apply constant pressure to teeth to move them where they ought to be.

The bottom line is that your jaws and teeth grow and fit together through many processes that involve more than just chewing forces, but a certain level of munching and crunching is necessary for the system to work properly. If you don't chew forcefully enough when you are young, your teeth won't be in the right position, and your jaws may not grow large enough to accommodate your wisdom teeth. Many people today therefore need orthodontists to straighten their teeth and oral surgeons to remove their impacted teeth because our genes haven't changed very much over the last few hundred years but our food has become so soft and processed that we don't chew hard enough and often enough. Think about what you ate today. It was probably highly processed: pureed, ground, mashed, whipped, or otherwise chopped into bite-sized pieces and then cooked to be soft and tender. Thanks to blenders, grinders, and other machines, you can go through a day eating wonderful food (oatmeal, soup, soufflé!) without having to chew at all. As chapter 5 reviewed, cooking and food processing were important innovations that allowed teeth to become smaller and thinner during the evolution of the genus *Homo,* but we have recently taken food processing to such extremes that children often don't chew as much as they need to for normal jaw growth. Try eating like a caveman for a few days: eat only roasted game, roughly chopped vegetables, and nothing that has been ground, pureed, boiled, or softened using modern technologies. Your jaw muscles will fatigue because they are not used to working that hard. Not surprisingly, the effects of modern, wimpy diets are abundantly clear wherever orthodontists look in people's mouths. For example, younger Australian Aborigines whose families recently transitioned to Western diets have smaller jaws and serious tooth crowding problems compared to their elders, who grew up eating more traditional foods.[23] In fact, over the last few thousand years human faces have become about 5 to 10 percent smaller after correcting for body size, about the same size reduction we see in the faces of animals fed cooked, softened food.[24]

Much as I think malocclusions and impacted wisdom teeth are mismatch conditions whose causes we fail to prevent, it would be absurd to abandon orthodontics and force children to chew mostly hard, tough food. I can only imagine the tantrums and other problems that parents would confront if they tried to save on orthodontic bills this way. I wonder, however, if we could reduce the incidence of orthodontic problems by encouraging children to chew more gum? Many grown-ups consider chewing gum to be unaesthetic and annoying, but dentists have long known that sugar-free gum reduces the incidence of cavities.[25] In addition, a few experiments have shown that children who chew hard, resinous gum grow larger jaws and have straighter teeth.[26] More research is needed, but I predict that chewing more gum would help the next generation to have its cake and more often eat it with their wisdom teeth too.

A Little Dirt Never Hurt

To many people, microbes are germs: invisible pests that cause disease and make things rot. The fewer of them the better! So we assiduously disinfect our homes, clothes, food, and bodies with an arsenal of germ-killing weapons, including soap, bleach, steam, and antibiotics. Many parents also try to prevent their children from putting all manner of filth into their mouths—an apparently natural instinct that seems impossible to halt (my daughter had a special penchant for gravel when she was a toddler). Few would question the assumption that being cleaner is healthier, and parents, advertisers, and others relentlessly remind us that the world is full of dangerous germs. And not without justification. Pasteurization, sanitation, and antibiotics have saved more lives than any other advances in medicine.

Yet from an evolutionary perspective, recent efforts to sterilize the body and everything that it contacts are abnormal and may sometimes carry possibly harmful consequences. One reason is that you are not entirely "you." Your body is host to a microbiome: the trillions of other organisms that naturally inhabit your gut, respiratory tract, skin, and other organs. According to some estimates, there are ten times as many foreign microbes in your body as there are of your own cells, and altogether these microbes weigh several pounds.[27] We coevolved with these microbes as well as with many species of worms over millions of years, which explains why most of your microbiome is either harmless or performs important functions, such as helping you digest and cleaning your skin and scalp.[28] You depend on these critters as much as they depend on you, and were you to eradicate them, you would suffer. Fortunately, antibiotics and antiparasitic drugs don't kill off your entire microbiome, but the overuse of these powerful medications does eliminate some helpful microbes and worms, whose absence may actually contribute to new diseases.

A related reason—relevant to this chapter—not to sterilize everything in sight or overuse antibiotics and other such medications is that certain microbes and worms appear to play a crucial role in helping to stress the immune system appropriately. Just as your bones need stress to grow, your immune system requires germs to mature properly. Like any other system of the body, the developing immune system needs to interact with the environment in order to match capacity appropriately with demand. An insufficient immune response to a harmful foreign invader can mean death, but an excessive response is also dangerous, either in the form of an allergic reaction or an autoimmune disease, when the immune system mistakenly attacks the body's own cells. Further, as with other systems, the first few years of life have especially important training effects on the immune system. When you first encountered the cruel world outside the relatively protected environment of your mother's womb, you were assaulted by a host of novel pathogens. Like other infants, you probably endured an endless series of little colds and gastrointestinal troubles. These colds caused distress, but they helped you develop your adaptive immune system, in which your white blood cells learn to recognize and then kill a wide array of foreign pathogens, such as harmful bacteria and viruses.[29] If you were breast-fed, you were also kept healthy by your mother's milk, which is loaded with antibodies and other protective factors, providing an immunological umbrella.[30] Hunter-gatherer children typically nurse for about three years, giving their immature immune systems lots of assistance as they grow amid a world of germs and worms. When farmers started weaning children at younger ages, they lessened their children's immune defenses even as they were creating environments with more harmful pathogens.

The idea that a certain amount of filth is both normal and necessary to develop a healthy immune system has come to be known as the hygiene hypothesis. The hypothesis, first articulated formally by David Strachan,[31] has created a revolution in how we think about a wide variety of diseases, ranging from inflammatory bowel disease and autoimmune disorders to some cancers and even autism.[32] Its original application was to hypothesize why the immune system sometimes causes allergies. Allergies, unlike the previous examples discussed in this chapter, don't occur from a lack of capacity in response to demand. Instead, allergies are harmful inflammatory responses that occur when the immune system overreacts to normally benign substances like peanuts, pollen, or wool. Many allergic reactions are mild, but as everyone knows, they can be severe and life-threatening. Some of the scariest allergic responses are asthma attacks, when the muscles surrounding the airways of the lungs contract and the linings of the airways swell, making it hard or impossible to breathe. Other allergic responses cause skin rashes, itchy eyes, runny noses, vomiting, and more. A particularly troubling trend, suggestive of dysevolution, is that allergies and asthma are on the rise in developed nations. The incidence of asthma and other immune-related disorders has more than tripled since the 1960s in high-income countries, while the rate of infectious diseases has fallen.[33] As an example, peanut allergies have doubled over the last two decades in the United States and other wealthy countries.[34] Since genetic changes and better diagnoses cannot explain these rapid and recent trends, their cause must be partly environmental. Could lack of exposure to certain germs and worms with which we coevolved be a culprit?

To explore how and why too much hygiene might cause otherwise innocuous substances like milk or pollen to trigger potentially deadly overreactions, let's begin with a brief review of how your immune system protects you. Whenever foreign substances enter your body, special cells digest the trespassers and then display the fragments (known as antigens) on their surfaces, like Christmas tree ornaments. Other immune cells, T-helper cells, which are present all over your body, are then attracted, bringing them into contact with the antigens. Usually, T-helper cells are tolerant of the antigens, and they do nothing. Occasionally, however, a T-helper cell decides that an antigen is harmful. When this happens, the T-helper cell has two options. One is to recruit giant white blood cells, which engulf and then digest anything with that antigen. This kind of cellular response works best to remove entire cells in your body that have been infected by viruses or bacteria. The other option, which works better to combat invaders that are swimming about in your bloodstream or other fluids, is for T-helper cells to activate cells that produce antibodies specific to the foreign antigen. There are several kinds of antibodies, but allergic reactions almost always involve IgE antibodies (also called IgE immunoglobulins). When these antibodies bind to an antigen they attract yet other immune cells, which launch an all-out attack on anything that displays the antigen. Among the weapons deployed are chemicals, such as histamines, that cause inflammation—rashes, runny noses, or clogged airways in the lungs. They also trigger muscle spasms, contributing to asthma, diarrhea, coughing, vomiting, and other unpleasant symptoms that help you eject the invaders.

Antibodies protect you from many deadly pathogens, but they cause allergies when they inappropriately target common, harmless substances. The first time this happens, the response is usually mild or moderate. However, your immune system has a memory, and when you encounter the same antigen a second time, antibody-producing cells specific to that antigen are lying in wait, ready to pounce. Activated cells rapidly clone themselves and produce staggering quantities of antibodies for just that antigen. Once this trigger is pulled, your attack cells then respond like a swarm of angry killer bees, causing a massive inflammatory response that can kill you. Viewed in this light, allergic reactions are therefore inappropriate immune responses caused by misguided T-helper cells. Why would T-helper cells wrongly decide that harmless substances are mortal enemies? And what might this response have to do with a lack of germs and worms?

Allergies have multiple causes, but there are several ways in which abnormally sterile conditions during early development could help explain why allergies are becoming more prevalent. The first hypothesis has to do with different T-helper cells. Most bacteria and viruses activate T-helper 1 cells, which recruit white blood cells that demolish infected cells like a big fish devouring a small one. In contrast, T-helper 2 cells stimulate the production of antibodies, which activate the inflammatory responses described above. When certain infections such as hepatitis A viruses stimulate T-helper 1 cells, they suppress the number of T-helper 2 cells.[35] The original hygiene hypothesis is that since people were constantly fighting off mild infections for much of human history, their immune systems were always moderately busy with bacteria and viruses, limiting the number of T-helper 2 cells. Ever since bleach, sterilization, and antibiotic soaps made our environments more germ-free, children's immune systems have had more unemployed T-helper 2 cells swimming about, increasing the likelihood that one of them will make a terrible mistake and wrongly target a harmless substance as an enemy. Once this happens, an allergy develops.

The original hygiene hypothesis has received much attention but does not fully explain why so many allergies are becoming more common. First, although T-helper 1 cells sometimes regulate T-helper 2 cells, these two cell types usually work together.[36] In addition, over the last few decades we have nearly eradicated many viral infections, such as measles, mumps, rubella, and chickenpox, all of which activate T-helper 1 cells. Yet having these diseases provides no protection against developing allergies.[37] An alternative idea, known as the "old friends" hypothesis, is that many allergies and other inappropriate immune responses are occurring more often because our microbiomes are seriously abnormal.[38] For millions of years, we have been living with countless microbes, worms, and other tiny critters that were omnipresent in our environments. These microorganisms are not always totally harmless, but it was probably adaptive to tolerate them, merely keeping them in check rather than fighting them with a full-blown immune response. Imagine how miserable and short life would be if you were always sick, fighting a massive battle against every bug in your microbiome! For good reason, our immune systems and the pathogens with which we live coevolved a sort of cold war–like equilibrium, just keeping each other in balance.

Viewed in this context, many inappropriate immune responses such as allergies may be becoming more prevalent in developed nations because we have upset the longstanding equilibrium that our immune systems coevolved with many "old friends." Thanks to antibiotics, bleach, mouthwash, water treatment plants, and other forms of hygiene, we no longer encounter a broad spectrum of little worms and bacteria. Freed from coping with worms and germs, our immune systems have become overactive, getting into trouble like vagrant youths with no constructive outlets for their pent-up energy. The "old friends" hypothesis explains why exposure to a wide range of germs from animals, dirt, water, and other sources is associated with lower rates of allergies.[39] In addition, the hypothesis might also help explain accumulating evidence that exposure to certain parasites sometimes helps treat autoimmune diseases like multiple sclerosis, inflammatory bowel disease, and more.[40] In the not-too-distant future, your doctor may prescribe you worms or feces.[41]

In short, there is good reason to believe that asthma and other allergies are mismatch diseases in which too little exposure to microorganisms contribute to an imbalance that, paradoxically, causes too much of a response to otherwise harmless foreign substances. The immune system, however, is far more complex than the above description, and there is no question that other factors— many of them genetic—also play key roles. Twins, for example, are more likely than not to share the same allergy.[42] Although it is unlikely that allergy-causing genes are increasingly rapidly in frequency, other environmental factors that disrupt the immune system are certainly more common, such as pollution and various toxic chemicals in our food, water, and air.

The hygiene and "old friends" hypotheses suggest that the way we have been treating some immune disorders is sometimes a case of dysevolution. It is vital, sometimes lifesaving, to focus on the symptoms of an allergic response, but we also need to better address the causes to prevent them in the first place. Maybe children would be less likely to develop life-threatening allergies and possibly certain autoimmune diseases if we made sure they have the right microbiomes. Just as children need the right kinds of food and exercise, it appears that they also need the right kinds of microorganisms in their guts and respiratory tracts. Further, when they get sick and require antibiotics (which do save lives), perhaps the *anti*biotic prescriptions should always be followed by *pro*biotic prescriptions to restore old friends and help keep their immune systems appropriately occupied.

No Strain, No Gain

Diseases of disuse in which too little stress causes inadequate or inappropriate capacity are widespread. I am sure you can think of other mismatch diseases that belong to the same general category: insufficient vitamins and other nutrients, too little sleep, weak back muscles, not enough sunlight, and more. Perhaps the most obvious example of the principle of "no strain, no gain" is the necessity to be physically active to be physically fit. Vigorous activities like running, hiking, and swimming require your muscles to use more oxygen, so you breathe harder, your heart rate increases, your blood pressure goes up, your muscles fatigue, and so on. These stresses trigger numerous adaptive responses in your cardiovascular, respiratory, and musculoskeletal systems that increase their capacity. Heart muscles strengthen and enlarge, arteries grow and become more elastic, muscles add fibers, bones thicken. The flip side of this highly adaptable system, however, are the problems caused by prolonged inactivity. Natural selection never adapted bodies to grow in pathologically abnormal conditions of low activity. In addition, adaptations to save energy by reducing unneeded capacity (muscles are very costly to maintain) lead to serious declines in fitness in couch potatoes, whose muscles atrophy, arteries stiffen, and more. Many studies show that people who are more physically active are more likely to live longer and age better than people who are inactive.[43]

Many mismatch diseases of disuse are also diseases of dysevolution because we have allowed them to remain prevalent or get worse by not addressing their causes. The examples discussed here—osteoporosis, impacted teeth, and allergies—all fit the characteristics of dysevolutionary mismatch diseases. First, we have become reasonably proficient at treating or coping with most of their symptoms, but we do little to prevent their causes, sometimes because of ignorance. Second, none of the mismatch diseases discussed above normally affect people's reproductive fitness (the one exception is an extreme untreated allergic reaction). One can live for years with osteoporosis, bad teeth, and certain allergies. Third, for all these diseases, the relationship between the environmental causes of the mismatch and the physiological effects are gradual, obscure, delayed, marginal, or indirect, and many of them are promoted to some extent by cultural factors we value, such as eating delicious processed food, minimizing toil, and being clean. In fact, many of these problems stem from a basic, common urge to avoid stress and mess. Children love to run around and play (often in filth), but as people age they typically cease to enjoy such pleasures. It is probably adaptive for adults to take it easy and be clean whenever possible. However, it is only recently that a fortunate few have been able to indulge these predilections to extreme degrees, creating environments of ease, comfort, and cleanliness that no caveman could ever imagine. Yet just because we *can* live lives of exceptional cleanliness and comfort doesn't mean they are good for us, especially children. To grow properly, almost every part of the body needs to be stressed appropriately by interactions with the outside world. Just as not requiring a child to reason critically will stunt her intellect, not stressing a child's bones, muscles, and immune systems will fail to match these organs' capacities to their demands.

The solution to diseases of disuse is not to go back to the Stone Age. Many recent inventions have made life better, more convenient, tastier, and more comfortable. Many readers of this book might not be alive were it not for antibiotics and modern sanitation. There is no sense in abandoning these and other advances, but we will benefit from reconsidering how much and when we use, permit, and prescribe them. The good news about the most common diseases of disuse is that efforts to deal with them are usually a question of kind rather than degree. This is especially true of physical activity. Most parents encourage their children to exercise, and most schools require a modest (though inadequate) level of physical education. What we haven't figured out is how much exercise is enough and how to be more effective at getting people to be active, especially as they age. But how much dirt is enough but not too much? Can you imagine public service announcements encouraging parents to let their children eat dirt? I can imagine, however, a world in which antibiotic treatments lead to follow-up visits to a gastroenterologist, who will prescribe worms, bacteria, or specially processed fecal matter to restore a patient's gut ecology.

In conclusion, human bodies were not engineered like the Brooklyn Bridge but instead evolved to grow by interacting with their environment. Because of millions of generations of natural selection on these interactions, every body needs appropriate, sufficient stresses to tune its capacities. The old adage "no strain, no gain" is profoundly true. Allowing our children to ignore this adage leads to a pernicious feedback loop in which problems like osteoporosis become more prevalent, especially as people live longer. Maybe someday we will invent miracle drugs that cure these problems, but I doubt it. In any case, we already know how to prevent or lessen their incidence and intensity through diet and exercise, which yield myriad other benefits and pleasures. How we might get people to change their habits, hence their bodies, is the subject of chapter 13, but before we move on to that, I'd like to consider a final category of mismatch conditions that lead to a host of troubles, in part because of the way we respond to them: diseases of novelty.

12

The Hidden Dangers of Novelty and Comfort

Why Everyday Innovations Can Damage Us

Consider any individual at any period of his life,
and you will always find him preoccupied with fresh
plans to increase his comfort.

—ALEXIS DE TOCQUEVILLE, *Democracy in America*

Danger is everywhere, but why do so many people knowingly engage in potentially harmful behaviors they can avoid? The archetypal example is tobacco. More than a billion people today have willingly become addicted to cigarettes despite being aware that smoking jeopardizes their health. For various reasons, millions engage in other obviously unnatural and potentially hazardous activities, like going to tanning booths, abusing narcotics, or bungee jumping. We also willingly suspend disbelief about many of the dangerous chemicals in our environment. I buy products such as paint and deodorant that are made with suspicious substances, some of which I suspect are toxic or cause cancer, but which I choose not to investigate and which I do not trust my government to regulate as stringently as I would prefer. An example is sodium nitrite, a chemical used to preserve foods (it prevents botulism) and to make meats look red, but which is also linked to cancer. After the U.S. government mandated lower sodium nitrite levels in the 1930s, stomach cancer rates dropped markedly, but why do we still allow small quantities in food?[1] Why also do we allow builders to construct houses using particleboard that contains formaldehyde, a known carcinogen? And why do we permit companies to pollute our air, water, and food with chemicals known to contribute to illness and death?

There are no simple answers to these conundrums, but one major, well-studied factor is the way we assess costs relative to benefits. We habitually value costs and benefits more highly in the near term than in the future (economists call this behavior hyperbolic discounting), allowing us to appear more rational about our long-term goals than our less rational immediate desires, actions, and pleasures. As a result, we tolerate or take pleasure in potentially harmful things because they enhance our lives now more than what we judge to be their eventual costs or risks. Dosage often plays a key role in these judgment calls. The U.S. government permits small amounts of sodium nitrite in food and formaldehyde in particleboard based on estimates of the long-term health risks relative to the short-term economic benefits of cheap meat and inexpensive wood. We accept other less subtle trade-offs all the time. Having a certain percentage of people die from automotive pollutants and car accidents is a price we are apparently willing to pay for the benefit of having cars. Most states sponsor gambling to generate revenue in spite of the social costs from gambling-related addiction and corruption.

I think there are other, deeper evolutionary explanations for why humans sometimes do novel things that are potentially harmful. Chief among them is that we don't actually consider many novel behaviors to be potentially harmful because we don't consider them novel and we are psychologically disposed to consider the world around us to be normal, hence benign. I grew up thinking that it is traditional and ordinary to go to school, to travel in cars and airplanes, to eat canned food, and to watch TV. I also grew up thinking it's normal for people to sometimes have car accidents, just as I think it is abnormal for people to die of the flu or from starvation. It's a habit to form habits, and questioning everything you do can lead to great unhappiness. As a result, I do not question my behaviors or environment in a way that a rational person should or might do. It is standard practice to paint the walls of a house, and we consider the potentially harmful chemicals in the paint as simply an unavoidable potential side effect of living in a house. History teaches us that ordinary people can grow accustomed to horrible, normally unthinkable acts—what the philosopher Hannah Arendt called "the banality of evil." Evolutionary logic suggests that humans become accustomed to novel, unhealthy behaviors and aspects of our environment when they become quotidian.

The inherent tendency to accept the world around us as normal (the banality of the everyday) can have insidious effects that lead to mismatches and dysevolution in surprising ways. Look around you. You are probably seated while reading this and using artificial light to see the words. Maybe you are wearing shoes, and the air in the room is either heated or cooled. Perhaps you are sipping a soda. Your grandmother might find these circumstances normal, but none of these conditions, including the fact that you are sitting and reading, are actually normal for a human being, and all of them are potentially harmful in excess. Why? Because our bodies aren't well adapted for novelties such as reading, sitting too much, and drinking soda. This is hardly news. Just as everyone knows that tobacco is harmful, we also know that too much alcohol is bad for your liver, too much sugar cause cavities, and being physically inactive causes your body to deteriorate. However, I think most people are surprised to learn that many other everyday things we do are also potentially harmful in excess for just the same reason: our bodies aren't well adapted to them.

Which leads to the second evolutionary explanation for why humans often do novel, potentially harmful things: we frequently mistake comfort for well-being. Who doesn't love a state of physical ease? It is pleasant to avoid toiling for long hours, sitting on the hard ground, or being too hot or too cold. Right now, I am sitting in a chair to write these words because it is more comfortable than standing, and the heat in my house is set to an idyllic 68 degrees Fahrenheit (20 degrees Celsius). Later this morning, I'll put on shoes and a coat to go to work, where I can take an elevator to my office's floor to avoid the stress of climbing the stairs. I can then sit in comfort for the rest of the day in another climate-controlled room. The foods I eat will require little effort to procure or consume, the water in my shower will be just the right temperature, and the bed in which I sleep tonight will be soft and warm. If, by chance, I get a headache, I might take some medicine to relieve the pain. Like most of my fellow human beings, I assume that anything comfortable must be good for me. And to some extent this is true. Shoes that hurt are usually bad, as are clothes that are too tight. But is more comfort better? Of course not. Most people suspect that overly soft mattresses can lead to back troubles, and everyone knows that avoiding physical exertion is unhealthy. Yet it is human nature to let one's instincts for comfort override better judgment (I'll take the elevator just this once), and we often fail to recognize that certain everyday, normal comforts are harmful when taken to extremes. Comfort is also profitable. All day long we see and hear advertisements for products that appeal to our apparently insatiable desire for more comfort.

There are plenty of examples of everyday, abnormal, comfortable things that are actually novel and that can lead to ill health. This chapter focuses on just three behaviors noted above that you are probably doing right now: wearing shoes, reading, and sitting. These activities can contribute to the vicious circle of dysevolution because the evolutionary mismatches they sometimes cause (abnormal feet, myopia, back pain) have stimulated the invention of remedies (orthotics, eyeglasses, spinal surgery) to treat their symptoms, but we do a miserable job of preventing the problems from occurring in the first place. As a result, these maladies have become so prevalent that most people think they, too, are normal and inevitable. But they needn't be, and the solution is not to abandon them, but instead to adopt an evolutionary perspective on what is normal to help us devise better shoes, books, and chairs.

The Sense and Sensibility of Shoes

I sometimes run barefoot and over the years have become accustomed to being shouted at: "Doesn't it hurt?" "Watch out for the dog crap!" "Don't step on glass!" I especially enjoy such reactions from people who are walking their dogs. For some reason, they think it acceptable to let their dogs walk and run unshod but abnormal for humans to do the same. These and other reactions highlight just how out of touch we have become with our bodies, leading to a warped perspective of novelty and normality. After all, humans have been walking and running on their bare feet for millions of years, and many people still do. Moreover, when people did start to wear shoes, probably around 45,000 years ago,[2] their footwear was minimal by today's standards, without thick, cushioned heels, arch supports, and other common features. The oldest known sandals, dated to 10,000 years ago, had thin soles that were tied onto the ankle with twine; the oldest preserved shoes, dated to 5,500 years ago, were basically moccasins.[3]

Shoes are now ubiquitous in the developed world, where being barefoot is often considered eccentric, vulgar, or unhygienic. Many restaurants and businesses won't serve barefoot customers, and it is commonly believed that comfortable, supportive shoes are healthy.[4] The mind-set that wearing shoes is more normal and better than being barefoot has been especially evident in the controversy over barefoot running. Interest in the topic was ignited in 2009 by the best-selling book *Born to Run,* which was about an ultramarathon in a remote region of northern Mexico, but which also argued that running shoes cause injury.[5] A year later, my colleagues and I published a study on how and why barefoot people can run comfortably on hard surfaces by landing in an impact-free way that requires no cushioning from a shoe (more on this below).[6] Ever since, there has been much passionate public debate. And, as is often the case, the most extreme views tend to get the most attention. At one extreme are enthusiasts of barefoot running, who decry shoes as unnecessary and injurious, and at the other extreme are vigorous opponents of barefoot running, who think that most runners should wear supportive shoes to avoid injury. Some critics have derided the barefoot running movement as nothing more than "just another passing fad in the running community."[7]

As an evolutionary biologist, I find both extreme views to be implausible and revealing. On the one hand, given that humans have been barefoot for millions of years, one must conclude that wearing shoes is a recent fad. On the other hand, people have also been using shoes to varying degrees for thousands of years, and often without apparent harm. In reality, shoes have benefits but also some costs, which we often fail to consider because wearing shoes has become as normal and commonplace as wearing underwear. In addition, most shoes, especially athletic shoes, are extremely comfortable. Most people assume that any comfortable shoe must also be healthy. But is this assumption true?

Beyond considerations of style, the most important function of shoes is to protect the soles of your feet. The feet of unshod people and other animals accomplish this function with calluses, which are made of keratin, a flexible hairlike protein that also makes up rhino horns and horse hooves. Your skin naturally generates calluses whenever you go barefoot. Every spring when it gets warm enough to spend more time barefoot, my calluses grow, and they recede every winter when I cease going unshod. Not wearing shoes thus creates a circle of dependency: it hurts to go barefoot without calluses, which leads you to wear shoes, which inhibit callus formation. There is no question that a shoe's soles can be more protective than calluses, but the drawback of thick-soled shoes is that they limit sensory perception. You have a rich, extensive network of nerves on the bottom of your feet that provides vital information to your brain about the ground beneath you and that activates key reflexes that help you avoid injury when you sense something sharp, uneven, or hot underfoot. Any shoe interferes with this feedback, and the thicker the sole, the less information you get. In fact, even socks lessen stability, which explains why martial artists, many dancers, and yoga practitioners prefer to go barefoot to enhance their sensory awareness.

Of all the parts of a shoe that cushion your feet, the heel does the most. The heel is the first part of the body (or shoe) to hit the ground when you walk, and sometimes when you run. This collision generates a rapid spike of force on the ground, shown in figure 27, known as an impact peak. Impact peaks can equal the force of your body's weight when walking and be as much as three times your body's weight when running.[8] Since every action has an equal and opposite reaction, impact peaks send a shock wave of force up your legs and spine that quickly reaches your head (within a hundredth of a second when running). Landing hard on your heel

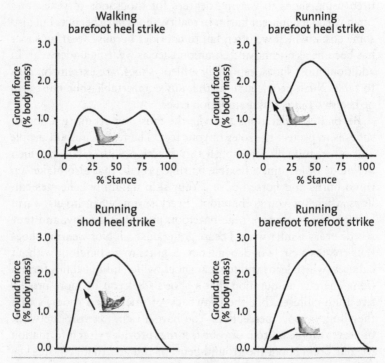

FIGURE 27. Forces on the ground during walking and running (barefoot and shod), measured as units of body weight. In walking, one normally heel strikes, which generates a small impact peak. During barefoot running, heel striking generates a much higher, faster impact peak. A cushioned shoe slows the rate of the impact peak considerably. A forefoot strike (shod or unshod) generates no impact peak.

can feel like being hit with a sledgehammer. Fortunately, the foot's heel pad absorbs these forces enough to make barefoot walking entirely comfortable, but it can be painful to run barefoot for long distances on hard surfaces like concrete or asphalt while landing on your heels. Most running shoes therefore have thick, cushioned heels made from elastic materials that slow down each impact peak, making heel striking comfortable and less injurious (as shown in figure 27). These shoes also make walking more comfortable.

What habitually barefoot people know, however, is that you don't need a shoe with a cushioned heel pad to avoid discomfort when walking and running on hard surfaces. When walking barefoot, one tends to land more gently on the heel, lessening the impact peak, and when running, you can actually avoid any impact peak at all if you land on the ball of the foot before bringing down the heel, in what is known as a forefoot strike.[9] You can demonstrate this yourself by simply jumping barefoot (go ahead, do it right now). I'll bet you naturally first touch down on the balls of your feet before your heels come down, making the landing soft, gentle, and quiet. However, if you forced yourself to land first on your heels, the impact would be loud, hard, and painful (be careful if you try it). The same principle applies to running, which is really just jumping from one leg to the other. By landing gently on the forefoot or sometimes on the midfoot you can run fast on hard surfaces without any cushioning because you don't generate any noticeable impact peak—as far as your foot is concerned, the landing is collision free. Since pain is an adaptation to avoid harmful behaviors, it should hardly be surprising that many experienced barefoot or minimally shod runners tend to forefoot or midfoot strike when running long distances on hard or uneven surfaces, and many habitually shod runners who usually heel strike switch to a forefoot strike when asked to run barefoot on a hard surface.[10] To be sure, some barefoot people land on their heels, especially when going slow, for short distances, or on soft surfaces, but they don't have to run this way if it becomes painful.[11] Many of the world's best and fastest runners forefoot strike even when they are wearing shoes.

To be clear, I am not arguing that it is unnatural or wrong to heel strike. On the contrary, there are several reasons why barefoot and shod people sometimes prefer to heel strike, especially on soft surfaces. Heel striking allows you to lengthen your stride easily, and it requires much less strength in your calf muscles (which have to contract forcefully while they lengthen to help bring down your heel gently when you forefoot strike). Heel striking is also easier on the Achilles tendon. The thick heels on many shoes also make it difficult not to heel strike. My point is that when you heel strike in a cushioned shoe, your body no longer gets the sensory feedback it expects to help modulate your gait to change the impact. As a result, if you run in cushioned shoes with poor form, it is easy to be a thumper, hitting the ground hard with every step.[12] Thanks to the shoe's cushioned heel, these impact peaks don't hurt. But if you run 40 kilometers (25 miles) a week this way, each leg experiences about a million forceful impacts per year. These impacts, in turn, may be damaging. Studies by Irene Davis and others have shown that runners who generate higher, more rapid impact peaks are significantly more likely to accumulate repetitive stress injuries in their feet, shins, knees, and lower back.[13] My students and I found that members of the Harvard cross country team who heel strike are injured more than twice as frequently as those who forefoot strike.[14] The bottom line is that whether you forefoot strike or heel strike, you should do so gently, and being barefoot gives you less choice.

Shoes have other features designed to increase comfort that also affect your body. Many shoes, including running shoes, have an arch support, which props up the arch of the foot. A normal foot arch looks like a half dome and naturally flattens a little when you walk, stretching to help stiffen the foot and transfer your weight to the ball of the big toe. When you run, the arch collapses much more, acting like a giant spring that stores and releases energy, helping push you into the air (see chapter 4). Your foot has about a dozen ligaments and four layers of muscles that hold the arch's bones together. Just as a neck brace relieves your neck muscles from supporting your head, an arch support in your shoe relieves the foot's ligaments and muscles from having to hold up the arch. Arch supports are therefore built into many shoes because they lessen how much work the foot's muscles have to do. Another labor-saving feature is a stiff sole, which allows the foot's muscles to work less hard to push your body forward and upward (this is why walking on a sandy beach can tire your feet). Most shoes also have a sole that curves upward toward the front. This curvature, called a toe spring, requires less muscular effort when your toes push off at the end of stance.

Arch supports and stiff, curved soles are unquestionably comfortable, but they can lead to several problems. One of the most common is flat feet, which occurs when the foot's arch either doesn't develop or it collapses permanently. About 25 percent of Americans have flat feet[15] and thus are more likely to suffer from discomfort and sometimes injury, because a fallen arch changes the way the foot works, causing improper movements in the ankle, knee, and even the hips. Some people's genes may predispose them to getting flat feet, but the problem is mostly caused by weak foot muscles, which otherwise help create and maintain the shape of the arch. Studies that compare habitually barefoot and shod people have found that barefoot people almost never have flat feet but instead have much more consistently shaped arches, neither low nor high.[16] I have examined vast numbers of feet, and I have almost never seen a flat arch in any habitually barefoot person, reinforcing my belief that flat feet are an evolutionary mismatch.

Another related and common problem that may occur from wearing shoes is plantar fasciitis. Have you had a sharp, searing pain in the bottom of your foot either when you get up in the morning or after a run? This pain comes from inflammation of the plantar fascia, a tendonlike sheet of tissue at the base of your foot, which works in conjunction with your muscles to stiffen the arch. Plantar fasciitis has multiple causes, but one way it develops is when the muscles of the foot's arch become weak and the fascia has to compensate for these weak muscles that are unable to maintain the arch. The fascia is not well designed for this much stress and becomes painfully inflamed.[17]

When your feet hurt, your whole body hurts, so most people with foot pain are desperate for treatment. Unfortunately, we too often help these unhappy souls by relieving their symptoms rather than remedying the causes of their problems. Strong, flexible feet are healthy feet, but instead of strengthening their patients' feet, many podiatrists prescribe orthotics and advise patients to wear comfortable shoes with arch supports and stiff soles. These treatments do effectively relieve the symptoms of flat feet and plantar fasciitis, but if their use is not discontinued they can create a pernicious feedback loop because they don't prevent the problem from occurring and instead eventually allow the muscles of the foot to become even weaker. Consequently, people who wear orthotics become increasingly reliant on them. In this regard, perhaps we should treat the foot more like other parts of the body. If you sprain or damage your neck or shoulder, you might use a brace to relieve the pain temporarily, but doctors rarely prescribe permanent braces. Instead you discontinue using the brace as soon as possible and often have physical therapy to regain your strength.

Since the forces that cause repetitive injuries result from the way your body moves, another underused form of prevention and treatment is to look at how people actually move when they walk and run and how well their muscles can control these movements. Although some doctors will examine the gait of a patient who suffers from a repetitive stress injury, too many just treat the symptoms of the problem by prescribing medications, orthotics, or cushioned shoes. Several studies have found that prescriptions of motion-control shoes, which limit how much the foot rolls in (pronation) or out (supination), have no effect on reducing injury rates among runners.[18] Another study found that runners are actually more likely to be injured in more expensive, cushioned shoes.[19] Sadly, between 20 percent and 70 percent of runners incur repetitive stress injuries every year, and there is no evidence that the rates have declined as shoe technology has become more sophisticated over the last thirty years.[20]

Other aspects of shoes also lead to mismatches. How often do you wear uncomfortable shoes because they look good? Millions, maybe even billions of people wear shoes with narrow toe boxes or high heels. Such shoes may be stylish, but they are unhealthy. Narrow toe boxes unnaturally scrunch the front of the foot and contribute to common problems such as bunions, misaligned toes, and hammertoes.[21] High heels show off a person's calves, but they disrupt normal posture, permanently shorten the calf muscles, and subject the ball of the foot, the arch, and even the knee to abnormal forces that cause injury.[22] Encasing feet all day in leather or plastic is commonly considered hygienic but actually creates a sweaty, warm, oxygen-free environment that is heaven for many funguses and bacteria that cause irritating infections, such as athlete's foot.[23]

In short, many people suffer from foot problems because our feet evolved to be bare. Minimal shoes have been around for many thousands of years, yet some modern shoes designed for a combination of comfort and style can interfere substantially with the foot's natural functions. I suspect that we don't need to abandon shoes entirely, and a growing number of shod consumers are responding to these mismatches by wearing minimal shoes that lack heels, stiff soles, arch supports, and narrow toe boxes. It will be interesting to see if they fare any better, and we urgently need to understand how to adapt people with weak feet to the greater muscular demands of wearing minimal shoes. I also suspect it is healthy to encourage infants and children to go barefoot and to ensure that children's shoes are minimal so their feet develop properly and become strong. Sadly, however, most people today with unhappy feet respond by treating the symptoms of their foot pain with orthotics, ever more comfortable shoes, surgery, medications, and a host of other products available in your local pharmacy's extensive foot care section. As long as we continue to encase our feet in comfortable, apparently normal shoes, podiatrists and others who care for aching modern feet will continue to be very busy.

Focusing on Focusing

Reading is to the mind what exercise is to the body and is such an ordinary and essential activity that we think little about the actual physical task of reading words. Even if you are reading this book the way Samuel Goldwyn used to read—"part of it, all the way through"[24]—you are nonetheless focusing for extended periods of time on a string of black and white letters that are probably an arm's length away from your eyes. As your eyes flit from word to word, they remain intently focused on the page. Sometimes when I am engrossed in a really good book, I lose conscious sense of my body and the world around me for hours at a time. But staring at words or anything else so close to your face for hours isn't natural. Writing was first invented about 6,000 years ago, printing presses were invented during the fifteenth century, and it was not until the nineteenth century that it became commonplace for an average person to spend long hours reading. Today, people in developed nations spend many hours staring intently at computer screens.

All of this focusing brings many benefits, but it may come at the cost of poor vision. If you are nearsighted, you have no problem focusing on anything close up, like a book or a computer screen, but everything distant, usually beyond 2 meters (6 feet), is blurry.

In the United States and Europe, nearly a third of children between the age of seven and seventeen become nearsighted (myopic) and need glasses to see properly; the percentage of myopic people is higher in some Asian countries.[25] Myopia is so common that wearing eyeglasses is utterly ordinary and even fashionable. Yet the evidence suggests that being nearsighted used to be very rare. Studies from all over the globe indicate that rates of myopia are less than 3 percent among hunter-gatherers and in populations that practice subsistence agriculture.[26] In addition, myopia among Europeans used to be uncommon except among the educated upper classes. In 1813, James Ware noted that "among the Queen's Guard many were myopic, while of the 10,000 footguards less than a half dozen were myopic."[27] In late-nineteenth-century Denmark, the incidence of myopia among unskilled laborers, seamen, and farmers was less than 3 percent but was 12 percent for craftsmen and 32 percent for university students.[28] Similar shifts in the prevalence of myopia have also been documented in hunter-gatherer populations who transition to Western lifestyles. One such study from the 1960s tested eyesight among Inuits on Barrow Island, in Alaska.[29] Although less than 2 percent of the elders had even mild myopia, a majority of the young adults and schoolchildren were nearsighted, some severely. Evidence that myopia is a modern disease makes sense because it is highly probable that being nearsighted was a serious disadvantage until recently. In the old days, people with poor distance vision probably were less able to hunt animals or gather food effectively, and they were less capable of spotting predators, snakes, and other perils. People with genes that contributed to myopia probably died younger and had fewer children, keeping the trait infrequent.

Nearsightedness is a complex trait caused by many interactions among a large number of genes and multiple environmental factors.[30] However, since people's genes haven't changed much in the last few centuries, the recent worldwide epidemic of myopia must result primarily from environmental shifts. Of all the factors identified, the most commonly identified culprit is close work: intent focusing for long periods of time on nearby images such as sewing and words on a page or screen.[31] One study of more than a thousand Singaporean children found that those who read more than two books a week were three times more likely to have strong myopia (after controlling for sex, race, school, and their parents' degree of myopia).[32] Some studies, however, have found that youngsters who spend less time outside are more likely to get myopia, regardless of how much they read. Therefore, a related but more important cause may be a lack of sufficiently intense and diverse visual stimuli during childhood and adolescence.[33] Additional factors whose causal roles are not as well supported but that merit further study include diets rich in starch and early adolescent growth spurts.[34]

To investigate what factors cause myopia and reevaluate how we treat the problem, let's first consider how the eye normally works to focus light. The process of focusing involves two main steps, summarized in figure 28. The first step happens in the cornea, the transparent outer covering of the front of the eye. Because the cornea is naturally curved like a magnifying glass, it bends light beams, redirecting them through the pupil and onto the lens. The next step, fine focusing, occurs in the lens, a transparent disk the size of a shirt button. Like the cornea, the lens is convex, which enables it to focus light coming from the cornea onto the retina, at the back of the eyeball. There, specialized nerve cells turn light into a stream of signals that are sent to your brain and transformed into a perceptible image. However, unlike the cornea, the lens can change its shape to alter its focus. These shape changes are achieved by hundreds of tiny fibers that suspend the lens behind the pupil.[35] A normal lens is very convex, but the fibers are like springs that constantly pull on the lens, flattening it like a trampoline. In this flattened state, the lens focuses light from distant objects onto the retina. However, in order to focus light beams from relatively larger nearby objects onto the retina, the lens needs to become more convex. This adjustment (termed *accommodation*) occurs when the tiny ciliary muscles that attach to each fiber contract, lessening the tension placed on the lens, allowing it to return to its natural, more convex shape. In other words, while you are reading these words, hundreds of tiny muscles are firing in each eyeball to slacken the fibers and keep your lenses curved, thereby focusing light from the nearby page or screen on your retinas. If you look up and gaze into the distance, those muscles will relax, and the fibers will tighten, flattening the lens so you can focus on faraway objects.

Many hundreds of millions of years of natural selection have

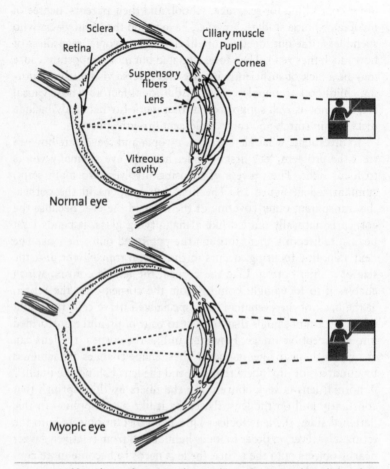

FIGURE 28. How the eye focuses on distant objects. In a normal eye, light is bent first by the cornea and then by the lens (which is relaxed by contractions of the ciliary muscles) to focus on the back of the retina. A myopic eye (bottom) is too long, causing the focus point of distant objects to fall short of the retina.

perfected the eyeball. Its focusing system usually works so well that most of us take clear vision for granted. But, as in any system with much complexity, small variations can impair function, and near-sightedness is no exception. Most cases of nearsightedness occur when the eyeball grows too long, as figure 28 depicts.[36] When this happens the lens can still focus on nearby objects by contracting the ciliary muscles, which allows the lens to become more convex. However, when someone with an overly long eyeball tries to focus on a distant object by relaxing the ciliary muscles, the focus point of the flattened lens falls short of the retina. As a result, everything distant (usually beyond 2 meters, or 6 feet) is out of focus, sometimes dreadfully. Unfortunately, people with myopia are also at greater risk of other eye problems, such as glaucoma, cataracts, detached retinas, and retinal degeneration.[37]

One might suppose that a problem as widespread and as important as myopia would be better understood, but the mechanisms by which prolonged close work or a lack of outdoor visual stimuli can cause eyeballs to grow too long are still uncertain. One longstanding hypothesis is that hours of focusing on nearby objects elongate the eyeball by increasing pressure inside the eye. The hypothesis goes as follows: When you stare at something close (like this page), the ciliary muscles have to contract continuously and other muscles rotate the eyeballs inward (*converge*) to maintain binocular vision. Because the ciliary and eye-rotating muscles are anchored in the outer wall of the eye (the sclera), they essentially squeeze the eyeball, raising pressures within the large posterior (vitreous) chamber, causing it to elongate.[38] Experiments that implanted sensors inside the posterior chamber of the eyeball in macaques measured elevations in pressure when the monkeys were forced to focus on nearby objects.[39] Although direct pressure measurements have not been made in humans, people's eyeballs elongate very slightly when they focus on nearby objects.[40] It has therefore been hypothesized that growing children whose eyeball walls have yet to fully strengthen and who stare persistently at nearby objects stretch the eyeball's walls so much that they permanently elongate, ever so slightly, but enough to cause myopia. Extreme and incessant close work might also cause this process in adults. People whose job requires them to spend long hours with their eyes pressed into microscope lenses often suffer from progressively worsening myopia.[41]

The close work hypothesis is controversial and has never been tested directly in humans. It also fails to explain the findings of other experiments on animals, which indicate that abnormal visual input can cause myopia independent of close work. This phenomenon was discovered by accident when a group of researchers studying how the brain perceives visual information noticed that monkeys whose eyelids had been stitched shut had abnormally elongated eyeballs, as much as 21 percent longer than normal.[42] Intrigued, the researchers followed up with further experiments that showed that the monkeys' myopia was not triggered by excessive close work but instead by a lack of normal visual input (if what a monkey sees in a lab can ever be considered normal).[43] More recent studies that experimentally blurred the vision of kittens and chickens confirmed that myopia can be caused by unfocused images, which somehow disrupt normal eyeball growth.[44] In addition, children who spend more time inside than outdoors are more likely to get myopia.[45] The mechanism by which this abnormal growth occurs is currently unknown, but these various lines of evidence have led to the hypothesis that normal eye elongation requires a mix of complex visual stimuli, such as varying intensities of light and different colors rather than the drab, muted colors typical of the inside of the house or the pages of a book.

Whatever environmental factors contribute to myopia, the problem has been around for a few millennia, albeit less frequently in the past than now. In fact, an inability to see distant objects is used as a metaphor in the New Testament: "But he that lacketh these things is blind, and cannot see afar off, and hath forgotten that he was purged from his old sins."[46] The condition was also diagnosed by the second-century doctor Galen, who purportedly coined the term "myopia." But until eyeglasses were invented in the Renaissance, nearsighted people had to endure their disability without much help. Eyeglasses have since been improved and refined through numerous innovations, including the development of bifocals by Benjamin Franklin in 1784. Today, people with overly long eyeballs can see distant objects just fine with the help of technology, and it is doubtful that myopia now has any negative effects on anyone's reproductive fitness. In this regard, eyeglasses buffer nearsighted people from natural selection. If anything, eyeglasses have been the focus of much cultural evolution themselves as they have become lighter, thinner, more multipurpose, and even invisible (contact lenses). Eyeglass styles change constantly, enticing nearsighted people to buy new frames every few years in order to see and be seen fashionably.

The recent cultural evolution of eyeglasses combined with the importance of being able to focus has led to the intriguing hypothesis that eyeglasses have caused coevolution. As a reminder, this kind of evolution occurs when cultural developments actually stimulate natural selection on genes, as in the classic case of how drinking farm animals' milk favors the prevalence of genes for lactase persistence (see chapter 8). The hypothesis that eyeglasses have caused coevolution is difficult to test, but it is conceivable that since eyeglasses became affordable and common over the last few hundred years, there has been a relaxation of selection against deleterious genes that contribute to myopia. If so, we could predict that the prevalence of myopia has increased gradually and independently of the environmental factors that also cause the problem. This hypothesis is unlikely given how rapidly the prevalence of myopia has increased. A more extreme and frankly troubling idea is that the invention of eyeglasses has been so beneficial for so many individuals that eyeglasses have actually permitted indirect selection for genes for intelligence that indirectly cause myopia. A much-discussed 1958 study found that myopic children in America had significantly higher intelligence quotients (IQs) than normal-sighted children, and this correlation has since been replicated elsewhere in places such as Singapore, Denmark, and Israel.[47] Correlation is not causation, but numerous hypotheses have been proposed to explain these correlations. One possibility is that because eyeball size and brain size are strongly correlated, eyeglasses have permitted selection for larger brains, hence larger eyeballs, which could be more prone to becoming myopic.[48] If so, the high incidence of myopia might be a by-product of selection for bigger brains. This hypothesis is likely wrong for many reasons, not the least of which is that brain size has actually diminished since the Ice Age (see chapter 5), and it's doubtful that bigger-brained Ice Age humans were nearsighted. Alternative hypotheses are that some of the genes that affect intelligence also affect eyeball growth, or that genes that contribute to intelligence are located on chromosomes near genes that cause myopia.[49] If so, then the invention of eyeglasses not only removed any negative selection on myopia but also permitted selection for intelligence that has resulted in a higher proportion of smart near-sighted people. I am skeptical of this hypothesis because children who read more frequently are simply more likely to get myopia, and it's also possible that children with myopia end up reading more and spending more time inside because they don't focus well on distant objects. In either case, nearsighted children end up reading more than children with normal vision and thus do better on IQ tests, which favor children who read more.

We have much to learn about myopia, but two facts are clear. First, myopia is a formerly rare evolutionary mismatch that is exacerbated by modern environments. Second, even though we don't entirely understand which factors cause children's eyeballs to elongate too much, we do know how to treat the symptoms of myopia effectively with eyeglasses. Eyeglasses are just simple lenses that bend waves of light before they hit the eyeball, moving the point of focus back onto the retina. Eyeglasses permit approximately one billion nearsighted people to see clearly, and as more countries undergo economic development, this number will surely rise. Eyeglasses, like shoes, are now so ubiquitous that they have gone from being unattractive—"men seldom make passes at girls who wear glasses"—to being either unnoticed or fashion accessories.

The high prevalence of nearsightedness, combined with the way we use eyeglasses to treat the problem's symptoms rather than its causes, raises several hypotheses about how we promote dysevolution of this disease. One controversial idea, based on the theory that close work causes myopia, is that eyeglasses actually exacerbate the problem. If contractions of the eye's muscles cause myopia in the first place, then giving corrective glasses, which cause all distant objects to appear is if they were close, sets up a positive feedback loop by causing everything to appear close.[50] As noted above, not all the evidence is consistent with this theory, but it has received some support from a few studies that apparently decreased the progression of myopia in children by giving them reading glasses.[51] An alternative idea, based on the visual deprivation hypothesis, is that eyeglasses neither prevent nor exacerbate myopia, but they may indirectly promote other factors that cause myopia by making it easier for children at risk of myopia to spend too many hours reading or doing other indoor activities that provide insufficient visual stimuli. One obvious solution is to encourage these children to spend more time outside. Another might. be to replace boring printed pages (such as this one) with exciting electronic books that are more visually stimulating with intense changes in color and brightness that would challenge young eyes. Wouldn't it be cool if children's books were projected brightly and dynamically on distant walls? Illuminating interior environments more brightly and colorfully might also help.

We have much to learn about myopia, but how and why people become nearsighted and how we help them highlights several typical characteristics of dysevolution. First, like many evolutionary mismatches, myopia is unwittingly transmitted by parents to their children in a non-Darwinian manner. Although certain genes may predispose some children to becoming nearsighted, the primary factors that cause myopia and that parents pass on to their children are environmental, and it is even possible that eyeglasses sometimes exacerbate the problem. Second, we arguably know enough to try to prevent nearsightedness from developing, but so far its prevention has received little attention. I suspect our efforts to prevent myopia would be much more intense if eyeglasses were less effective and less attractive.

Fetch the Comfy Chair

In the late 1920s, two enterprising young men from Michigan held a contest to name the upholstered reclining chair they had invented. From the many submissions, they chose La-Z-Boy (other entries were Sit-N-Snooze and Slack-Back), and the company is still producing luxury chairs of the same name. Today's models feature eighteen "comfort levels" with independently moving backrests and footrests, plus "total lumbar support in all positions." If you pay extra you can add features such as vibrating motors to massage you, a tilting seat that helps you in and out of the chair, cup holders, and more. Yet for the same price as some La-Z-Boy chairs, you could buy a round-trip airplane ticket to the Kalahari Desert or some other remote part of the world, where you'll be hard-pressed to find chairs, let alone ones with cushioning, reclining backs, and leg rests. But this doesn't mean you won't find anyone sitting. Hunter-gatherers and subsistence farmers work hard to obtain every calorie they eat, and they rarely have an energy surplus. When hardworking people with limited food have the chance, they sensibly sit or lie, which costs much less energy than standing. However, when they sit, they usually squat, or they rest on the ground with their legs folded or straight out. Chairs, when they exist, tend to be stools, and the only backrests are trees, rocks, and walls.

To those of us reading this book, sitting in a comfy chair is an utterly normal and pleasant activity, but an evolutionary perspective teaches us that this kind of sitting is unusual. But are chairs unhealthy? Should I abandon the office chair in which I am writing these words and instead write this standing up, perhaps using a treadmill desk? Should you read these words while squatting? And for that matter, should we throw out our mattresses and sleep like our ancestors on hard mats?

Don't worry! I am not going to make you feel bad about sitting in chairs, and, for the record, I have no intention of getting rid of the chairs in my house. But there may be reasons to be concerned about the amount of time you spend in chairs, especially if you are inactive for the rest of the day. One major concern relates to energy balance (see chapter 10). For every hour you sit at a desk, you spend about 20 fewer calories than if you were to stand, because you are no longer tensing muscles in your legs, back, and shoulder, as you support and shift your weight.[52] Standing for eight hours a day adds up to 160 calories, the equivalent of a half-hour walk. Over weeks and years, the energetic difference between mostly sitting and standing is staggering.

A different problem caused by sitting for hours upon hours in comfortable seats is muscle atrophy, especially in the core muscles of the back and abdomen that stabilize the trunk. In terms of muscle activity, sitting in a chair is not much different from lying in bed. It is commonly appreciated that prolonged bed rest has many deleterious effects on the body, including a weaker heart, muscle degeneration, bone loss, and elevated levels of tissue inflammation.[53] Prolonged chair rest has almost the same effect because you also don't use any leg muscles to support your weight, and if the chair has a backrest, a headrest, and armrests, you may not be using as many muscles in your upper body either. This is why La-Z-Boy chairs are so comfortable. Slumping forward or slouching back in a chair also requires less muscle effort than sitting up straight.[54] But there is a price to pay for such comfort. Muscles deteriorate in response to prolonged periods of inactivity by losing muscle fibers, especially the slow-twitch fibers that provide endurance.[55] Months and years of sitting with poor posture in comfortable chairs combined with other sedentary habits therefore allow trunk and abdominal muscles to be weak and to fatigue rapidly. In contrast, squatting and sitting on the ground or even on a stool require more postural control from a variety of muscles in the back and abdomen, helping to maintain their strength.[56]

Another kind of atrophy caused by endless hours of sitting is muscle shortening. When you immobilize joints for lengthy periods, muscles that are no longer stretched can become shorter, which accounts for why wearing high-heeled shoes shortens calf muscles. Chairs are no exception. When you sit in a standard chair, your hips and knees are flexed at right angles, a position which shortens the hip flexor muscles that cross the front of your hip. As a result, many hours of sitting can permanently shorten the hip flexors. Then, when you stand, your shortened hip flexors are tight, so they tilt the pelvis forward leading to an exaggerated lumbar curve. Your hamstring muscle along the back of the thigh then must contract to counter this curvature, tilting your pelvis backward, leading to a flat-back posture, which hunches your shoulders forward. Fortunately, stretching effectively increases muscle length and flexibility, making it a good idea for anyone spending long hours in a chair to get up and stretch regularly.[57]

Muscle imbalances caused by hours of sitting in chairs have also been hypothesized to contribute to one of the most common health problems on the planet: lower back pain. Depending on where you live and what you do, your chances of getting lower back pain are between 60 and 90 percent.[58] Some cases of lower back pain are caused by structural failures like a collapsed disk or by a traumatic accident that damages the spine; however, the majority of lower back pain is diagnosed as "nonspecific," a medical euphemism for problems whose causes are poorly understood. Despite decades of intense research, we remain woefully ineffective at diagnosing, preventing, and treating lower back pain. Many experts have therefore concluded that lower back pain is a nearly inevitable consequence of evolution's unintelligent design of the human lumbar curve, which has cursed the human lineage ever since we stood up about 6 million years ago.

But is this conclusion true? Lower back pain is the most common cause of disability today, costing billions of dollars a year. Today we have painkillers, heat pads, and other largely ineffective ways to alleviate back pain, but imagine how a serious back injury would have affected a Paleolithic hunter-gatherer. Even if our ancestors simply suffered through the pain, back troubles would surely have lessened their ability to forage, hunt, evade predators, provision offspring, and do other tasks that affect reproductive success. Natural selection is therefore likely to have selected for individuals whose backs were less susceptible to injury. As chapter 2 reviewed, selection in response to the biomechanical demands of pregnancy likely explains why women have adaptations that spread their lumbar curve over more vertebrae and have more strongly reinforced joints than men. Selection to strengthen the spine may also explain why humans today tend to have five lumbar vertebrae, one fewer than early hominins such as *H. erectus*. Perhaps the lumbar spine is a much better adapted structure than we realize. If so, then is the high incidence of lower back pain today an example of an evolutionary mismatch in which our bodies are not well adapted to the way we use them? Could it be that we are simply poorly adapted to sitting and other forms of inactivity?

Unfortunately, lower back pain is such a complex, multifactorial problem that intensive efforts to find simple answers about why it occurs and how to prevent it have been (and will remain) frustratingly inconclusive. Studies designed to associate lower back pain with specific causal factors in developed countries have mostly failed to reveal any smoking guns, such as genes, height, weight, time spent sitting, bad posture, exposure to vibrations, participation in sports, or even frequent lifting.[59] However, comprehensive analyses of the incidence of back pain around the world consistently find that back pain is twice as high in developed versus less developed countries; further, within low-income countries, the incidence is roughly twice as high in urban versus rural areas.[60] For example, lower back pain afflicts about 40 percent of farmers in rural Tibet but 68 percent of sewing machine operators in India, many of whom describe their pain as "persistent and unbearable."[61] Neither of these populations lounges about in La-Z-Boys, but a general trend is that people who frequently carry heavy loads and do other "back-breaking" work get fewer back injuries than those who sit in chairs for hours bent over a machine.

If one considers cross-cultural patterns of back pain injury in conjunction with an understanding of how the back evolved to function, there are clues that lower back pain is partly an evolutionary mismatch, albeit one with multiple causes. The key point to consider is that, from an evolutionary perspective, none of the populations so far studied use their backs in a normal way. No one has yet quantified the incidence of lower back pain among hunter-gatherers, but foragers rarely sit in chairs, they never sleep on soft mattresses,[62] they often walk while carrying moderate loads, and they also dig, climb, prepare food, and run. They also don't engage in hours of strenuous work such as hoeing or lifting that repetitively load the back. In other words, hunter-gatherers use their backs moderately—neither as intensively as subsistence farmers nor as minimally as sedentary office workers. They fall generally near the middle of an important model for the risk of lower back pain proposed by Michael Adams and colleagues,[63] illustrated in figure 29. According to this model, a healthy back requires an appropriate balance between how much you use your back and how well your back functions. A normal, fit back needs to have a considerable degree of flexibility, strength, and endurance, as well as some degree of coordination and balance. Since people who are mostly sitters tend to have weak and inflexible backs, they are more likely to experience muscle strains, torn ligaments, stressed joints, bulging disks, and other causes of pain if and when they subject their backs to unusual, stressful movements. As predicted, people in developed countries who suffer from back pain tend to have a lower percentage of slow twitch fibers, which means that their backs fatigue more rapidly, and they also have lower core muscle strength, reduced flexibility in the hip and spine, and more abnormal patterns of motion.[64] At the other end of the spectrum are people whose livelihoods require lots

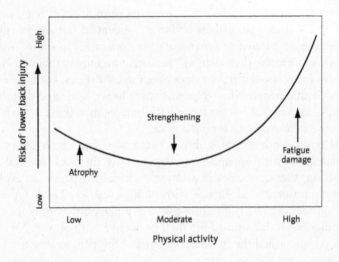

FIGURE 29. Model of the relationship between physical activity levels and back injury. Individuals with very low and high levels of activity have higher risk of injury but for different reasons. Modified from figure 6.4 in M. A. Adams et al. (2002). *The Biomechanics of Back Pain*. Edinburgh: Churchill-Livingstone.

of heavy lifting and other stressful activities that cause repetitive stress damage to the back's muscles, bones, ligaments, disks, and nerves. For this reason, subsistence farmers in Tibet, who dig their fields and harvest crops for weeks on end, and furniture movers, who carry enormous loads, both suffer from back injuries, but their injuries have a different set of causes than those suffered by people who sit all day long hunched over computers or sewing machines.

In short, there is probably a balance between how you use your back and how healthy your back is. A normal back doesn't get pampered by chairs but instead is used with varying degrees of moderate intensity all day long, even during sleep. The adoption of agriculture was probably bad news for human backs. Now we face the opposite problem, thanks to comfy chairs, as well as shopping carts, rolling suitcases, elevators, and thousands of other labor-saving devices. Liberated from overstressing our backs, we suffer from weak and inflexible backs. The resulting scenario is all too common: for months or even years, you may be pain-free, but your back is weak, hence susceptible to injury. Then one day you reach down to pick up a bag, sleep in an awkward position, or fall on the street, and—WHAM—your back gets injured. A visit to the doctor's office usually results in a diagnosis of nonspecific back pain, plus a handful of medicines to alleviate your suffering. The problem is that once lower back pain begins, a vicious circle often ensues. A natural instinct is to rest when following a back injury and then to avoid activities that stress the back. However, too much rest only weakens the muscles, making you more vulnerable to another injury. Fortunately, therapies that improve back strength, including low-impact aerobic exercise, appear to be effective ways to improve back health.[65]

Beyond Comfort

Just about every airplane seat pocket in the United States offers a magazine, *SkyMall,* which sells a bizarre array of products, many designed to enhance your comfort, including shock-absorbing shoes, inflatable cushions, and outdoor heaters that warm you on cold evenings by the pool. Sometimes at the end of a long flight, my daughter and I have a contest to find the most absurd product, and the winner usually comes from the large selection of comfort-enhancing inventions for pets. My favorite is the elevated food bowl so your poor dog doesn't need to strain its neck by having to eat and drink food on the floor. These and countless other products testify to our species' seemingly insatiable desire to increase not just our own comfort but also our pets'. It is generally assumed and widely advertised that anything that makes you feel more at ease must be good, and people pay vast sums of money to avoid having to get too hot or too cold, climb stairs, lift, twist, stand, and more. Over the last few generations, our cravings for comfort and physical pleasure have inspired many new, remarkable inventions, making some entrepreneurs rich. But at the same time, some of these innovations promote disability, especially among those of us unable to temper our urge to take it easy.

Machines that enhance comfort are, of course, just the tip of the iceberg when it comes to the remarkable range of innovations humans have devised since the Paleolithic that have created novel stimuli for the human body. Imagine transporting a caveman to a modern city and trying to explain the new technologies we take for granted, like telephones, showers, motorcycles, guns, and more. Just as natural selection weeds out deleterious mutations and promotes adaptations, cultural evolution eventually sorts out the better innovations from those that are less useful or harmful. Gone are the days of hand axes, astrolabes, and grainy black-and-white TVs, not to mention whalebone corsets and head binding. But cultural selection does not always operate with the same criteria as natural selection. Whereas natural selection only favors novel mutations that enhance an organism's abilities to survive and reproduce, cultural selection can promote novel behaviors simply because they are popular, lucrative, or otherwise beneficial. Wearing shoes, reading, and sitting in chairs have obviously been selected this way because they bring many benefits and pleasures, but the evolutionary mismatches they also cause easily fit the characteristics of dysevolution. In particular, we are adept at treating the symptoms of having bad feet, bad vision, and bad backs, but we do little to prevent their causes. In addition, none of these problems affect people's abilities to live long and happy lives, or have lots of children. In addition, these mismatches have remained prevalent or are becoming worse, in part because they bring many benefits.

Recognizing that many innovations, including those designed for comfort and convenience, are not always beneficial to human health doesn't mean one needs to avoid all new products and technologies. However, an evolutionary perspective on the human body teaches us that some novelties can lead to evolutionary mismatches. Our bodies were simply not adapted by millions of years of evolution to handle many modern technologies, at least not in extreme quantities or degrees. Consider the three examples highlighted in this chapter: wearing shoes, reading, and sitting in chairs. In and of themselves, these everyday behaviors that were unknown until recently are harmless and often beneficial. However, in excess they can cause a variety of problems that we often fail to recognize as harmful because any damage they cause accrues extremely gradually over extended periods of time, obscuring any relationship between cause and effect. They are also comfortable, convenient, pleasurable, and normal.

It is an interesting exercise to attempt to tally up the everyday things you eat, wear, or otherwise employ that are totally novel and that might lead to mismatch diseases or injuries when used in excess. Here are just a few. Your mattress, which is soft and comfortable, may weaken your back if it is too soft and too comfortable. Lightbulbs, which allow you to spend more time indoors, may also deprive you of sufficient bright sunlight, affecting your vision and your mood. Antibacterial soaps, which kill germs in your bathroom, may also be promoting the evolution of new bacteria that could make you even sicker. The earbuds that you use to listen to music may cause hearing loss if you don't keep the volume down. Even more insidious dangers are those that superficially make your life easier but that actually make you weaker: escalators, elevators, suitcases with wheels, shopping carts, automatic can openers, and more. These devices are wondrous aids to bodies that are already damaged but potentially deleterious to those that are still healthy. Years of unnecessarily relying too much on these labor-saving devices can contribute to decrepitude.

The solution to diseases of novelty and comfort is not to rid ourselves of modern conveniences but to halt the cycle of dysevolution in which we treat the symptoms of the problems they create rather than addressing their causes. Returning to the arguments made earlier in this chapter, there is no need to abandon shoes altogether, but instead we might be able to avoid some foot problems by encouraging people—especially children—to go barefoot more often and to wear more minimal shoes (this hypothesis has yet to be tested). Reading, too, is obviously a wonderful modern invention, which we neither can nor should discourage. However, we might prevent or lessen some cases of myopia by getting children to read in a different way (and to get outside more). And there is no need to throw out every chair in your house and office and resort only to standing or squatting, but maybe standing desks should be more common for sedentary office workers.

Of course, these and other shifts will not be easy to achieve for many reasons. For one, who doesn't like comfort and convenience?

There are billions of dollars to be made by creating products that make life easier and more enjoyable and then convincing one another to buy and use them. We don't need to abandon everything novel, but an evolutionary approach to what is normal and comfortable may help inspire more informed skepticism to help us build better shoes and chairs, not to mention mattresses, books, glasses, lightbulbs, houses, towns, and cities. How evolutionary logic might help us achieve such a transformation is the focus of the next and final chapter.

13

Survival of the Fitter

Can Evolutionary Logic Help Cultivate a Better Future for the Human Body?

> When we reflect on this struggle, we may console
> ourselves with the full belief that the war of nature
> is not incessant, that no fear is felt, that death is
> generally prompt, and that the vigorous, the healthy,
> and the happy survive and multiply.

—CHARLES DARWIN, *On the Origin of Species*

There's a popular joke about a group of octogenarians discussing their health problems. "My eyes are so bad I can no longer see clearly." "The arthritis in my neck hurts so much, I can't turn my head." "My heart medication makes me dizzy." "Yes, that's the price we pay for living so long, but at least we can still drive!"

In more ways than one, the joke is obviously recent. The last few thousand years of cultural evolution have significantly altered the human body's condition, sometimes for the worse (especially initially), but eventually and mostly for the better. Because of farming, industrialization, sanitation, new technologies, improved social institutions, and other cultural developments, we have more food, more energy, less work, and additional blessings that immeasurably enrich and improve our existence. Billions of people now take for granted a long life and good health. In fact, if you have the good fortune of being born in a wealthy, well-governed country, you can expect to live into your seventies or eighties, rarely if ever suffer from a serious communicable disease, never have to do hard physical labor, always have plenty of tasty food, and beget similarly healthy, pampered children. To those less lucky, such a prognosis must sound like an advertisement for a lifelong vacation.

To be honest, the most marked improvements to human health and well-being have occurred from the intense surge of scientific progress, still ongoing, that started in the last few hundred years. Many of these advances solved problems that were deleterious consequences of the Agricultural Revolution. As we have seen, although farmers have more food and can have more children than foragers, they have to labor more intensively, and they experience more famine, malnutrition, and infectious disease. Over the last few generations, we have figured out how to conquer many of the contagions that arose or became epidemic after farming took root. Diseases like smallpox, measles, plague, and even malaria have either been eradicated or can now be cured or prevented with proper measures. Likewise, diseases of malnutrition and poor sanitation that proliferated after people settled into permanent towns and cities exist today in some parts of the world chiefly because of poor government, social inequality, and ignorance. As democracy, information, and economic progress sweep across the globe, people are becoming taller, living longer, and otherwise thriving. Yet of course there are inevitable trade-offs, because everyone must die from something. Not dying young from diarrhea, pneumonia, or malaria means a greater likelihood of dying in old age from cancer or heart disease. Similarly, as bodies accumulate wear and tear over the years, aging inevitably brings increasing decrepitude, even when cars and other technologies permit us to still get around.

Our body's evolutionary journey is also far from over. Natural selection didn't stop when farming started but instead has continued and continues to adapt populations to changing diets, germs, and environments. Yet the rate and power of cultural evolution has vastly outpaced the rate and power of natural selection, and the bodies we inherited are still adapted to a significant extent to the various and diverse environmental conditions in which we evolved over millions of years. The end product of all that evolution is that we are big-brained, moderately fat bipeds who reproduce relatively rapidly but take a long time to mature. We are also adapted to be physically active endurance athletes who regularly walk and run long distances and who frequently climb, dig, and carry things. We evolved to eat a diverse diet that includes fruits, tubers, wild game, seeds, nuts, and other foods that tend to be low in sugar, simple carbohydrates, and salt but high in protein, complex carbohydrates, fiber, and vitamins. Humans are also marvelously adapted to make and use tools, to communicate effectively, to cooperate intensively, to innovate, and to use culture to cope with a wide range of challenges. These extraordinary cultural capacities enabled *Homo sapiens* to spread rapidly across the planet and then, paradoxically, cease being hunter-gatherers.

The principal trade-off between the novel environments we have created and the bodies we inherited has been mismatch diseases. Adaptation is a tricky concept, and there is no one environment to which the human body was adapted, but our biology remains imperfectly adapted to living at high population densities in permanent settlements amid the filth we create. We are also inadequately adapted to being too physically idle, too well fed, too comfortable, too clean, and more. Despite recent progress in medicine and sanitation, too many of us are getting sick from a wide range of diseases that used to be rare or unknown. Increasingly, these diseases are chronic noninfectious illnesses, many of which arise from having made too much progress. For millions of years, humans struggled to stay in energy balance, but billions of people are now obese from eating more calories (especially from massive doses of sugar) as well as from less physical activity. As we accumulate excess fat in our bellies while fitness dwindles, diseases of affluence are on the rise, especially heart disease, type 2 diabetes, osteoporosis, breast cancer, and colon cancer. In the United States, the rate of type 2 diabetes is rising even among teenagers, with nearly 25 percent now classified as having either prediabetes, diabetes, or other risk factors for cardiovascular disease.[1] Economic progress has also brought more pollution and other potentially harmful environmental changes (too much, too little, too new) that are contributing to rising rates of mismatch diseases, such as certain cancers, allergies, asthma, gout, celiac disease, depression, and more. The next generation of Americans risks being the first generation to live shorter lives than their parents.[2]

The ongoing epidemiological transition that is bringing lower mortality and higher morbidity is not just a problem for wealthy nations. The rest of the world is heading in the same direction.[3] India, for example, has achieved dramatic improvements in life expectancy but is now facing a tsunami of type 2 diabetes among the middle class, with the number of cases expected to grow from 50 million in 2010 to more than 100 million by 2030.[4] Economically developed countries are already having problems paying for the rising costs of chronic illness among the young and middle-aged (for example, diabetes doubles the average cost of a person's health care)[5]. How will less wealthy countries, such as India, cope?

The big picture we now confront is a paradoxical situation in which the human body is simultaneously doing better in many respects but worse in others. To understand this paradox and what to do requires using the lens of evolution to consider two related processes. The first, summarized above, is that changing environments have made us increasingly prone to diseases from evolutionary mismatches. Understanding why mismatches occur is vital to figuring out how to prevent or treat them, which highlights the importance of the second process, the pernicious feedback loop of dysevolution. Even though many (though not all) mismatch diseases are preventable, we too often fail to address their environmental causes, allowing the diseases to remain prevalent or to intensify when we pass on the same disease-inducing environmental conditions to our children through our culture. The obvious and important exceptions to this feedback loop have been infectious diseases, which we have become fairly skilled at preventing since the development of microbiology and modern sanitation. Diseases caused by malnutrition are also now uncommon when people have good government. But for various reasons outlined in chapters 10 through 12, we seem to be unable to apply the same preventive logic to a wide range of diseases caused by too much energy intake, not enough physiological stress, and other novel aspects of our environments. These mismatch diseases are the ones most likely to disable you, kill you, and cost you money. The United States, for example, spends more than two trillion dollars a year on health care, nearly 20 percent of the country's gross domestic product, and it is estimated that approximately 70 percent of the illnesses we treat are preventable.[6]

In conclusion, although the human body has come a long way over the last 6 million years, its journey is far from over. But what is that future? Will we just muddle along? Will we succeed in developing new technologies to finally cure cancer, solve the obesity epidemic, and make people otherwise healthier and happier? Or are we headed to a future like the one described in the movie *WALL-E*, in which we balloon into a race of fat, chronically ill weaklings who are dependent on medications, machines, and big corporations to survive? How can an evolutionary perspective help chart a better future for the human body? There is obviously no single approach to this Gordian knot, so let's look at each of the options using the lens of evolution.

Approach 1: Let Natural Selection Sort the Problem Out

In 1209, a Catholic army massacred between ten and twenty thousand people in the city of Béziers, France, in an effort to stamp out heresy. Since it was not possible to distinguish the faithful from the heretics, the slaughterers were reportedly told to "kill them all and let God sort them out." Such heartless attitudes are fortunately rare, but I am often asked if natural selection will solve the health problems we now confront in a similarly ruthless way. Will natural selection weed out those whose bodies can't handle modern environments, making our species better adapted to junk food and physical inactivity?

It bears repeating from earlier chapters that natural selection hasn't ceased to operate today. This is because natural selection is basically the inevitable outcome of two phenomena that still exist:

heritable variation and differential reproductive success. Just as selection must be acting on people with less immunity to certain infectious diseases, presumably there are people who are less well adapted genetically to today's environment of plenty and physical inactivity. If they have fewer surviving children, won't their genes be removed from the gene pool? By the same token, won't those better able to resist getting sick from inactivity, modern diets, and various pollutants be more likely to pass on those beneficial genes?

We cannot entirely discount these ideas. According to one 2009 study, American women who are shorter and stouter have very slightly higher fertility, hinting that future generations might become plumper and less tall if these selective trends continue for a very long time (which is far from clear).[7] In addition, infectious diseases can still be strong selective forces. When the next deadly pandemic eventually does arise, anyone whose immune system confers some resistance will have a major fitness advantage. Perhaps selection will also favor individuals with genes that help them resist common toxins, skin cancer, or other environmental causes of disease. It is also hypothetically possible that genetic screening technologies will allow parents of the future to artificially select characteristics in their offspring that provide some benefit.

Human evolution is not over, but the chances of natural selection adapting our species in dramatic, major ways to common noninfectious mismatch diseases are remote unless conditions change dramatically. One reason is that many of these diseases have little to no effect on fertility. Type 2 diabetes, for example, generally develops after people have reproduced, and even then, it is highly manageable for many years.[8] Another consideration is that natural selection can act only on variations that affect reproductive success and that are also genetically passed from parent to offspring. Some obesity-related illnesses can hinder reproductive function, but these problems have strong environmental causes.[9] Finally, although culture sometimes spurs selection, it is also a powerful buffer. Every year new products and therapies are being developed that allow people with common mismatch diseases to cope better with their symptoms. Whatever selection is operating is probably occurring at a pace too slow to measure in our lifetimes.

Approach 2: Invest More in Biomedical Research and Treatment

In 1795, the marquis de Condorcet predicted that medicine would eventually extend human life indefinitely, and intelligent people still make rashly optimistic predictions about dazzling new breakthroughs to halt aging, defeat cancer, and cure other diseases.[10] A friend of mine, for example, proposes that one day we will genetically modify foods with compounds that inhibit fat cells. He imagines specially bioengineered muffins you can eat for breakfast that will prevent obesity. Even if such a muffin could be developed, and even if it didn't have dangerous side effects (which is almost impossible), I predict it would do more harm than good, because people who ate the muffin would lose the incentive to be physically active and eat sensibly. As a result, they would not reap the many physical and mental benefits that come from good diet and exercise.

Quick fixes for complex diseases may be a dangerous form of science fiction, but decades of modern medical science have led (with plenty of mistakes along the way) to countless beneficial treatments for mismatch diseases that save lives and alleviate suffering. It should go without saying that we must keep investing in fundamental biomedical research to promote further advances. But we should expect little more than slow, incremental progress. Most currently available drugs have limited effectiveness as well as nasty side effects, and among treatments for noninfectious diseases, few offer real cures but instead only mitigate symptoms or lessen the risks of death or illness. For example, there are no pharmaceuticals or surgical procedures that permanently cure type 2 diabetes, osteoporosis, or heart disease. Many of the drugs that help adults with type 2 diabetes are less effective in adolescents who acquire the disease.[11] Despite significant investments, the mortality rate for many types of cancer has barely budged since the 1950s (after adjusting for the age and size of the population).[12] Autism, Crohn's disease, allergies, and a host of other diseases are still difficult to treat. We have a long, long way to go.

Another reason not to expect big biomedical breakthroughs in the near future for chronic mismatch diseases, especially those unrelated to pathogens, is that the causes of these diseases are not easy to target effectively. Harmful germs and worms can be defeated through sanitation, vaccination, and antibiotics, but diseases caused by poor diet, physical inactivity, and aging have complex origins involving many causal factors that defy simple remedies. The genes that have been identified as factors for many of these chronic illnesses turn out to be astonishingly numerous and diverse, and few of them have strong effects on any given disease.[13] Practically, this means that any genetic mutations that make your neighbor more susceptible to diabetes, heart disease, or cancer are rare, and unlikely to be the same as the mutations that might affect you or your children. In addition, even if we could design drugs to target these uncommon genes, those drugs will often have only limited effects. As a result, we cannot expect science to devise a few highly effective treatments to cure most noninfectious mismatch diseases. There will be no Pasteur for such diseases.

And therein lies a quandary, because many of these diseases are preventable to some extent—sometimes greatly—through environmental changes that are challenging to enact and through behavioral changes that are difficult to adhere to. Good old-fashioned diet and exercise are not panaceas, but dozens of studies unambiguously prove they substantially reduce the rate of most common mismatch diseases. To bring up one example among many, a study of thirty thousand elderly people in fifty-two countries found that switching to an overall healthy lifestyle—eating a diet rich in fruits and vegetables, not smoking, exercising moderately, and not drinking too much alcohol—lowered heart disease rates by approximately 50 percent.[14] Reducing exposure to carcinogens, such as tobacco and sodium nitrite, have been shown to decrease the incidence of lung and stomach cancers, and it is likely (more evidence is needed) that lowering exposures to other known carcinogens, such as benzene and formaldehyde, will reduce the incidence of other cancers. Prevention really is the most powerful medicine, but we as a species consistently lack the political or psychological will to act preventively in our own best interests.

It is worthwhile to ask to what extent efforts to treat the symptoms of common mismatch diseases have the effect of promoting dysevolution by taking attention and resources away from prevention. On an individual level, am I more likely to eat unhealthy foods and exercise insufficiently if I know I'll have access to medical care to treat the symptoms of the diseases these choices cause many years later? More broadly within our society, is the money we allocate to treating diseases coming at the expense of money to prevent them?

I don't know the answers to these questions, but by any objective measure, we are paying insufficient attention and devoting too few resources to prevention. To appreciate the scale of this point, consider that a large, carefully controlled, long-term intervention study showed that adult Americans who were unfit but then improved their level of fitness *halved* their rates of cardiovascular disease.[15] Because it costs an extra $18,000 a year to treat an American with heart disease, one can estimate that persuading just 25 percent more of the population to become fit could save in excess of $58 billion per year for just heart disease care alone.[16] To put this number into perspective, $58 billion is roughly twice the entire annual research budget of the National Institutes of Health (NIH). Only 5 percent of that NIH budget goes to research on disease prevention.[17] No one knows how much it would actually cost to get 25 percent more Americans to become fit (or how to do it), but a 2008 study estimated that spending $10 per year per person in community-based programs that increase physical activity, prevent smoking, and improve nutrition would save the United States more than $16 billion per year in health-care costs within five years.[18] The precise numbers are debatable, but my point is that no matter how you look at the issue, prevention is a fundamentally preferable and more cost-effective way to promote health and longevity.

Most people agree that we invest insufficiently in prevention, but they would also surmise that it is difficult to get young, healthy people to avoid behaviors that increase their risk of future illness. Consider smoking, which causes more preventable deaths than any major risk factor (the other big ones being physical inactivity, poor diet, and alcohol abuse). After prolonged legal battles, public health efforts to discourage smoking have managed to halve the percentage of Americans who smoke since the 1950s.[19] Yet 20 percent of Americans still smoke, causing 443,000 premature deaths in 2011 at a direct cost of $96 billion per year. Likewise, most Americans know they should be physically active and eat a healthy diet, yet only 20 percent of Americans meet the government's recommendations for physical activity, and fewer than 20 percent meet government dietary guidelines.[20]

There are many, diverse reasons we are bad at persuading, nudging, or otherwise encouraging people to use their bodies more as they evolved to be used (more on this later), but one contributing factor could be that we are still following in the footsteps of the marquis de Condorcet, waiting for the next promised breakthrough. Scared of death and hopeful about science, we spend billions of dollars trying to figure out how to regrow diseased organs, hunting for new drugs, and designing artifical body parts to replace the ones we wear out. I am in no way suggesting that we cease investing in these and other areas. Quite the contrary: let's spend more! But let's not do so in a way that promotes the pernicious feedback loop of just treating mismatch diseases rather than preventing them. In practical terms this means that health insurance plans should spend more on prevention (which, ultimately, will save money on treatment). In addition, public health budgets should not fund research on treatments for disease at the expense of funding research on preventive medicine. Unfortunately, the tiny percentage of funds that the NIH devotes to preventive medicine suggests that the United States is doing just that.

Another relevant factor is money. In the United States and many other countries, health care is partly a for-profit industry.[21] Consequently, there is a strong incentive to invest in or promote treatments such as antacids and orthotics that alleviate the symptoms of diseases and that people have to buy frequently and for many years. Another way to make lots of money is to favor costly procedures like surgery instead of less expensive preventive treatments like physical therapy. Preventive medicine is also distorted by profit. Dieting, for example, is a multibillion-dollar industry in America and elsewhere, largely because most diets are ineffective, and overweight people are willing to keep spending lots of money on new diet plans, many of which are literally too good to be true.

In the final analysis, we have no choice but to keep investing in and focusing on treating mismatch diseases, thus diverting time, money, and effort from prevention, because so many people are currently sick and because efforts to promote prevention don't work very well. This depressing assessment forces us to ask the question: can we do a better job of changing people's behaviors?

Approach 3: Educate and Empower

Knowledge is power. People therefore need and deserve useful, credible information about how their bodies work, and they require the right tools to achieve their goals. Consequently, a cornerstone of public health efforts is to devise ways to educate and empower people so they can better use and care for their bodies and make more rational decisions.

Research and much trial and error have caused public health strategies to evolve rapidly in the last few decades. Prior to the 1990s, most efforts focused on providing basic health education, on the assumption that people make more rational decisions when they have good information. When I was in high school, we were given scary statistics about smoking, drugs, and unprotected sex, and our teachers showed us gruesome images of smokers' lungs. Not surprisingly, studies of the efficacy of these programs revealed that providing such information is necessary but usually not sufficient to produce lasting behavioral change.[22] Now public health programs advocate a full-court-press approach that supplies not just information but also the skills people need to make changes within their social environment.[23] Effective public health intervention also requires programs that operate at multiple levels: between individuals such as doctors and patients, within communities such as schools and churches, and through governments via public media campaigns, regulations, and taxes.[24] Yet other competing factors limit the efficacy of these efforts. For example, advertisers in the United States spend billions of dollars yearly to market tasty, desirable, but unhealthy food to children. In 2004, the average American child between the ages of two and seven saw more than 4,400 advertisements on TV for children's foods but only about 164 public service announcements for fitness or nutrition—a twenty-seven-fold difference![25]

The effectiveness of many educational efforts is also depressingly modest. One study at a large American university required nearly two thousand students to take a fifteen-week course on health and wellness, which included information on the benefits of physical activity and diet. Half the students attended live lectures, and the other half took the course online. Behavioral assessments after the course revealed that the students increased their daily level of moderate intensity activity by 8 percent, but vigorous activity declined; they also ate 4 percent more fruits and vegetables and 8 to 11 percent more whole grains.[26] Those who took the course online changed their habits less than those who attended the lecture. Other studies have yielded similar results.[27] Education is essential, but it can only do so much.

It does not take a multimillion-dollar study to know that we should not have unrealistic expectations about behavioral changes even if we improve the quality and reach of health education. If I am hungry and have to choose between a piece of chocolate cake or celery, there is no question that I'm almost always going to prefer the cake. There is no wisdom of the body that naturally guides people to select foods that are healthy in the context of today's abundance.[28] Instead, experiments repeatedly reveal that children and adults instinctively prefer foods that we evolved to crave (sweet, starchy, salty, and fatty) and that factors such as advertising, range of available choices, peer pressure, and cost strongly affect modern foraging decisions.[29] The same is true for physical activity. When I can choose between taking an escalator or the stairs, I almost always prefer the escalator. I am in the majority. Moreover, banners and posters in malls designed to encourage shoppers to take stairs instead of escalators increase stair climbing by only 6 percent, which is about as effective as mass media campaigns that try to promote physical activity.[30]

Why we behave irrationally with regard to our health is increasingly the subject of innovative research. Numerous experiments have proved that humans behave in many ways that are beyond our conscious control. We react through instinct. These snap judgments tend to be for common, repetitive, instantaneous decisions such as whether to eat the chocolate cake or the celery, or whether to take the stairs or the elevator.[31] Although it is possible to suppress these instincts with slower, more deliberative kinds of thinking, such behavioral overrides are challenging. For example, we consistently discount the value of rewards in the present (such as one more cookie) relative to rewards in the distant future (such as health during old age) in proportion to the length of the delay. These and other unhealthy instincts are presumably ancient adaptations that used to benefit the chances of surviving and having more offspring during times of scarcity, and it is only recently that they have become perversely adaptive in an environment of plenty. Put differently, we constantly make irrational decisions through no fault of our own. These natural tendencies then make us vulnerable to manufacturers and marketers who easily exploit our basic urges to eat too much, eat the wrong foods, and exercise too little. Because these unhealthy behaviors are deep instincts they are very difficult to overcome.

The bottom line is that knowledge is power, but not enough. Most of us need information and skills, but we also require motivation and reinforcement to overcome basic urges in order to make healthier choices in environments replete with plentiful food and labor-saving devices.

Approach 4: Change the Environment

If you are concerned about the obesity epidemic, the global surge of chronic noninfectious diseases, rising costs of health care, or your family's health, then ask yourself if you agree with the following three statements:

1. For the foreseeable future, people will continue to get sick from mismatch diseases.
2. Future advancements in medical science will continue to improve our ability to diagnose and treat the symptoms of mismatch diseases but will not devise many actual cures.
3. Efforts to educate people about diet, nutrition, and other ways to promote health will have limited effects on their behavior in current environments.

If you concur, then the last remaining option is to change people's environments in ways that promote health through prevention. But how?

As a thought experiment, imagine that a tyrant who is both a health freak and obsessed by the cost of health care seizes control of your country and enforces radical changes to people's daily lives. Soda, fruit juice, candy, and other sugar-laden foods are banned, as are potato chips, white rice, white bread, and other simple carbohydrates. Fast-food restaurant owners are sent to prison, as are smokers, drunkards, and anyone who pollutes food, air, or water with a known carcinogen or toxin. Farmers are no longer subsidized to grow corn, and cows must be fed grass or hay. Everyone is mandated a regime of daily push-ups, 150 minutes of vigorous exercise per week, 8 hours of sleep per night, and regular tooth flossing.

Salubrious as it may seem, this sort of fascist national health camp is fortunately impossible (there would be an uprising or a coup d'état) and ethically wrong, because human beings have the right to decide what to do with their bodies. But it is almost certain that many common mismatch diseases would become rarer, as would the incidence of some cancers. Freedom is more precious than good health, but can we change our environments effectively in a way that also respects people's rights?

An evolutionary perspective, I think, provides a useful framework based on two principles. The first is that since all diseases result from gene-environment interactions, and we cannot reengineer our genes, the most effective way to prevent mismatch diseases is to reengineer our environments. The second principle is that the human body was adapted by millions of generations of what Darwin called "the struggle for existence" in conditions that differ substantially from today's. Until recently, humans had little choice but to behave in ways that natural selection dictated. Your ancestors were generally compelled by circumstance to eat a naturally healthy diet, to get plenty of physical activity and sleep, and to avoid chairs, and they were rarely if ever able to live in crowded, permanent, filthy settlements that promoted infectious diseases. It therefore follows that humans did not always evolve to choose to behave in ways that promoted health but instead were coerced by nature. Put differently, an evolutionary perspective suggests that we sometimes need help from external forces in order to help ourselves.

The logic that humans need to be encouraged and sometimes even obliged to act in their own best interests is not controversial when applied to children, who cannot be counted on to make rational decisions and who should not necessarily be penalized by circumstances beyond their control (including bad parents). For this reason, governments ban the sale of alcohol and tobacco to minors, require parents to immunize their children, and make physical education compulsory in school (albeit to varying extents). Many schools now ban soda or other unhealthy foods. Governments also prohibit children from being forced to work long hours in factories.[32] These and many other laws are widely considered acceptable for ethical, social, and practical reasons, but they also make sense from an evolutionary perspective. Certain types of coercion—inconceivable during the Paleolithic—safeguard children from novel, harmful aspects of the environment from which they are unable to protect themselves.

What about adults? I am not a philosopher, a lawyer, or a politician, but allow me to share my opinion, which is essentially an evolutionarily informed version of "libertarian paternalism," or "soft paternalism."[33] Like many, I think that adults have the right to do as they wish as long as they don't harm others. I have the right to smoke as long as you don't have to breathe my fumes or pay for the medical cost of my lung cancer treatment. I also have the right to eat as many donuts and drink as many sodas as I can tolerate and afford. At the same time, we humans (myself included) sometimes behave in ways that are not in our best interests because we lack sufficient information, we cannot control our environments, we are unfairly manipulated by others, and—crucially—because we are poorly adapted to control deep cravings for comforts and calories that used to be rare. Consequently, a sensible role of government that benefits everyone is to help one another make choices that we would rationally judge to be in our own self-interest. In other words, government has the right and even duty to nudge or sometimes push us to behave rationally while preserving our right to still behave irrationally if we so choose. Government also has the duty to ensure that we have the information we need to make rational decisions and to protect us from unfair manipulations. An uncontroversial example of this principle is that food producers should not be allowed to prevent customers from knowing what harmful chemicals are present in their food. In addition, the government shouldn't prevent me from smoking, but it should inform me of the dangers, give me incentives not to smoke, and tax me heavily to pay for the burden my smoking places on you. (As the adage goes, "You are free to do as you wish as long as I don't have to pay for it.")

If you agree that society should promote health through soft paternalism by using its influence to modify the evolutionarily unnatural environments in which we live, then the question is not whether to act, but how much, and in what ways.

Let's start with children because, as noted above, it is relatively uncontroversial to regulate the environments of children, who often cannot make rational decisions in their own best interests. Further, poor fitness, obesity, and exposure to harmful chemicals during childhood have strong negative effects on later health outcomes. Therefore, one obvious place to start would be to mandate more physical education in schools, with an emphasis on fitness over sports. The U.S. surgeon general recommends one hour of physical activity per day for children and teenagers, but only a minority of American students are this active.[34] For example, a study of more than five hundred American high schools found that only about half of the students participate in physical education at all, and few got even half as much as the surgeon general recommends.[35] What about college? Most colleges used to require physical education but rarely do today. The school where I teach, Harvard, for example, dropped its physical education requirement in 1970, and surveys of Harvard students indicate that only a minority exercise vigorously more than three times a week.

A more contentious realm of regulations to consider regarding children is junk food. There is nearly universal consensus that we should prohibit selling and serving alcohol to minors because wine, beer, and spirits can be addictive and, when used to excess, ruinous for their health. Is excess sugar any different? From an evolutionary perspective, how different would it be to limit the sale to children of soda, sweetened beverages, and other sugar-laden foods, which are also addictive and unhealthy in large quantities?[36] We evolved to crave sugar, but most wild fruits have very little sugar, and the only very sweet food that hunter-gatherer kids ever got to enjoy was honey. What about fast foods? These industrially engineered foods pose little risk in small and infrequent quantities, but they slowly cause ill health when overconsumed, and we crave them to the point of addiction.[37] Consequently, is banning or limiting the consumption of french fries and soda in schools different from requiring children to wear seat belts? For that matter, is limiting the sale of these foods outside of school different from limiting what kinds of movies they can attend?

Regulating what children do may be acceptable, albeit unpopular (especially with the food industry and their many lobbyists), but adults are a different matter, because they have the right to get sick. In addition, we usually afford companies the right to sell consumers products that they want, such as cigarettes and chairs, regardless of whether they are healthful. But, in actual fact, there are plenty of exceptions to these rights. In the United States it is illegal to sell not only LSD and heroin, but also unpasteurized milk and imported haggis (the national dish of Scotland). In the spirit of soft paternalism, a more sensible and fairer tactic is to enact regulations to help people make choices they can rationally judge to be in their self-interest. Since taxing things is less coercive than banning them, perhaps a first step is to tax or charge individuals for the unhealthy choices they knowingly make that affect others. In this regard, is taxing soda or fast food different from taxing cigarettes and alcohol? I am sure you can think of many nudges (or shoves) that could help make modern environments better promote prevention. One might be to regulate advertising for junk foods, as we do for cigarettes and alcohol. (Every large soda could come with a label that says "SURGEON GENERAL'S WARNING: Consuming too much sugar causes obesity, diabetes, and heart disease.") Another is to require packaged foods to label the content and quantity of food portions unambiguously and without deception, and to cease marketing highly fattening sugar-laden food as "fat free." Perhaps we can require buildings to make stairs more accessible than elevators. Yet more nudges would be to stop rewarding or incentivizing individuals and companies to act in ways that promote disease. Such logic suggests that we cease subsidizing farmers to grow so much corn that gets turned into high fructose corn syrup, corn-fed beef, and other unhealthy foods.

In short, if cultural evolution got us into this mess, then shouldn't cultural evolution be able to get us out? For millions of years, our ancestors relied on innovation and cooperation to get enough food, to help care for one another's children, and to survive in hostile environments, such as deserts, tundras, and jungles. Today we need to innovate and cooperate in new ways to avoid eating too much food, especially excess sugar and processed industrial foods, and to survive in cities, suburbs, and other unnatural environments. We therefore need government and other social institutions on our side, because we never evolved to choose healthy lifestyles. Most people don't get sick through any fault of their own, but instead they acquire chronic illnesses as they age because they grew up in an environment that encourages, entices, and sometimes even forces them to become sick. For many of these diseases, we can then only treat the symptoms. Unless we want to end up as a species ever more dependent on medicines and expensive technologies to cope with the symptoms of preventable diseases, we need to change our environments. In fact, it is questionable whether we can continue to afford the cost of our current trajectory of increased longevity and population sizes combined with increased chronic morbidity.

I think it is reasonable to conclude that cultural evolutionary processes today are gradually replacing one form of coercion with another. For millions of years, our ancestors were required to consume a naturally healthy diet and to be physically active. Cultural evolution, especially since humans began farming, has transformed how our bodies interact with the environment. Many people today still live in poverty and suffer from diseases caused by poor sanitation, contagion, and malnutrition that were much less common in the Paleolithic. Those of us fortunate enough to live in the developed world have escaped those miseries, and we can now choose to be inactive as much as we want and eat whatever we crave. In fact, for some, such habits are the default setting. Those choices or urges, however, often make us sick in other ways, which then compel us to treat our symptoms. Right now, we are generally satisfied with the system we have created, thanks to long life spans and overall decent health. But we could do better. And as the mismatch environments we have created and pass on to our children through the pernicious feedback loop of dysevolution intensify, we increase our risk of suffering from needless, preventable diseases.

Last Words: Marching Backward into the Future

Some people erroneously think that natural selection means "survival of the fittest." Darwin never used that phrase (it was coined in 1864 by Hebert Spencer), nor would he have, because natural selection is better described as "survival of the fitter." Natural selection doesn't produce perfection; it only weeds out those unlucky enough to be less fit than others. Does "survival of the fitter" have any useful meaning in today's world, in which so many of us believe we have put evolution behind us?

A common answer to that question is that evolution still matters because it explains why our bodies are the way they are, including why we get sick. Remember, "Nothing in biology makes sense except in the light of evolution." Our evolutionary history thus accounts for how and why our skeletons, hearts, intestines, and brains work the way they do. Evolution also explains how and why in the course of a mere 6 million years we changed from being apes in an African forest to being upright, striding bipeds who peer through telescopes into distant galaxies searching for other forms of life. It's been an amazing 6 million years, but our species' evolution occurred through just a few transformations. None of these shifts were drastic, all of them were chance events contingent on previous changes, and, more often than not, they were driven by climate change.

In the grand scheme of things, if there is any one most transformative human adaptation that we evolved it must be our ability to evolve through culture rather than just natural selection. Today, cultural evolution is outpacing and sometimes outwitting natural selection. Many recent human inventions were adopted because they helped our ancestors produce more food, harness more energy, and have more children. Unintended by-products of these cultural innovations, however, were increased levels of infectious disease from larger, denser populations, inadequate sanitation, and less nutritious food. Civilization also brought extreme famines, dictatorships, war, slavery, and other modern misfortunes. In recent years we have made much progress to redress these man-made problems, and arguably people in the developed world are now better off than hunter-gatherers ever were.

Evolution, or the survival of the fitter, has thus brought us to where we are, and it explains much that is good and bad about being a twenty-first-century human being. But what about our future? Will our infinitely inventive minds allow us to continue to make progress with new technologies? Or are we headed for collapse? Can thinking about evolution help us improve the human condition?

If there is any one most useful lesson to learn from our species' rich and complex evolutionary history, it is that culture does not allow us to transcend our biology. Human evolution never was a triumph of brains over brawn, and we should be skeptical of the science fiction that the future will be any different. Clever as we are, we cannot alter the bodies we inherited in more than superficial ways, and it is dangerously arrogant to think we can engineer feet, liver cells, brains, or other body parts any better than nature already does. Like it or not, we are slightly fat, furless, bipedal primates who crave sugar, salt, fat, and starch, but we are still adapted to eating a diverse diet of fibrous fruits and vegetables, nuts, seeds, tubers, and lean meat. We enjoy rest and relaxation, but our bodies are still those of endurance athletes evolved to walk many miles a day and often run, as well as dig, climb, and carry. We love many comforts, but we are not well adapted to spend our days indoors in chairs, wearing supportive shoes, staring at books or screens for hours on end. As a result, billions of people suffer from diseases of affluence, novelty, and disuse that used to be rare or unknown. We then treat the symptoms of these diseases because it is easier, more profitable, and more urgent than treating their causes, many of which we don't understand anyway. In doing so, we perpetuate a pernicious feedback loop—dysevolution—between culture and biology.

Maybe this feedback loop isn't so bad. Perhaps we'll reach a sort of steady state in which we perfect the science of treating preventable diseases of affluence, disuse, and novelty. I doubt it, and it's foolish to wait around hoping that scientists of the future will finally conquer cancer, osteoporosis, or diabetes. There is a better way, and it is available immediately by paying better attention to how and why our bodies got to be the way they are. We don't yet know how to cure most of the major diseases that kill or disable people, but we do know how to lessen their likelihood and sometimes prevent them by using the bodies we inherited more as they evolved to be used. Just as cultural innovations have caused many of these mismatch diseases, other cultural innovations can also help us prevent them. Doing so will take a mixture of science, education, and intelligent collective action.

Just as this is not the best of all possible worlds, your body is not the best of all possible bodies. But it's the only one you'll ever have, and it's worth enjoying, nurturing, and protecting. The human body's past was molded by the survival of the fitter, but your body's future depends on how you use it. At the end of *Candide*, Voltaire's critique of complacent optimism, the hero finds peace, declaring: "We must cultivate our garden." To that I would add: We must cultivate our bodies.

Acknowledgments

I am especially grateful to my wife, Tonia, who read every page (often many times) and to my daughter, Eleanor. Both of them have been supportive and tolerant of my long hours of work and they have put up with far too many conversations about australopiths, exercise, diet, and a host of illnesses (many of which, thankfully, never even made it into the book). Several wonderful friends and colleagues have helped me edit and revise portions of the book. Special thanks go to David Pilbeam, Carole Hooven, Alan Garber, and Tucker Goodrich, who read multiple chapters. I received critical help from Ofer Bar-Yosef, Rachel Carmody, Steve Corbett, Irene Davis, Jeremy DeSilva, Peter Ellison, David Haig, Katie Hinde, Pam Johnson, Benjamin Lieberman, Charlie Nunn, David Raichlen, and Chet Sherwood.

For additional help, collaborations, wonderful conversations, and other forms of support, I am grateful to Brian Addison, Meir Barak, Caroline Bleeke, Mark Blumenkrantz, Dennis Bramble, Eric Castillo, Fuzz Crompton, Adam Daoud, Chris Dean, Maureen Devlin, Pierre d'Hemecourt, Heather Dingwall, Carolyn Eng, Brenda Frazier, Michael and Dorothy Hintze, Jean-Jacques Hublin, Soumya James, Farish A. Jenkins Jr., Yana Kamberov, Karen Kramer, Kristi Lewton, Philip Lieberman, David Ludwig, Meg Lynch, Zarin Machanda, Mickey Mahaffey, Chris McDougall, Richard Meadow, Bruce Morgan, Yannis Pitsiladis, John Polk, Herman Pontzer, Anne Prescott, Philip Rightmire, Neil Roach, Craig Rodgers, Campbell Rolian, Maryellen Ruvolo, Pardis Sabeti, Lee Saxby, John Shea, Tanya Smith, Cliff Tabin, Noreen Tuross, Madhusudhan Venkadesan, Anna Warrener, William Werbel, Katherine Whitcome, Richard Wrangham, and Katie Zink. I apologize to anyone I inadvertently left out.

I am also grateful to my agent, Max Brockman, for his unceasing support and help, and to my extraordinary and very helpful editor, Erroll McDonald, with whom I am fortunate to have worked.

Finally, I thank all the many students I have had the privilege to teach and learn from.

Notes

1 Introduction: *What Are Humans Adapted For?*

1. Haub, C., and O. P. Sharma (2006). India's population reality: Reconciling change and tradition. *Population Bulletin* 61: 1–20; http://data.worldbank.org/indicator/SP.DYN.LE00.IN.
2. I'll return to these issues in chapter 9. A comprehensive summary of the evidence for the epidemiological transition is summarized in a special issue on the global burden of disease published in December 2012 in *The Lancet*.
3. Hayflick, N. (1998). How and why we age. *Experimental Gerontology* 33: 639–53.
4. Khaw, K.-T., et al. (2008). Combined impact of health behaviours and mortality in men and women: The EPIC-Norfolk Prospective Population Study. *PLoS Medicine* 5: e12.
5. OECD (2011). *Health at a Glance 2011*. Paris: Organization of Economic Cooperation and Development Publishing; http://dx.doi.org/10.1787/health_glance-2011-en.
6. Alfred Russel Wallace also came up with the same basic theory, which Darwin and Wallace jointly presented in 1858 to the Linnaean Society of London. Wallace deserves more credit than he often gets, but Darwin had a much more complete and documented theory, which he published the next year in *On the Origin of Species*.
7. Sometimes natural selection is termed "survival of the fittest," a phrase Darwin never used, and which really ought to be "survival of the fitter."
8. The ENCODE Project Consortium (2012). An integrated encyclopedia of DNA elements in the human genome. *Nature* 489: 57–74.
9. Biologists often refer to these features as "spandrels" because of a famous essay by Stephen J. Gould and Richard Lewontin that argued that many features are not adaptations but instead are emergent properties of development or structure. The analogy they used were spandrels, the space between two

adjoining arches that is often used in churches for decoration. Gould and Lewontin argued that just as spandrels were by-products of the way arches are built and not intentional design features, many features of organisms that apparently have some function were not originally adaptations. To read the article, see Lewontin, R. C., and S. J. Gould (1979). The spandrels of San Marcos and the Panglossian paradigm: A critique of the adaptationist programme. *Proceedings of the Royal Society of London B* 205: 581–98.

10. There are many excellent discussions of this issue. A classic, still worth reading, is Williams, G. C. (1966). *Adaptation and Natural Selection*. Princeton, NJ: Princeton University Press.

11. Although Darwin first wrote about the Galápagos finches, most of what we know about selection on these finches comes from the work of Peter and Rosemary Grant. For very readable summaries of their research, see Grant, P. R. (1991). Natural selection and Darwin's finches. *Scientific American* 265: 81–87; Weiner, J. (1994). *The Beak of the Finch: A Story of Evolution in Our Time*. New York: Knopf.

12. Jablonski, N. G. (2006). *Skin: A Natural History*. Berkeley: University of California Press.

13. For a wonderful big-picture review of these events, I recommend Shubin, N. (2008). *Your Inner Fish: A Journey into the 3.5-Billion-Year History of the Human Body*. New York: Vintage Books.

14. For a thoughtful analysis of how scientists tell human evolutionary history using stories, and how analyzing the structure of those stories tells us something about science, see Landau, M. (1991). *Narratives of Human Evolution*. New Haven, CT: Yale University Press.

15. Dobzhansky, T. (1973). Nothing in biology makes sense except in the light of evolution. *The American Biology Teacher* 35: 125–29.

16. Primates in zoos who eat overly processed diets and get insufficient physical activity acquire type 2 diabetes through similar mechanisms as their human counterparts. See Rosenblum, I. Y., T. A. Barbolt, and C. F. Howard Jr. (1981). Diabetes mellitus in the chimpanzee (*Pan troglodytes*). *Journal of Medical Primatology* 10: 93–101.

17. For an introduction to the field of evolutionary medicine, see Williams, G. C., and R. M. Nesse (1996). *Why We Get Sick: The New Science of Darwinian Medicine*. New York: Vintage Books. Other excellent reviews are also available: Stearns, S. C., and J. C. Koella (2008). *Evolution in Health and Disease,* 2nd ed. Oxford: Oxford University Press; Gluckman, P., and M. Hanson (2006). *Mismatch: The Lifestyle Diseases Timebomb*. Oxford: Oxford University Press; Trevathan, W. R., E. O. Smith, and J. J. McKenna (2008). *Evolutionary Medicine and Health*. Oxford: Oxford University Press; Gluckman, P., A. Beedle, and M. Hanson (2009). *Principles of Evolutionary Medicine*. Oxford: Oxford University Press; Trevathan, W. R. (2010). *Ancient Bodies, Modern Lives: How Evolution Has Shaped Women's Health*. Oxford: Oxford University Press.

2 Upstanding Apes—*How We Became Bipeds*

1. Experiments that try to measure chimp strength are hard to evaluate because of factors such as motivation and inhibition. The first such study, from 1926, suggested that chimps had five times more strength than humans, but more recent studies by Finch (1943), Edwards (1965), and Scholz et al. (2006) suggest that chimps may be only twice as powerful as the strongest humans. Even so, the difference is striking. For references, see Bauman, J. E. (1926). Observations on the strength of the chimpanzee and its implications. *Journal of Mammalology* 7: 1–9; Finch, G. (1943). The bodily strength of chimpanzees. *Journal of Mammalogy* 24: 224–28; Edwards, W. E. (1965). *Study of monkey, ape and human morphology and physiology relating to strength and endurance. Phase IX: The strength testing of five chimpanzee and seven human subjects.* Holloman Air Force Base, NM, 6571st Aeromedical Research Laboratory, Holloman, New Mexico; Scholz, M. N., et al. (2006). Vertical jumping performance of bonobo (*Pan paniscus*) suggests superior muscle properties. *Proceedings of the Royal Society B: Biological Sciences* 273: 2177–84.

2. Darwin, C. (1871). *The Descent of Man.* London: John Murray, 140–42.

3. There are hundreds of fossil apes from dozens of extinct species that lived during the period between about 20 and 10 million years ago. However, the relationships between these species and their relationships to chimps, gorillas, and the LCA are unclear and highly debated. For a review of these fossils, see Fleagle, J. (2013). *Primate Adaptation and Evolution,* 3rd ed. New York: Academic Press.

4. The term used to be "hominid." However, according to the complex rules of Linnaean classification, the fact that humans are more closely related to chimps than to gorillas requires the term "hominin" because we belong to the tribe Homininae.

5. Shea, B. T. (1983). Paedomorphosis and neoteny in the pygmy chimpanzee. *Science* 222: 521–22; Berge, C., and X. Penin (2004). Ontogenetic allometry, heterochrony, and interspecific differences in the skull of African apes, using tridimensional Procrustes analysis. *American Journal of Physical Anthropology* 124: 124–38; Guy, F., et al. (2005). Morphological affinities of the *Sahelanthropus tchadensis* (Late Miocene hominid from Chad) cranium. *Proceedings of the National Academy of Sciences* 102: 18836–41.

6. Lieberman, D. E., et al. (2007). A geometric morphometric analysis of heterochrony in the cranium of chimpanzees and bonobos. *Journal of Human Evolution* 52: 647–62; Wobber, V., R. Wrangham, and B. Hare (2010). Bonobos exhibit delayed development of social behavior and cognition relative to chimpanzees. *Current Biology* 20: 226–30.

7. The chief proponent of this idea was the great British anatomist Sir Arthur Keith, which he championed in his classic book: Keith, A. (1927). *Concerning Man's Origin.* London: Watts.

8. White, et al. (2009). *Ardipithecus ramidus* and the paleobiology of early hominids. *Science* 326: 75–86.

9. For the original descriptions of the cranial material, see Brunet, M., et al. (2002). A new hominid from the upper Miocene of Chad, central Africa.

Nature 418: 145–51; Brunet, M., et al. (2005). New material of the earliest hominid from the Upper Miocene of Chad. *Nature* 434: 752–55. The postcrania have yet to be described. For popular accounts of these remains and how they were found, see Reader, J. (2011). *Missing Links: In Search of Human Origins*. Oxford: Oxford University Press; Gibbons, A. (2006). *The First Human*. New York: Doubleday.

10. One method of dating compares fossils from the site with similar dated fossils from eastern Africa. Another method used a new technique based on beryllium isotopes. See Vignaud, P., et al. (2002). Geology and palaeontology of the Upper Miocene Toros-Menalla hominid locality, Chad. *Nature* 418: 152–55; Lebatard, A. E., et al. (2008). Cosmogenic nuclide dating of *Sahelanthropus tchadensis* and *Australopithecus bahrelghazali* Mio-Pliocene early hominids from Chad. *Proceedings of the National Academy of Sciences USA* 105: 3226–31.

11. Pickford, M., and B. Senut (2001). "Millennium ancestor," a 6-million-year-old bipedal hominid from Kenya. *Comptes rendus de l'Académie des Sciences de Paris, série 2a*, 332: 134–44.

12. Haile-Selassie, Y., G. Suwa, and T. D. White (2004). Late Miocene teeth from Middle Awash, Ethiopia, and early hominid dental evolution. *Science* 303: 1503–5; Haile-Selassie, Y., G. Suwa, and T. D. White (2009). Hominidae. In *Ardipithecus kadabba: Late Miocene Evidence from the Middle Awash, Ethiopia*, ed. Y. Haile-Selassie and G. WoldeGabriel. Berkeley: University of California Press, 159–236.

13. White, T. D., G. Suwa, and B. Asfaw (1994). *Australopithecus ramidus*, a new species of early hominid from Aramis, Ethiopia. *Nature* 371: 306–12; White, T. D., et al. (2009). *Ardipithecus ramidus* and the paleobiology of early hominids. *Science* 326: 75–86; Semaw, S., et al. (2005). Early Pliocene hominids from Gona, Ethiopia. *Nature* 433: 301–5.

14. For details, see Guy, F., et al. (2005). Morphological affinities of the *Sahelanthropus tchadensis* (Late Miocene hominid from Chad) cranium. *Proceedings of the National Academy of Sciences USA* 102: 18836–41; Suwa, G., et al. (2009). The *Ardipithecus ramidus* skull and its implications for hominid origins. *Science* 326: 68e1–7; Suwa, G., et al. (2009). Paleobiological implications of the *Ardipithecus ramidus* dentition. *Science* 326: 94–99; Lovejoy, C. O. (2009). Reexamining human origins in the light of *Ardipithecus ramidus*. *Science* 326: 74e1–8.

15. Wood, B., and T. Harrison (2012). The evolutionary context of the first hominins. *Nature* 470: 347–52.

16. The best predictor of when animals walk is their rate of brain development (starting the clock at conception), and in this regard humans are just where they should be compared to other animals from mice to elephants. See Garwicz, M., M. Christensson, and E. Psouni (2009). A unifying model for timing of walking onset in humans and other mammals. *Proceedings of the National Academy of Sciences USA* 106: 21889–93.

17. Lovejoy, C. O., et al. (2009). The pelvis and femur of *Ardipithecus ramidus*: The emergence of upright walking. *Science* 326: 71e1–6.

18. Richmond, B. G., and W. L. Jungers (2008). *Orrorin tugenensis* femoral morphology and the evolution of hominin bipedalism. *Science* 319: 1662–65.

19. Lovejoy, C. O., et al. (2009). The pelvis and femur of *Ardipithecus ramidus:* The emergence of upright walking. *Science* 326: 71e1–6.

20. Zollikofer, C. P., et al. (2005). Virtual cranial reconstruction of *Sahelanthropus tchadensis*. *Nature* 434: 755–59.

21. Lovejoy, C. O., et al. (2009). Combining prehension and propulsion: The foot of *Ardipithecus ramidus*. *Science* 326: 72e1–8; Haile-Selassie, Y., et al. (2012). A new hominin foot from Ethiopia shows multiple Pliocene bipedal adaptations. *Nature* 483: 565–69.

22. DeSilva, J. M., et al. (2013). The lower limb and mechanics of walking in *Australopithecus sediba*. *Science* 340: 1232999.

23. Lovejoy, C. O. (2009). Careful climbing in the Miocene: The forelimbs of *Ardipithecus ramidus* and humans are primitive. *Science* 326: 70e1–8.

24. Brunet, M., et al. (2005). New material of the earliest hominid from the Upper Miocene of Chad. *Nature* 434: 752–55; Haile-Selassie, Y., G. Suwa, and T. D. White (2009). Hominidae. In *Ardipithecus kadabba: Late Miocene Evidence from the Middle Awash, Ethiopia,* ed. Y. Haile-Selassie and G. WoldeGabriel. Berkeley: University of California Press, 159–236; Suwa, G., et al. (2009). Paleobiological implications of the *Ardipithecus ramidus* dentition. *Science* 326: 94–99.

25. Guy, F., et al. (2005). Morphological affinities of the *Sahelanthropus tchadensis* (Late Miocene hominid from Chad) cranium. *Proceedings of the National Academy of Sciences USA* 102: 18836–41; Suwa, G., et al. (2009). The *Ardipithecus ramidus* skull and its implications for hominid origins. *Science* 326: 68e1–7.

26. Haile-Selassie, Y., G. Suwa, and T. D. White (2004). Late Miocene teeth from Middle Awash, Ethiopia, and early hominid dental evolution. *Science* 303: 1503–5.

27. Some researchers have suggested that smaller canines are signs of a social system with less fighting between males and perhaps even pair-bonding. However, differences between male and female canine size in other primate species do not predict very well how much males compete with one another, and body size estimates of later species hint that early hominin males were about 50 percent bigger than females—a sign that males were engaged in intense competition with one another. An alternative hypothesis is that canine length limits gape hence bite force. In order to have large canines, one has to have a wide gape and position the jaw-closing muscles farther back, making these muscles less efficient at generating chewing force. For this reason, smaller canines are associated with smaller gapes and more powerful chewing. For more details of these hypotheses, see Lovejoy, C. O. (2009). Reexamining human origins in the light of *Ardipithecus ramidus*. *Science* 326: 74e1–8; Plavcan, J. M. (2000). Inferring social behavior from sexual dimorphism in the fossil record. *Journal of Human Evolution* 39: 327–44; Hylander, W. L. (2013). Functional links between canine height and jaw gape in catarrhines with special reference to early hominins. *American Journal of Physical Anthropology* 150: 247–59.

28. These data come from many sources, but the best evidence comes from the shells of tiny sea creatures, foraminifera, which form shells of calcium carbonate ($CaCO_3$) and then sink to the ocean's floor when they die. When

the oceans are warmer, the oxygen atoms incorporated into the shells have a higher percentage of the heaver oxygen isotope (O_{18} versus O_{16}). Thus, by excavating and analyzing the ratio of O_{18} versus O_{16} in long cores from the ocean floor, one can measure how the ocean's temperature changed over time. Figure 4 is from an especially comprehensive study of oxygen isotopes: Zachos, J., et al. (2001). Trends, rhythms, and aberrations in global climate 65 Ma to present. *Science* 292: 686–93.

29. Kingston, J. D. (2007). Shifting adaptive landscapes: Progress and challenges in reconstructing early hominid environments. *Yearbook of Physical Anthropology* 50: 20–58.

30. Laden, G., and R. W. Wrangham (2005). The rise of the hominids as an adaptive shift in fallback foods: Plant underground storage organs (USOs) and the origin of the Australopiths. *Journal of Human Evolution* 49: 482–98.

31. For a description of how orangutans cope, see Knott, C. D. (2005). Energetic responses to food availability in the great apes: Implications for Hominin evolution. In *Primate Seasonality: Implications for Human Evolution*, ed. D. K. Brockman and C. P. van Schaik. Cambridge: Cambridge University Press, 351–78.

32. Thorpe, S. K. S., R. L. Holder, and R. H. Crompton (2007). Origin of human bipedalism as an adaptation for locomotion on flexible branches. *Science* 316: 1328–31.

33. Hunt, K. D. (1992). Positional behavior of *Pan troglodytes* in the Mahale Mountains and Gombe Stream National Parks, Tanzania. *American Journal of Physical Anthropology* 87: 83–105.

34. Carvalho, S., et al. (2012). Chimpanzee carrying behaviour and the origins of human bipedality. *Current Biology* 22: R180–81.

35. Sockol, M. D., D. Raichlen, and H. D. Pontzer (2007). Chimpanzee locomotor energetics and the origin of human bipedalism. *Proceedings of the National Academy of Sciences USA* 104: 12265–69.

36. Pontzer, H. D., and R. W. Wrangham (2006). The ontogeny of ranging in wild chimpanzees. *International Journal of Primatology* 27: 295–309.

37. Lovejoy, C. O. (1981). The origin of man. *Science* 211: 341–50; Lovejoy, C. O. (2009). Reexamining human origins in the light of *Ardipithecus ramidus*. *Science* 326: 74e1–8.

38. To be honest, there are not enough fossils to figure out male versus female body size in any of the earliest hominins. The best evidence for size differences between males and females comes from later australopiths, in which males are about 50 percent more dimorphic than females. See Plavcan, J. M., et al. (2005). Sexual dimorphism in *Australopithecus afarensis* revisited: How strong is the case for a human-like pattern of dimorphism? *Journal of Human Evolution* 48: 313–20.

39. Mitani, J. C., J. Gros-Louis, and A. Richards (1996). Sexual dimorphism, the operational sex ratio, and the intensity of male competition among polygynous primates. *American Naturalist* 147: 966–80.

40. Pilbeam, D. (2004). The anthropoid postcranial axial skeleton: Comments on development, variation, and evolution. *Journal of Experimental Zoology Part B* 302: 241–67.

41. Whitcome, K. K., L. J. Shapiro, and D. E. Lieberman (2007). Fetal load and the evolution of lumbar lordosis in bipedal hominins. *Nature* 450: 1075–78.

3 Much Depends on Dinner—*How the Australopiths Partly Weaned Us Off Fruit*

1. Raw foodists deem it harmful to cook their food above normal body temperature, based on the logic that humans originally evolved to eat raw food, and because they believe that heating destroys natural vitamins and enzymes. Although it is true that our ancestors ate only raw food and that overly processed food can be unhealthy, the other claims are generally untrue. Cooking actually increases the availability of nutrients from most foods. In addition, humans have been cooking their food long enough to have made cooking a biological necessity and a human universal. Raw foodism has only become possible recently through processing highly domesticated foods that are much less fibrous and more energy rich than the wild foods that used to be available. Even so, raw foodists often lose weight, suffer from low fertility, and increase their risk of getting sick from bacteria and other pathogens that are otherwise destroyed by heat. For more, see Wrangham, R. W. (2009). *Catching Fire: How Cooking Made Us Human.* New York: Basic Books. For comparative data on feeding times, see Organ, C., et al. (2011). Phylogenetic rate shifts in feeding time during the evolution of *Homo. Proceedings of the National Academy of Sciences USA* 108: 14555–59.
2. Wrangham, R. W. (1977). Feeding behaviour of chimpanzees in Gombe National Park, Tanzania. In *Primate Ecology,* ed. T. H. Clutton-Brock. London: Academic Press, 503–38.
3. McHenry, H. M., and K. Coffing (2000). *Australopithecus* to *Homo:* Transitions in body and mind. *Annual Review of Anthropology* 29: 145–56.
4. Haile-Selassie, Y., et al. (2010). An early *Australopithecus afarensis* postcranium from Woranso-Mille, Ethiopia. *Proceedings of the National Academy of Sciences USA* 107: 12121–26.
5. Dean, M. C. (2006). Tooth microstructure tracks the pace of human life-history evolution. *Proceedings of the Royal Society B* 273: 2799–808.
6. In truth, there are no reasonably complete partial skeletons of robust australopiths. Thus, although we know much about their distinctive skulls, we are less sure what the rest of their bodies were like.
7. DeSilva, J. M., et al. (2013). The lower limb and walking mechanics of *Australopithecus sediba. Science* 340: 1232999.
8. Cerling, T. E., et al. (2011). Woody cover and hominin environments in the past 6 million years. *Nature* 476: 51–56; deMenocal, P. B. (2011). Anthropology. Climate and human evolution. *Science* 331(6017): 540–42; Passey, B. H., et al. (2010). High-temperature environments of human evolution in East Africa based on bond ordering in paleosol carbonates. *Proceedings of the National Academy of Sciences USA* 107: 11245–49.

9. As chapter 1 discussed, an especially well documented example of selection for fallback foods comes from the Galápagos finches, first studied by Darwin and more recently by Peter and Rosemary Grant. During lengthy droughts, many finches perish from starvation, because preferred foods, such as cactus fruits, become rare. However, finches with thicker beaks are more likely to survive these droughts because they are better able to eat harder foods like seeds. Under such conditions, thicker-beaked finches have more surviving progeny, and since beak thickness is inherited, the percentage of thick-beaked birds increases in the next generation. For a superb description of this research, see Weiner, J. (1994). *The Beak of the Finch: A Story of Evolution in Our Time*. New York: Knopf.

10. Grine, F. E., et al. (2012). Dental microwear and stable isotopes inform the paleoecology of extinct hominins. *American Journal of Physical Anthropology* 148: 285–317; Ungar, P. S. (2011). Dental evidence for the diets of Plio-Pleistocene hominins. *Yearbook of Physical Anthropology* 54: 47–62; Ungar, P., and M. Sponheimer (2011). The diets of early hominins. *Science* 334: 190–93.

11. Wrangham, R. W. (2005). The delta hypothesis. In *Interpreting the Past: Essays on Human, Primate, and Mammal Evolution,* eds. D. E. Lieberman, R. J. Smith, and J. Kelley. Leiden: Brill Academic, 231–43.

12. Wrangham, R. W., et al. (1999). The raw and the stolen: Cooking and the ecology of human origins. *Current Anthropology* 99: 567–94.

13. Wrangham, R. W., et al. (1991). The significance of fibrous foods for Kibale Forest chimpanzees. *Philosophical Transactions of the Royal Society, Part B Biological Science* 334: 171–78.

14. Laden, G., and R. Wrangham (2005). The rise of the hominids as an adaptive shift in fallback foods: Plant underground storage organs (USOs) and australopith origins. *Journal of Human Evolution* 49: 482–98.

15. Wood, B. A., S. A. Abbott, and H. Uytterschaut (1988). Analysis of the dental morphology of Plio-Pleistocene hominids IV. Mandibular postcanine root morphology. *Journal of Anatomy* 156: 107–39.

16. Lucas, P. W. (2004). *How Teeth Work*. Cambridge: Cambridge University Press.

17. Efficient force production takes advantage of simple principles of Newtonian physics. Like all muscles, the chewing muscles create rotational forces, called torques, that move the jaw. Just as a longer wrench handle enables you to generate more torque with the same amount of applied force, moving the insertions of the chewing muscles away from the jaw joint increases how much torque, hence bite force, these muscles can generate. This principle explains much about the configuration of australopith skulls. For example, as you can see from figure 6, the cheekbones in the australopiths were impressively long, emerging far forward on the face, and they spread far out to the side. Wide and forwardly positioned cheekbones enabled the australopith's masseter muscles to generate high vertical and sideways forces when they chewed. By summing up how much force every chewing muscle could produce, we can estimate that an *Au. boisei* would have been able to bite down with about 2.5 times as much force as a human. It would have been very unwise to stick your finger in an australopith's mouth. For more details, see Eng, C. M., et

al. (2013). Bite force and occlusal stress production in hominin evolution. *American Journal of Physical Anthropology* online. 10.1002/ajpa.22296 http://www.ncbi.nlm.nih.gov/pubmed/23754526.

18. Currey, J. D. (2002). *Bones: Structure and Mechanics*. Princeton: Princeton University Press.

19. Rak, Y. (1983). *The Australopithecine Face*. New York: Academic Press; Hylander, W. L. (1988). Implications of in vivo experiments for interpreting the functional significance of "robust" australopithecine jaws. In *Evolutionary History of the "Robust" Australopithecines*, ed. F. Grine. New York: Aldine De Gruyter, 55–83; Lieberman, D. E. (2011). *The Evolution of the Human Head*. Cambridge, MA: Harvard University Press.

20. Climate change therefore explains the general trend we see among the australopiths toward thicker, bigger teeth, larger faces, and more massive jaws, culminating in robust species, such as *Au. boisei* and *Au. robustus*, all of which evolved after 2.5 million years ago.

21. Pontzer, H., and R. W. Wrangham. The ontogeny of ranging in wild chimpanzees. *International Journal of Primatology* 27: 295–309.

22. The cost of Groucho-walking was measured in Gordon, K. E., D. P. Ferris, and A. D. Kuo (2009). Metabolic and mechanical energy costs of reducing vertical center of mass movement during gait. *Archives of Physical Medicine and Rehabilitation* 90: 136–44. The comparison of chimps and humans derives from data in Sockol, M. D., D. A. Raichlen, and H. D. Pontzer (2007). Chimpanzee locomotor energetics and the origin of human bipedalism. *Proceedings of the National Academy of Sciences USA* 104: 12265–69. This important study found that a walking chimp spends 0.20 milliliters of oxygen per kilogram per meter, whereas a walking human spends 0.05 milliliters of oxygen per kilogram per meter. During aerobic respiration, a liter of oxygen converts to 5.13 kilocalories.

23. Schmitt, D. (2003). Insights into the evolution of human bipedalism from experimental studies of humans and other primates. *Journal of Experimental Biology* 206: 1437–48.

24. Latimer, B., and C. O. Lovejoy (1990). Hallucal tarsometatarsal joint in *Australopithecus afarensis*. *American Journal of Physical Anthropology* 82: 125–33; McHenry, H. M., and A. L. Jones (2006). Hallucial convergence in early hominids. *Journal of Human Evolution* 50: 534–39.

25. Harcourt-Smith, W. E., and L. C. Aiello (2004). Fossils, feet and the evolution of human bipedal locomotion. *Journal of Anatomy* 204: 403–16; Ward, C. V., W. H. Kimbel, and D. C. Johanson (2011). Complete fourth metatarsal and arches in the foot of *Australopithecus afarensis*. *Science* 331: 750–53; DeSilva, J. M., and Z. J. Throckmorton (2010). Lucy's flat feet: The relationship between the ankle and rearfoot arching in early hominins. *PLoS One* 5(12): e14432.

26. Latimer, B., and C. O. Lovejoy (1989). The calcaneus of *Australopithecus afarensis* and its implications for the evolution of bipedality. *American Journal of Physical Anthropology* 78: 369–86.

27. Zipfel, B., et al. (2011). The foot and ankle of *Australopithecus sediba*. *Science* 333: 1417–20.

28. Aiello, L. C., and M. C. Dean (1990). *Human Evolutionary Anatomy*. London: Academic Press.

29. We lack complete femurs from older hominins, so we don't know if this feature is unique to australopiths or evolved earlier in hominins such as *Ardipithecus*.

30. Been, E., A. Gómez-Olivencia, and P. A. Kramer (2012). Lumbar lordosis of extinct hominins. *American Journal of Physical Anthropology* 147: 64–77; Williams, S. A., et al. (2013). The vertebral column of *Australopithecus sediba*. *Science* 340: 1232996

31. Raichlen, D. A., H. Pontzer, and M. D. Sockol (2008). The Laetoli footprints and early hominin locomotor kinematics. *Journal of Human Evolution* 54: 112–17.

32. Churchill, S. E., et al. (2013). The upper limb of *Australopithecus sediba*. *Science* 340: 1233447.

33. Wheeler, P. E. (1991). The thermoregulatory advantages of hominid bipedalism in open equatorial environments: The contribution of increased convective heat loss and cutaneous evaporative cooling. *Journal of Human Evolution* 21: 107–15.

34. Tocheri, M. W., et al. (2008). The evolutionary history of the hominin hand since the last common ancestor of *Pan* and *Homo*. *Journal of Anatomy* 212: 544–62.

35. Goodall, J. (1986). *The Chimpanzees of Gombe: Patterns of Behavior.* Cambridge, MA: Harvard University Press; Boesch, C., and H. Boesch (1990). Tool use and tool making in wild chimpanzees. *Folia Primatologica* 54: 86–99.

4 The First Hunter-Gatherers—*How Nearly Modern Bodies Evolved in the Human Genus*

1. Zachos, J., et al. (2001). Trends, rhythms, and aberrations in global climate 65 Ma to present. *Science* 292: 686–93.

2. For a review of climate change and its effects on human evolution, I recommend Potts, R. (1986). *Humanity's Desert: The Consequences of Ecological Instability.* New York: William Morrow and Co.

3. Trauth, M. H., et al. (2005). Late Cenozoic moisture history of East Africa. *Science* 309: 2051–53.

4. Bobe, R. (2006). The evolution of arid ecosystems in eastern Africa. *Journal of Arid Environments* 66: 564–84; Passey, B. H., et al. (2010). High-temperature environments of human evolution in East Africa based on bond ordering in paleosol carbonates. *Proceedings of the National Academy of Sciences USA* 107: 11245–49.

5. For an engaging biography, see Shipman, P. (2001). *The Man Who Found the Missing Link: The Extraordinary Life of Eugene Dubois.* New York: Simon & Schuster.

6. It was actually a specialist on birds, Ernst Mayr, who restored sense to this taxonomic mess in a famous essay: Mayr, E. (1951). Taxonomic categories in fossil hominids. *Cold Spring Harbor Symposia on Quantitative Biology* 15: 109–18.

7. Ruff, C. B., and A. Walker (1993). Body size and body shape. In *The Nariokotome* Homo erectus *Skeleton*, ed. A. Walker and R. E. F. Leakey. Cambridge, MA: Harvard University Press, 221–65; Antón, S. C. (2003). Natural history of *Homo erectus*. *Yearbook of Physical Anthropology* 46: 126–70; Lordkipanidze, D., et al. (2007). Postcranial evidence from early *Homo* from Dmanisi, Georgia. *Nature* 449: 305–10; Graves, R. R., et al. (2010). Just how strapping was KNM-WT 15000? *Journal of Human Evolution* 59(5): 542–54.

8. Leakey, M. G., et al. (2012). New fossils from Koobi Fora in northern Kenya confirm taxonomic diversity in early *Homo*. *Nature* 488: 201–4.

9. Wood, B., and M. Collard (1999). The human genus. *Science* 284: 65–71.

10. Kaplan, H. S., et al. (2000). Theory of human life history evolution: Diet, intelligence, and longevity. *Evolutionary Anthropology* 9: 156–85.

11. Marlowe, F. W. (2010). *The Hadza: Hunter-Gatherers of Tanzania*. Berkeley: University of California Press.

12. The oldest clear evidence is 2.6 million years old and comes from several sites. For references, see de Heinzelin, J., et al. (1999). Environment and behavior of 2.5-million-year-old Bouri hominids. *Science* 284: 625–29; Semaw, S., et al. (2003). 2.6-million-year-old stone tools and associated bones from OGS-6 and OGS-7, Gona, Afar, Ethiopia. *Journal of Human Evolution* 45: 169–77. Bones dated to 3.4 million years old with putative cut marks have also been found, but these finds have been controversial. See McPherron, S. P., et al. (2010). Evidence for stone-tool-assisted consumption of animal tissues before 3.39 million years ago at Dikika, Ethiopia. *Nature* 466: 857–60.

13. Kelly, R. L. (2007). *The Foraging Spectrum: Diversity in Hunter-Gatherer Lifeways*. Clinton Corners, NY: Percheron Press.

14. Marlowe, F. W. (2010). *The Hadza: Hunter-Gatherers of Tanzania*. Berkeley: University of California Press.

15. Hawkes, K., et al. (1998). Grandmothering, menopause, and the evolution of human life histories. *Proceedings of the National Academy of Sciences USA* 95: 1336–39.

16. Hrdy, S. B. (2009). *Mothers and Others*. Cambridge, MA: The Belknap Press.

17. Wrangham, R. W., and N. L. Conklin-Brittain (2003). Cooking as a biological trait. *Comparative Biochemistry and Physiology—Part A: Molecular & Integrative Physiology* 136: 35–46.

18. Zink, K. D. (2013). Hominin food processing: material property, masticatory performance and morphological changes associated with mechanical and thermal processing techniques. Doctoral thesis, Harvard University, Cambridge, MA.

19. Carmody, R. N., G. S. Weintraub, and R. W. Wrangham (2011). Energetic consequences of thermal and nonthermal food processing. *Proceedings of the National Academy of Sciences USA* 108: 19199–203.

20. Meegan, G. (2008). *The Longest Walk: An Odyssey of the Human Spirit*. New York: Dodd Mead.

21. Marlowe, F. W. (2010). *The Hadza: Hunter-Gatherers of Tanzania*. Berkeley: University of California Press.

22. Pontzer, H., et al. (2010). Locomotor anatomy and biomechanics of the Dmanisi hominins. *Journal of Human Evolution* 58: 492–504.

23. Pontzer, H. (2007). Predicting the cost of locomotion in terrestrial animals: A test of the LiMb model in humans and quadrupeds. *Journal of Experimental Biology* 210: 484–94; Steudel-Numbers, K. (2006). Energetics in *Homo erectus* and other early hominins: The consequences of increased lower limb length. *Journal of Human Evolution* 51: 445–53.

24. Bennett, M. R., et al. (2009). Early hominin foot morphology based on 1.5-million-year-old footprints from Ileret, Kenya. *Science* 323: 1197–201; Dingwall, H. L., et al. (2013). Hominin stature, body mass, and walking speed estimates based on 1.5-million-year-old fossil footprints at Ileret, Kenya. *Journal of Human Evolution* 2013.02.004.

25. Ruff, C. B., et al. (1999). Cross-sectional morphology of the SK 82 and 97 proximal femora. *American Journal of Physical Anthropology* 109: 509–21; Ruff, C. B., et al. (1993). Postcranial robusticity in *Homo*. I: Temporal trends and mechanical interpretation. *American Journal of Physical Anthropology* 91: 21–53.

26. Ruff, C. B. (1988). Hindlimb articular surface allometry in Hominoidea and *Macaca,* with comparisons to diaphyseal scaling. *Journal of Human Evolution* 17: 687–714; Jungers, W. L. (1988). Relative joint size and hominoid locomotor adaptations with implications for the evolution of hominid bipedalism. *Journal of Human Evolution* 17: 247–65.

27. Wheeler, P. E. (1991). The thermoregulatory advantages of hominid bipedalism in open equatorial environments: The contribution of increased convective heat loss and cutaneous evaporative cooling. *Journal of Human Evolution* 21: 107–15.

28. See Ruff, C. B. (1993). Climatic adaptation and hominid evolution: The thermoregulatory imperative. *Evolutionary Anthropology* 2: 53–60; Simpson, S. W., et al. (2008). A female *Homo erectus* pelvis from Gona, Ethiopia. *Science* 322: 1089–92; Ruff, C. B. (2010). Body size and body shape in early hominins: Implications of the Gona pelvis. *Journal of Human Evolution* 58: 166–78.

29. Franciscus, R. G., and E. Trinkaus (1988). Nasal morphology and the emergence of *Homo erectus*. *American Journal of Physical Anthropology* 75: 517–27.

30. You can demonstrate this with a simple experiment on a cold day. Have a friend breathe out though his or her nose and then mouth. You'll notice much more steam when your friend exhales orally than nasally, because nasal turbulence recaptures more water vapor.

31. Van Valkenburgh, B. (2001). The dog-eat-dog world of carnivores: A review of past and present carnivore community dynamics. In *Meat-Eating and Human Evolution*, ed. C. B. Stanford and H. T. Bunn. Oxford: Oxford University Press, 101–21.

32. Wilkins, J., et al. (2012). Evidence for early Hafted hunting technology. *Science* 338: 942–46; Shea, J. J. (2006). The origins of lithic projectile point technology: Evidence from Africa, the Levant, and Europe. *Journal of Archaeological Science* 33: 823–46.

33. O'Connell, J. F., et al. (1988). Hadza scavenging: Implications for Plio-Pleistocene hominid subsistence. *Current Anthropology* 29: 356–63.

34. Potts, R. (1988). Environmental hypotheses of human evolution. *Yearbook of Physical Anthropology* 41: 93–136; Dominguez-Rodrigo, M. (2002). Hunting and scavenging by early humans: The state of the debate. *Journal of World Prehistory* 16: 1–54; Bunn, H. T. (2001). Hunting, power scavenging, and butchering by Hadza foragers and by Plio-Pleistocene *Homo*. In *Meat-Eating and Human Evolution*, ed. C. B. Stanford and H. T. Bunn. Oxford: Oxford University Press, 199–218; Braun, D. R., et al. (2010). Early hominin diet included diverse terrestrial and aquatic animals 1.95 Myr ago in East Turkana, Kenya. *Proceedings of the National Academy of Sciences USA* 107: 10002–7.

35. An untipped spear, unless it is very heavy, just bounces off an animal's hide. In addition, it is not the hole that a spear creates that usually kills an animal; instead, the lacerations caused by the jagged, sharp edges of a spearhead cause internal bleeding, hence death. Even today, hunters armed with tipped metal spears need to get within a few yards of their prey to have a chance of killing them. For details, see Churchill, S. E. (1993). Weapon technology, prey size selection and hunting methods in modern hunter-gatherers: Implications for hunting in the Palaeolithic and Mesolithic. In *Hunting and Animal Exploitation in the Later Palaeolithic and Mesolithic of Eurasia*, ed. G. L. Peterkin, H. M. Bricker, and P. A. Mellars. Archeological Papers of the American Anthropological Association no. 4, 11–24.

36. Carrier, D. R. (1984). The energetic paradox of human running and hominid evolution. *Current Anthropology* 25: 483–95; Bramble, D. M., and D. E. Lieberman (2004). Endurance running and the evolution of *Homo*. *Nature* 432: 345–52.

37. The explanation for this constraint is that galloping is a seesaw gait that causes an animal's guts to slosh back and forth with every stride, rhythmically slamming like a piston into the diaphragm. A galloping quadruped must therefore synchronize each stride with a single breath, preventing it from panting (which involves lots of rapid, short, shallow breaths). For more, see Bramble, D. M., and F. A. Jenkins Jr. (1993). Mammalian locomotor-respiratory integration: Implications for diaphragmatic and pulmonary design. *Science* 262: 235–40.

38. Hunters often pursue the biggest prey they can, because bigger animals overheat more rapidly. The reason for this is that body heat increases in proportion to body size, a cubic function, but the ability to lose heat increases linearly.

39. Liebenberg, L. (2006). Persistence hunting by modern hunter-gatherers. *Current Anthropology* 47: 1017–26.

40. Montagna, W. (1972). The skin of nonhuman primates. *American Zoologist* 12: 109–24.

41. It takes 531 kilocalories for a liter of water (a little more than a quart) to evaporate, and because of the law of conservation of energy, this change in state cools the skin by the same amount.

42. Schwartz, G. G., and L. A. Rosenblum (1981). Allometry of hair density and

the evolution of human hairlessness. *American Journal of Physical Anthropology* 55: 9–12.

43. Recall from chapter 3 that this is the opposite of walking, in which the center of mass rises during the first half of each step. Walking mostly uses pendular mechanics to move the body, whereas running uses mass-spring mechanics.

44. The same phenomenon has also been documented in kangaroos. For a full explanation, see Alexander, R. M. (1991). Energy-saving mechanisms in walking and running. *Journal of Experimental Biology* 160: 55–69.

45. Ker, R. F., et al. (1987). The spring in the arch of the human foot. *Nature* 325: 147–49.

46. Lieberman, D. E., D. A. Raichlen, and H. Pontzer (2006). The human gluteus maximus and its role in running. *Journal of Experimental Biology* 209: 2143–55.

47. Spoor, F., B. Wood, and F. Zonneveld (1994). Implications of early hominid labyrinthine morphology for evolution of human bipedal locomotion. *Nature* 369: 645–48.

48. Lieberman, D. E. (2011). *Evolution of the Human Head*. Cambridge, MA: Harvard University Press.

49. For a complete list of these features and their function, see Bramble, D. M., and D. E. Lieberman (2004). Endurance running and the evolution of *Homo*. *Nature* 432: 345–52.

50. Rolian, C., et al. (2009). Walking, running and the evolution of short toes in humans. *Journal of Experimental Biology* 212: 713–21.

51. Humans have a highly mobile torso that can twist independently relative to the hips and head. This twisting is important during running, because, unlike in walking, a runner spends part of every stride in the air, swinging one leg forward and the other back. This scissorlike motion creates an angular momentum that, unchecked, would rotate the runner's body to the left or right. A runner therefore needs to simultaneously swing the arms and rotate the trunk opposite the legs to generate equal angular momentum in the opposite direction. In addition, the torso's independent twisting helps keep the head from yawing from side to side. For a further explanation, see Hinrichs, R. N. (1990). Upper extremity function in distance running. In *Biomechanics of Distance Running*, ed. P. R. Cavanagh. Champaign, IL: Human Kinetics, 107–34; Pontzer, H., et al. (2009). Control and function of arm swing in human walking and running. *Journal of Experimental Biology* 212: 523–34.

52. Muscles have two major kinds of fibers: fast and slow twitch. Fast-twitch fibers contract more rapidly and forcefully than slow-twitch fibers, but they fatigue quickly and use more energy. Slow-twitch fibers are thus better for economy but limit speed. Most animals, including apes and monkeys, have high percentages of fast-twitch fibers in their legs, helping them run fast in short bursts, but human legs are dominated by slow-twitch fibers, which give us endurance. For example, the calf muscle is about 60 percent slow-twitch fibers in humans, but about 15 percent to 20 percent in macaques and chimps. We can only guess that *H. erectus* legs were also dominated by slow-twitch fibers. For references, see Acosta, L., and R. R. Roy (1987). Fiber-type

composition of selected hindlimb muscles of a primate (cynomolgus monkey). *Anatomical Record* 218: 136–41; Dahmane, R., et al. (2005). Spatial fiber type distribution in normal human muscle: Histochemical and tensiomyographical evaluation. *Journal of Biomechanics* 38: 2451–59; Myatt, J. P., et al. (2011). Distribution patterns of fiber types in the triceps surae muscle group of chimpanzees and orangutans. *Journal of Anatomy* 218: 402–12.

53. Goodall, J. (1986). *The Chimpanzees of Gombe*. Cambridge, MA: Harvard University Press.

54. Napier, J. R. (1993). *Hands*. Princeton, NJ: Princeton University Press.

55. Marzke, M. W., and R. F. Marzke (2000). Evolution of the human hand: Approaches to acquiring, analysing and interpreting the anatomical evidence. *Journal of Anatomy* 197 (pt. 1): 121–40.

56. Rolian, C., D. E. Lieberman, and J. P. Zermeno (2012). Hand biomechanics during simulated stone tool use. *Journal of Human Evolution* 61: 26–41.

57. Susman, R. L. (1998). Hand function and tool behavior in early hominids. *Journal of Human Evolution* 35: 23–46; Tocheri, M. W., et al. (2008). The evolutionary history of the hominin hand since the last common ancestor of *Pan* and *Homo. Journal of Anatomy* 212: 544–62; Alba, D., et al. (2003). Morphological affinities of the *Australopithecus afarensis* hand on the basis of manual proportions and relative thumb length. *Journal of Human Evolution* 44: 225–54.

58. Roach, N. T., et al. (2013). Elastic energy storage in the shoulder and the evolution of high-speed throwing in *Homo. Nature* 498: 483–86.

59. Another important feature that helps humans throw is low "torsion" of the humerus. In most people, as in chimps, the humerus has a twist, so that the elbow joint naturally faces inward, but people such as professional baseball players who throw often develop as much as 20 degrees less humeral torsion in their throwing versus nonthrowing arm. This configuration is an advantage because the less torsion you have, the more you can cock your arm back and store elastic energy. The two known *H. erectus* skeletons have humeral torsion values below those of most professional baseball players. For details, see Roach, N. T., et al. (2012). The effect of humeral torsion on rotational range of motion in the shoulder and throwing performance. *Journal of Anatomy* 220: 293–301; Larson, S. G. (2007). Evolutionary transformation of the hominin shoulder. *Evolutionary Anthropology* 16: 172–87.

60. The oldest archaeological evidence for fire is from Wonderwerk Cave in southern Africa. Whether the fire was used for cooking and when cooking became commonplace is unclear (this is discussed further in chapter 5). See Berna, F., et al. (2012). Microstratigraphic evidence of in situ fire in the Acheulean strata of Wonderwerk Cave, Northern Cape province, South Africa. *Proceedings of the National Academy of Sciences USA* 109: 1215–20.

61. Carmody, R. N., G. S. Weintraub, and R. W. Wrangham (2011). Energetic consequences of thermal and nonthermal food processing. *Proceedings of the National Academy of Sciences USA* 108: 19199–203.

62. Brace, C. L., S. L. Smith, and K. D. Hunt (1991). What big teeth you had, grandma! Human tooth size, past and present. In *Advances in Dental Anthropology*, ed. M. A. Kelley and C. S. Larsen. New York: Wiley-Liss, 33–57.

63. For an excellent review, see Alexander, R. M. (1999). *Energy for Animal Life*. Oxford: Oxford University Press.

64. For brain size, see Martin, R. D. (1981). Relative brain size and basal metabolic rate in terrestrial vertebrates. *Nature* 293: 57–60; for data on gut size, see Chivers, D. J., and C. M. Hladik (1980). Morphology of the gastrointestinal tract in primates: Comparisons with other mammals in relation to diet. *Journal of Morphology* 166: 337–86.

65. Aiello, L. C., and P. Wheeler (1995). The expensive-tissue hypothesis: The brain and the digestive system in human and primate evolution. *Current Anthropology* 36: 199–221.

66. Lieberman, D. E. (2011). *The Evolution of the Human Head*. Cambridge, MA: Harvard University Press.

67. See Hill, K. R., et al. (2011). Co-residence patterns in hunter-gatherer societies show unique human social structure. *Science* 331: 1286–89; Apicella, C. L., et al. (2012). Social networks and cooperation in hunter-gatherers. *Nature* 481: 497–501.

68. For a detailed description and analysis of these skills, see L. Liebenberg (2001). *The Art of Tracking: The Origin of Science*. Claremont, South Africa: David Philip Publishers.

69. Kraske, R. (2005). *Marooned: The Strange but True Adventures of Alexander Selkirk*. New York: Clarion Books.

70. There are several accounts of her deeds, the most famous being Marguerite de Navarre's rather pious version in the *Heptameron*. Http://digital.library .upenn.edu/women/navarre/heptameron/heptameron.html.

5 Energy in the Ice Age—*How We Evolved Big Brains Along with Large, Fat, Gradually Growing Bodies*

1. For a review of evolutionary theory behind these alternative strategies, see Stearns, S. C. (1992). *The Evolution of Life Histories*. Oxford: Oxford University Press.

2. Under the best of circumstances it is difficult to define fossil species precisely. Some experts consider *H. erectus* to be one highly variable species, but others see variants in East Africa, Georgia, and elsewhere as different but closely related species. For the purposes of this book, we'll consider *H. erectus* in the broadest sense (*sensu lato*) without worrying about the precise taxonomy.

3. Rightmire, G. P., D. Lordkipanidze, and A. Vekua (2006). Anatomical descriptions, comparative studies, and evolutionary significance of the hominin skulls from Dmanisi, Republic of Georgia. *Journal of Human Evolution* 50: 115–41; Lordkipanidze, D., et al. (2005). The earliest toothless hominin skull. *Nature* 434: 717–18.

4. Antón, S. C. (2003). Natural history of *Homo erectus*. *Yearbook of Physical Anthropology* 46: 126–70.

5. Some scholars classify the first Europeans as a separate species, *H. antecessor,* but the evidence to distinguish these fossils from *H. erectus* is very subtle. Bermúdez de Castro, J., et al. (1997). A hominid from the Lower Pleis-

tocene of Atapuerca, Spain: Possible ancestor to Neandertals and modern humans. *Science* 276: 1392–95.

6. This admittedly crude estimate assumes an annual growth rate of 0.004, a mean distance between territory centers of 24 kilometers (15 miles), and the establishment of a new territory northward every 500 years.

7. See Shreeve, D. C. (2001). Differentiation of the British late Middle Pleistocene interglacials: The evidence from mammalian biostratigraphy. *Quaternary Science Reviews* 20: 1693–705.

8. deMenocal, P. B. (2004). African climate change and faunal evolution during the Pliocene-Pleistocene. *Earth and Planetary Science Letters* 220: 3–24.

9. Rightmire, G. P., D. Lordkipanidze, and A. Vekua (2006). Anatomical descriptions, comparative studies and evolutionary significance of the hominin skulls from Dmanisi, Republic of Georgia. *Journal of Human Evolution* 50: 115–41; Lordkipanidze, D. T., et al. (2007). Postcranial evidence from early *Homo* from Dmanisi, Georgia. *Nature* 449: 305–10.

10. Ruff, C. B., and A. Walker (1993). Body size and body shape. In *The Nariokotome* Homo erectus *Skeleton*, ed. A. Walker and R. E. F. Leakey. Cambridge, MA: Harvard University Press, 221–65; Graves, R. R., et al. (2010). Just how strapping was KNM-WT 15000? *Journal of Human Evolution* 59(5): 542–54.; Spoor, F., et al. (2007). Implications of new early *Homo* fossils from Ileret, east of Lake Turkana, Kenya. *Nature* 448: 688–91; Ruff, C. B., E. Trinkaus, and T. W. Holliday (1997). Body mass and encephalization in Pleistocene *Homo*. *Nature* 387: 173–76.

11. Rightmire, G. P. (1998). Human evolution in the Middle Pleistocene: The role of *Homo heidelbergensis*. *Evolutionary Anthropology* 6: 218–27.

12. Arsuaga, J. L., et al. (1997). Size variation in Middle Pleistocene humans. *Science* 277: 1086–88.

13. Reich, D., et al. (2010). Genetic history of an archaic hominin group from Denisova Cave in Siberia. *Nature* 468: 1053–60; Scally, A., and R. Durbin (2012). Revising the human mutation rate: Implications for understanding human evolution. *Nature Reviews Genetics* 13: 745–53.

14. Reich, D., et al. (2011). Denisova admixture and the first modern human dispersals into Southeast Asia and Oceania. *American Journal of Human Genetics* 89: 516–28.

15. Klein, R. G. (2009). *The Human Career*, 3rd ed. Chicago: University of Chicago Press.

16. The oldest spears so far unearthed come from a 400,000-year-old site in Germany. These impressive javelins were seven and a half feet long, made of very dense wood, and were probably used as lances to kill horses, deer, and maybe even elephants. See Thieme, H. (1997). Lower Palaeolithic hunting spears from Germany. *Nature* 385: 807–10.

17. This way of making stone tools is called the Levallois technique, named after a suburb of Paris where these sorts of tools were discovered and named in the nineteenth century. However, the oldest evidence for this technique comes from the South African site of Kathu Pan. See Wilkins, J., et al. (2012). Evidence for early hafted hunting technology. *Science* 338: 942–46.

18. Berna, F. P., et al. (2012). Microstratigraphic evidence of in situ fire in the Acheulean strata of Wonderwerk Cave, Northern Cape province, South

Africa. *Proceedings of the National Academy of Sciences USA* 109: 1215–20; Goren-Inbar, N., et al. (2004). Evidence of hominin control of fire at Gesher Benot Ya'aqov, Israel. *Science* 304: 725–27. For a discussion of the limitations of the interpretations, see also Roebroeks, W., and P. Villa (2011). On the earliest evidence for habitual use of fire in Europe. *Proceedings of the National Academy of Sciences USA* 108: 5209–14.

19. Karkanas, P., et al. (2007). Evidence for habitual use of fire at the end of the Lower Paleolithic: Site-formation processes at Qesem Cave, Israel. *Journal of Human Evolution* 53: 197–212.

20. Green, R. E., et al. (2008). A complete Neandertal mitochondrial genome sequence determined by high-throughput sequencing. *Cell* 134: 416–26.

21. Green, R. E., et al. (2010). A draft sequence of the Neandertal genome. *Science* 328: 710–22; Langergraber, K. E., et al. (2012). Generation times in wild chimpanzees and gorillas suggest earlier divergence times in great ape and human evolution. *Proceedings of the National Academy of Sciences USA* 109: 15716–21.

22. Evidence for interbreeding does not mean that Neanderthals and modern humans are the same species. Many species can and do interbreed (the technical term is "hybridize"), but if the hybridization is minimal and the species remain very different, then it is more confusing than helpful to classify them as a single species.

23. Chemical analyses of their bones have suggested that they ate as much meat as other carnivores like wolves and foxes. See Bocherens, H. D., et al. (2001). New isotopic evidence for dietary habits of Neandertals from Belgium. *Journal of Human Evolution* 40: 497–505; Richards, M. P., and E. Trinkaus (2009). Out of Africa: Modern human origins special feature: Isotopic evidence for the diets of European Neanderthals and early modern humans. *Proceedings of the National Academy of Sciences USA* 106: 16034–39.

24. To be precise, brain mass scales to body mass to the power of 0.75. Expressed as an equation, brain mass = body mass$^{0.75}$. See Martin, R. D. (1981). Relative brain size and basal metabolic rate in terrestrial vertebrates. *Nature* 293: 57–60.

25. For a summary of these data, and all the equations to do the calculations yourself, see Lieberman, D. E. (2011). *Evolution of the Human Head.* Cambridge, MA: Harvard University Press.

26. Ruff, C. B., E. Trinkaus, and T. W. Holliday (1997). Body mass and encephalization in Pleistocene *Homo*. *Nature* 387: 173–76.

27. Vrba, E. S. (1998). Multiphasic growth models and the evolution of prolonged growth exemplified by human brain evolution. *Journal of Theoretical Biology* 190: 227–39; Leigh, S. R. (2004). Brain growth, life history, and cognition in primate and human evolution. *American Journal of Primatology* 62: 139–64.

28. DeSilva, J., and J. Lesnik (2006). Chimpanzee neonatal brain size: Implications for brain growth in *Homo erectus*. *Journal of Human Evolution* 51: 207–12.

29. Human brains have about 11.5 billion neurons, whereas chimps average 6.5 billion neurons. Haug, H. (1987). Brain sizes, surfaces, and neuronal sizes of the cortex cerebri: A stereological investigation of man and his variability

and a comparison with some mammals (primates, whales, marsupials, insectivores, and one elephant). *American Journal of Anatomy* 180: 126–42.

30. Changizi, M. A. (2001). Principles underlying mammalian neocortical scaling. *Biological Cybernetics* 84: 207–15; Gibson, K. R., D. Rumbaugh, and M. Beran (2001). Bigger is better: Primate brain size in relationship to cognition. In *Evolutionary Anatomy of the Primate Cerebral Cortex*, ed. D. Falk and K. R. Gibson. Cambridge: Cambridge University Press, 79–97.

31. She would need about 2,000 calories plus an extra 15 percent for her fetus; a typical three-year-old who exercises moderately needs 990 calories, and a seven-year-old needs 1,200 calories, assuming moderate levels of physical activity.

32. One way the human brain protects itself is with extra-thick membranes, which divide the brain into compartments (left and right, top and bottom). These bands act like cardboard divisions in a case of wine that prevent bottles from banging into one another. The brain is also housed in a large bath of pressurized fluid that absorbs impacts. In addition, the human braincase is especially thick-walled.

33. Leutenegger, W. (1974). Functional aspects of pelvic morphology in simian primates. *Journal of Human Evolution* 3: 207–22.

34. Rosenberg, K. R., and W. Trevathan (1996). Bipedalism and human birth: The obstetrical dilemma revisited. *Evolutionary Anthropology* 4: 161–68.

35. Tomasello, M. (2009). *Why We Cooperate.* Cambridge, MA: MIT Press.

36. The one exception is meat, which males sometimes share among other members of the hunting party. Muller, M. N., and J. C. Mitani (2005). Conflict and cooperation in wild chimpanzees. *Advances in the Study of Behavior* 35: 275–331.

37. Dunbar, R. I. M. (1998). The social brain hypothesis. *Evolutionary Anthropology* 6: 178–90.

38. Liebenberg, L. (1990). *The Art of Tracking: The Origin of Science.* Cape Town: David Philip.

39. Some experts consider adolescence a unique human stage, primarily defined by the growth spurt. However, almost all large-bodied mammals have a growth spurt (especially in body mass) that long precedes the end of skeletal growth.

40. Bogin, B. (2001). *The Growth of Humanity.* Cambridge: Cambridge University Press.

41. Smith, T. M., et al. (2013). First molar eruption, weaning, and life history in living wild chimpanzees. *Proceedings of the National Academy of Sciences USA* 110: 2787–91.

42. Although it takes more total energy for humans to mature compared to apes, the cost of each baby is lower for human mothers. In an important, insightful paper, Leslie Aiello and Cathy Key pointed out that producing milk is especially costly for larger-bodied mothers, increasing a mother's energy demands by 25 to 50 percent. An early human mother who weighed 50 kilograms (110 pounds) and was nursing an infant would need an average of 2,300 calories per day, 50 percent more than a 30-kilogram (66-pound) mother who was also nursing. One can thus calculate that a 50-kilogram human mother who weans her children like an ape after five years will expend an astonishing

4.2 million calories per child, 1.7 million more calories than if she weans her infants at three. Therefore, any mother with reliable access to enough high-quality food such as meat, marrow, and processed plants would gain a serious reproductive benefit if she managed to wean her infants when they were still immature. For more, see Aiello, L. C., and C. Key (2002). The energetic consequences of being a *Homo erectus* female. *American Journal of Human Biology* 14: 551–65.

43. Kramer, K. L. (2011). The evolution of human parental care and recruitment of juvenile help. *Trends in Ecology and Evolution* 26: 533–40.

44. These estimates are possible because in all mammals, including humans and other primates, the brain reaches adult size about the same age as the first permanent molar erupts. Further, because teeth have microscopic structures like tree rings that preserve a record of time, anatomists can use teeth to estimate at what age an animal's first molar erupted, hence when its brain stopped growing. For details, see Smith, B. H. (1989). Dental development as a measure of life history in primates. *Evolution* 43: 683–88; Dean, M. C. (2006). Tooth microstructure tracks the pace of human life-history evolution. *Proceedings of the Royal Society B Biological Sciences* 273: 2799–2808.

45. Dean, M. C., et al. (2001). Growth processes in teeth distinguish modern humans from *Homo erectus* and earlier hominins. *Nature* 414: 628–31.

46. Smith, T. M., et al. (2007). Rapid dental development in a Middle Paleolithic Belgian Neanderthal. *Proceedings of the National Academy of Sciences USA* 104: 20220–25.

47. Dean, M. C., and B. H. Smith (2009). Growth and development in the Nariokotome youth, KNM-WT 15000. In *The First Humans: Origin of the Genus Homo*, ed. F. E. Grine, J. G. Fleagle, and R. F. Leakey. New York: Springer, 101–20.

48. Smith, T. M., et al. (2010). Dental evidence for ontogenetic differences between modern humans and Neanderthals. *Proceedings of the National Academy of Sciences USA* 107: 20923–28.

49. A fat molecule is technically a triglyceride comprised of three fatty acids plus one glycerol. Fatty acids are basically long chains of carbon and hydrogen atoms; glycerol is a colorless, odorless, and sweet-tasting form of alcohol.

50. Kuzawa, C. W. (1998). Adipose tissue in human infancy and childhood: An evolutionary perspective. *Yearbook of Physical Anthropology* 41: 177–209.

51. Pond, C. M., and C. A. Mattacks (1987). The anatomy of adipose tissue in captive *Macaca* monkeys and its implications for human biology. *Folia Primatologica* 48: 164–85.

52. Clandinin, M. T., et al. (1980). Extrauterine fatty acid accretion in infant brain: Implications for fatty acid requirements. *Early Human Development* 4: 131–38.

53. Glycogen (the form of carbohydrate you store in your muscles and liver) burns faster than fat but it is much heavier and denser, and the body can store only limited amounts. Unless you run really fast, you burn mostly fat. See chapter 10 for more details.

54. Ellison, P. T. (2003). *On Fertile Ground.* Cambridge, MA: Harvard University Press.

55. This general relationship is known at Kleiber's law, the observation that as

organisms get bigger in mass, their metabolism increases to the exponent of 0.75 (the equation is BMR = body mass$^{0.75}$).

56. Leonard, W. R., and M. L. Robertson (1997). Comparative primate energetics and hominoid evolution. *American Journal of Physical Anthropology* 102: 265–81; Froehle, A. W., and M. J. Schoeninger (2006). Intraspecies variation in BMR does not affect estimates of early hominin total daily energy expenditure. *American Journal of Physical Anthropology* 131: 552–59.

57. For data, see Leonard, W. R., and M. L. Robertson (1997). Comparative primate energetics and hominoid evolution. *American Journal of Physical Anthropology* 102: 265–81; Pontzer, H., et al. (2010). Metabolic adaptation for low energy throughput in orangutans. *Proceedings of the National Academy of Sciences USA* 107: 14048–52; Dugas, L. R., et al. (2011). Energy expenditure in adults living in developing compared with industrialized countries: A meta-analysis of doubly labeled water studies. *American Journal of Clinical Nutrition* 93: 427–41; Pontzer, H., et al. (2012). Hunter-gatherer energetics and human obesity. *PLoS One* 7(7): e40503.

58. Kaplan, H. S., et al. (2000). A theory of human life history evolution: diet, intelligence, and longevity. *Evolutionary Anthropology* 9: 156–85.

59. This is true not just for humans but mammals in general. See Pontzer, H. (2012). Relating ranging ecology, limb length, and locomotor economy in terrestrial animals. *Journal of Theoretical Biology* 296: 6–12.

60. For a review, see chapter 5 of Wrangham, R. W. (2009). *Catching Fire: How Cooking Made Us Human*. New York: Basic Books.

61. For some key theories and references, see Charnov, E. L., and D. Berrigan (1993). Why do female primates have such long lifespans and so few babies? Or life in the slow lane. *Evolutionary Anthropology* 1: 191–94; Kaplan, H. S., J. B. Lancaster, and A. Robson (2003). Embodied capital and the evolutionary economics of the human lifespan. In *Lifespan: Evolutionary, Ecology and Demographic Perspectives*, ed. J. R. Carey and S. Tuljapakur. *Population and Development Review* 29, supp. 2003, 152–82; Isler, K., and C. P. van Schaik (2009). The expensive brain: A framework for explaining evolutionary changes in brain size. *Journal of Human Evolution* 57: 392–400; Kramer, K. L., and P. T. Ellison (2010). Pooled energy budgets: Resituating human energy-allocation trade-offs. *Evolutionary Anthropology* 19: 136–47.

62. Several human "pygmy" populations (people whose height does not exceed 150 centimeters, or 4.9 feet) have evolved in energy-limited places like rain forests or islands. Perhaps the small size of the Dmanisi hominins from Georgia also reflected selection to save energy among the first colonists of Eurasia.

63. Morwood, M. J., et al. (1998). Fission track age of stone tools and fossils on the east Indonesian island of Flores. *Nature* 392: 173–76.

64. Brown, P., et al. (2004). A new small-bodied hominin from the Late Pleistocene of Flores, Indonesia. *Nature* 431: 1055–61.

65. Morwood, M. J., et al. (2005). Further evidence for small-bodied hominins from the Late Pleistocene of Flores, Indonesia. *Nature* 437: 1012–17.

66. Falk, D., et al. (2005). The brain of LB1, *Homo floresiensis*. *Science* 308: 242–45; Baab, K. L., and K. P. McNulty (2009). Size, shape, and asymmetry in fossil hominins: The status of the LB1 cranium based on 3D morphometric

analyses. *Journal of Human Evolution* 57: 608–22; Gordon, A. D., L. Nevell, and B. Wood (2008). The *Homo floresiensis* cranium (LB1): Size, scaling, and early *Homo* affinities. *Proceedings of the National Academy of Sciences USA* 105: 4650–55.

67. Martin, R. D., et al. (2006). Flores hominid: new species or microcephalic dwarf? *Anatomical Record A* 288: 1123–45.

68. Argue, D., et al. (2006). *Homo floresiensis*: Microcephalic, pygmoid, *Australopithecus*, or *Homo*? *Journal of Human Evolution* 51: 360–74; Falk, D., et al. (2009). The type specimen (LB1) of *Homo floresiensis* did not have Laron syndrome. *American Journal of Physical Anthropology* 140: 52–63.

69. Weston, E. M., and A. M. Lister (2009). Insular dwarfism in hippos and a model for brain size reduction in *Homo floresiensis*. *Nature* 459: 85–88.

6 A Very Cultured Species—*How Modern Humans Colonized the World with a Combination of Brains plus Brawn*

1. Sahlins, M. D. (1972). *Stone Age Economics*. Chicago: Aldine.

2. Scally, A., and R. Durbin (2012). Revising the human mutation rate: Implications for understanding human evolution. *Nature Reviews Genetics* 13: 745–53.

3. Laval, G. E., et al. (2010). Formulating a historical and demographic model of recent human evolution based on resequencing data from noncoding regions. *PLoS ONE* 5(4): e10284.

4. Lewontin, R. C. (1972). The apportionment of human diversity. *Evolutionary Biology* 6: 381–98; Jorde, L. B., et al. (2000). The distribution of human genetic diversity: A comparison of mitochondrial, autosomal, and Y-chromosome data. *American Journal of Human Genetics* 66: 979–88.

5. Gagneux, P., et al. (1999). Mitochondrial sequences show diverse evolutionary histories of African hominoids. *Proceedings of the National Academy of Sciences USA* 96: 5077–82; Becquet, C., et al. (2007). Genetic structure of chimpanzee populations. *PLoS Genetics* 3(4): e66.

6. Green, R. E. (2008). A complete Neandertal mitochondrial genome sequence determined by high-throughput sequencing. *Cell* 134: 416–26; Green, R. E., et al. (2010). A draft sequence of the Neandertal genome. *Science* 328: 710–22; Langergraber, K. E., et al. (2012). Generation times in wild chimpanzees and gorillas suggest earlier divergence times in great ape and human evolution. *Proceedings of the National Academy of Sciences USA* 109: 15716–21.

7. For estimates of the date, see Sankararaman, S. (2012). The date of interbreeding between neandertals and modern humans. *PLoS Genetics* 8: e1002947.

8. Reich D., et al. (2010). Genetic history of an archaic hominin group from Denisova Cave in Siberia. *Nature* 468: 1053–60; Krause, J. (2010). The complete mitochondrial DNA genome of an unknown hominin from southern Siberia. *Nature* 464: 894–97.

9. The fossil, designated Omo I, is from southern Ethiopia. McDougall, I., F. H. Brown, and J. G. Fleagle (2005). Stratigraphic placement and age of modern humans from Kibish, Ethiopia. *Nature* 433: 733–36.

10. For example, the Herto sample includes three individuals dated to 160,000 years old, the site of Djebel Irhoud includes several fossils dated to 160,000 years ago, and the Singa cranium from Sudan is 133,000 years old. A few modern human fossils might even be a little older, such as a partial cranium from Florisbad, South Africa, which could predate 200,000 years. See White, T. D., et al. (2003). Pleistocene *Homo sapiens* from Middle Awash, Ethiopia. *Nature* 423: 742–47; McDermott, F., et al. (1996). New Late-Pleistocene uranium-thorium and ESR ages for the Singa hominid (Sudan). *Journal of Human Evolution* 31: 507–16.

11. Bar-Yosef, O. (2006). Neanderthals and modern humans: A different interpretation. In *Neanderthals and Modern Humans Meet*, ed. N. J. Conard. Tübingen: Tübingen Publications in Prehistory, Kerns Verlag, 165–87.

12. Bowler, J. M., et al. (2003). New ages for human occupation and climatic change at Lake Mungo, Australia. *Nature* 421: 837–40; Barker, G., et al. (2007). The "human revolution" in lowland tropical Southeast Asia: The antiquity and behavior of anatomically modern humans at Niah Cave (Sarawak, Borneo). *Journal of Human Evolution* 52: 243–61.

13. Genetic data and most archaeological evidence shows that human occupation of the New World occurred less than 30,000 years ago, probably less than 22,000 years ago. For a thorough overall review, see Meltzer, D. J. (2009). *First Peoples in a New World: Colonizing Ice Age America*. Berkeley, CA: University of California Press. For more information, see Goebel, T., M. R. Waters, and D. H. O'Rourke (2008). The late Pleistocene dispersal of modern humans in the Americas. *Science* 319: 1497–1502; Hamilton, M. J., and B. Buchanan (2010). Archaeological support for the three-stage expansion of modern humans across northeastern Eurasia and into the Americas. *PLoS One* 5(8): e12472. A few very old sites, notably Monte Verde, in Chile, are claimed to support an earlier initial colonization, but the evidence is controversial. See Dillehay, T. D., and M. B. Collins (1998). Early cultural evidence from Monte Verde in Chile. *Nature* 332: 150–52.

14. Hublin, J. J., et al. (1995). The Mousterian site of Zafarraya (Granada, Spain): Dating and implications on the palaeolithic peopling processes of Western Europe. *Comptes Rendus de l'Académie des Sciences, Paris*, 321: 931–37.

15. Lieberman, D. E., C. F. Ross, and M. J. Ravosa (2000b). The primate cranial base: Ontogeny, function and integration. *Yearbook of Physical Anthropology* 43: 117–69; Lieberman, D. E., B. M. McBratney, and G. Krovitz (2002). The evolution and development of cranial form in *Homo sapiens*. *Proceedings of the National Academy of Sciences USA* 99: 1134–39.

16. Weidenreich, F. (1941). The brain and its rôle in the phylogenetic transformation of the human skull. *Transactions of the American Philosophical Society* 31: 328–442; Lieberman, D. E. (2000). Ontogeny, homology, and phylogeny in the Hominid craniofacial skeleton: The problem of the browridge. *In Development, Growth and Evolution*, ed. P. O'Higgins and M. Cohn. London: Academic Press, 85–122.

17. Bastir, M., et al. (2008). Middle cranial fossa anatomy and the origin of modern humans. *Anatomical Record* 291: 130–40; Lieberman, D. E. (2008). Speculations about the selective basis for modern human cranial form. *Evolutionary Anthropology* 17: 22–37.

18. One unlikely idea is that chins function to strengthen the jaw, but why do humans, who cook their food, need extra strengthening? Other poorly supported speculations are that they help orient our lower incisors properly, they help us speak, or that they are attractive. For a review of these and other ideas, see Lieberman, D. E. (2011). *The Evolution of the Human Head.* Cambridge, MA: Harvard University Press.

19. Rak, Y., and B. Arensburg (1987). Kebara 2 Neanderthal pelvis: First look at a complete inlet. *American Journal of Physical Anthropology* 73: 227–31; Arsuaga, J. L., et al. (1999). A complete human pelvis from the Middle Pleistocene of Spain. *Nature* 399: 255–58; Ruff, C. B. (2010). Body size and body shape in early hominins: Implications of the Gona pelvis. *Journal of Human Evolution* 58: 166–78.

20. Ruff, C. B., et al. (1993). Postcranial robusticity in *Homo*. I: Temporal trends and mechanical interpretation. *American Journal of Physical Anthropology* 91: 21–53.

21. McBrearty, S., and A. S. Brooks (2000). The revolution that wasn't: A new interpretation of the origin of modern human behavior. *Journal of Human Evolution* 39: 453–563.

22. Brown, K. S., et al. (2012). An early and enduring advanced technology originating 71,000 years ago in South Africa. *Nature* 491: 590–93; Yellen, J. E., et al. (1995). A middle stone age worked bone industry from Katanda, Upper Semliki Valley, Zaire. *Science* 268: 553–56; Wadley, L., T. Hodgskiss, and M. Grant (2009). Implications for complex cognition from the hafting of tools with compound adhesives in the Middle Stone Age, South Africa. *Proceedings of the National Academy of Sciences USA* 106: 9590–94; Mourre, V., P. Villa, and C. S. Henshilwood (2010). Early use of pressure flaking on lithic artifacts at Blombos Cave, South Africa. *Science* 330: 659–62.

23. Henshilwood, C. S., et al. (2001). An early bone tool industry from the Middle Stone Age at Blombos Cave, South Africa: Implications for the origins of modern human behaviour, symbolism and language. *Journal of Human Evolution* 41: 631–78; Henshilwood, C. S., F. d'Errico, and I. Watts (2009). Engraved ochres from the Middle Stone Age levels at Blombos Cave, South Africa. *Journal of Human Evolution* 57: 27–47.

24. For a review of this debate, see D'Errico, F., and C. Stringer (2011). Evolution, revolution, or saltation scenario for the emergence of modern cultures? *Philosophical Transactions of the Royal Society, London, Part B, Biological Science* 366: 1060–69.

25. Jacobs, Z., et al. (2008). Ages for the Middle Stone Age of southern Africa: Implications for human behavior and dispersal. *Science* 322: 733–35.

26. For historical reasons, archaeologists use the term "Later Stone Age" for the Upper Paleolithic in sub-Saharan Africa. I am using the term "Upper Paleolithic" for both.

27. Stiner, M. C., N. D. Munro, and T. A. Surovell (2000). The tortoise and

the hare. Small-game use, the broad-spectrum revolution, and paleolithic demography. *Current Anthropology* 41: 39–79.

28. Weiss, E., et al. (2008). Plant-food preparation area on an Upper Paleolithic brush hut floor at Ohalo II, Israel. *Journal of Archaeological Science* 35: 2400–14; Revedin, A., et al. (2010). Thirty-thousand-year-old evidence of plant food processing. *Proceedings of the National Academy of Sciences USA* 107: 18815–19.

29. This enigmatic industry, known as the Châtelperronian, is found in just a few sites dated to between 35,000 and 29,000 years ago. It contains some typical Middle Paleolithic tools but also has Upper Paleolithic tools and some decorative pieces, like carved pendants and rings made from ivory. Some believe this industry is mixed, but others think it is a Neanderthal version of the Upper Paleolithic. For more information, and differing views, see Bar-Yosef, O., and J. G. Bordes (2010). Who were the makers of the Châtelperronian culture? *Journal of Human Evolution* 59: 586–93; Mellars, P. (2010). Neanderthal symbolism and ornament manufacture: The bursting of a bubble? *Proceedings of the National Academy of Sciences USA* 107: 20147–48; Zilhão, J. (2010). Did Neandertals think like us? *Scientific American* 302: 72–75; Caron, F., et al. (2011). The reality of Neandertal symbolic behavior at the Grotte du Renne, Arcy-sur-Cure, France. *PLoS One* 6: e21545.

30. This is a thorny issue for several reasons. First, brain size needs to be scaled by body size (bigger people tend to have bigger brains), but the relationship is not very tight within species, making such corrections imprecise. Second, how do you define, let alone measure, intelligence? Most studies have found very slight correlations (of 0.3–0.4) between brain size and test-based measures of intelligence, but one needs to exercise caution about drawing strong conclusions from these studies, because it is impossible to measure intelligence without some preconceived biases about what intelligence actually is. Is it the ability to solve math problems and use proper grammar, or is it the ability to track a kudu and figure out what others are thinking? In addition, it is not possible to correct for all the myriad effects of environment on measures of intelligence. Even so, people try. For an example, see Witelson, S. F., H. Beresh, and D. L. Kigar (2006). Intelligence and brain size in 100 postmortem brains: Sex, lateralization and age factors. *Brain* 129: 386–98.

31. Please don't think these studies have anything to do with phrenology, the nineteenth-century pseudoscience that assumed that subtle variations in the external form of the skull reflect meaningful differences in the brain relevant to personality, intellect, and other functions.

32. Lieberman, D. E., B. M. McBratney, and G. Krovitz (2002). The evolution and development of cranial form in *Homo sapiens*. *Proceedings of the National Academy of Sciences USA* 99: 1134–39; Bastir, M., et al. (2011). Evolution of the base of the brain in highly encephalized human species. *Nature Communications* 2: 588. For scaling studies, see Rilling, J., and R. Seligman (2002). A quantitative morphometric comparative analysis of the primate temporal lobe. *Journal of Human Evolution* 42: 505–34; Semendeferi, K. (2001). Advances in the study of hominoid brain evolution: Magnetic resonance imaging (MRI) and 3-D imaging. In *Evolutionary Anatomy of the*

Primate Cerebral Cortex, ed. D. Falk and K. Gibson. Cambridge: Cambridge University Press, 257–89.

33. Damage to a region of the temporal lobe known as Wernicke's area actually renders language meaningless.

34. Persinger, M. A. (2001). The neuropsychiatry of paranormal experiences. *Journal of Neuropsychiatry and Clinical Neurosciences* 13: 515–24.

35. Bruner, E. (2004). Geometric morphometrics and paleoneurology: Brain shape evolution in the genus *Homo. Journal of Human Evolution* 47: 279–303.

36. Culham, J. C., and K. F. Valyear (2006). Human parietal cortex in action. *Current Opinions in Neurobiology* 16: 205–12.

37. Semendeferi, K., et al. (2001). Prefrontal cortex in humans and apes: A comparative study of area 10. *American Journal of Physical Anthropology* 114: 224–41; Schenker, N. M., A. M. Desgouttes, and K. Semendeferi (2005). Neural connectivity and cortical substrates of cognition in hominoids. *Journal of Human Evolution* 49: 547–69.

38. The most famous case of damage to the prefrontal region is the case of Phineas Gage, a railway worker who was injured in a freak explosion that sent an iron bar rocketing through his eye socket and brain. Amazingly, Gage survived, but he then became irascible and impatient. For more information, see Damasio, A. R. (2005). *Descartes' Error: Emotion, Reason, and the Human Brain.* New York: Penguin.

39. For an explanation of these processes, see Lieberman, D. E., K. M. Mowbray, and O. M. Pearson (2000). Basicranial influences on overall cranial shape. *Journal of Human Evolution* 38: 291–315. For evidence that this happens differently in modern humans versus Neanderthals in the first few years of life, see Gunz, P., et al. (2012). A uniquely modern human pattern of endocranial development. Insights from a new cranial reconstruction of the Neandertal newborn from Mezmaiskaya. *Journal of Human Evolution* 62: 300–13. Note also that another factor that leads to more flexed cranial bases and rounder brains are smaller faces. Just as brains grow on top of the cranial base, faces grow downward and forward from the cranial base. Therefore, the length of the face also affects how much the cranial base flexes. Animals that have relatively longer faces have flatter skull bases, which enables more of the face to project in front of the braincase.

40. Miller, D. T., et al. (2012). Prolonged myelination in human neocortical evolution. *Proceedings of the National Academy of Sciences USA* 109: 16480–85; Bianchi, S., et al. (2012). Dendritic morphology of pyramidal neurons in the chimpanzee neocortex: Regional specializations and comparison to humans. *Cerebral Cortex.*

41. For a summary, see Lieberman, P. (2013). *The Unpredictable Species: What Makes Humans Unique.* Princeton, NJ: Princeton University Press.

42. Kandel, E. R., J. H. Schwartz, and T. M. Jessel (2000). *Principles of Neural Science,* 4th ed. New York: McGraw-Hill; Giedd, J. N. (2008). The teen brain: Insights from neuroimaging. *Journal of Adolescent Health* 42: 335–43.

43. A study by Tanya Smith and colleagues compared two juvenile Neanderthals who were not children with a large sample of modern human juveniles. One

of these Neanderthals (from the Belgian site of Scladina) died at the age of eight but was as mature as a human ten-year-old. Another Neanderthal (Le Moustier 1) was about twelve years old at death but had the skeleton of a modern human sixteen-year-old boy. Analyses of more fossils are needed to confirm these differences, but if they hold up it would mean that archaic humans had a shorter period of juvenile and adolescent growth before they became adults. See Smith, T., et al. (2010). Dental evidence for ontogenetic differences between modern humans and Neanderthals. *Proceedings of the National Academy of Sciences USA* 107: 20923–28.

44. Kaplan, H. S., et al. (2001). The embodied capital theory of human evolution. In *Reproductive Ecology and Human Evolution*, ed. P. T. Ellison. Hawthorne, NY: Aldine de Gruyter; Yeatman, J. D., et al. (2012). Development of white matter and reading skills. *Proceedings of the National Academy of Sciences USA* 109: 3045–53; Shaw, P., et al. (2005). Intellectual ability and cortical development in children and adolescents. *Nature* 44: 676–79; Lieberman, P. (2010). *Human Language and Our Reptilian Brain.* Cambridge, MA: Harvard University Press.

45. Klein, R. G., and B. Edgar (2002). *The Dawn of Human Culture.* New York: Nevreaumont Publishing.

46. Enard, W., et al. (2009). A humanized version of *Foxp2* affects cortico-basal ganglia circuits in mice. *Cell* 137: 961–71.

47. Krause, J., et al. (2007). The derived *FOXP2* variant of modern humans was shared with Neandertals. *Current Biology* 17: 1908–12; Coop, G., et al. (2008). The timing of selection at the human *FOXP2* gene. *Molecular Biology and Evolution* 25: 1257–59.

48. Lieberman, P. (2006). *Toward an Evolutionary Biology of Language.* Cambridge, MA: Harvard University Press.

49. This reshaping largely occurs because the size of the tongue scales very closely with body mass in primates, so when the human face became shorter, it did not make the tongue smaller. Instead, the human tongue became shorter but taller, with its base positioned lower in the human throat than in other primates.

50. This property of human speech is known as quantal speech. It was first proposed by Kenneth Stevens and Arthur House. See Stevens, K. N., and A. S. House (1955). Development of a quantitative description of vowel articulation. *Journal of the Acoustical Society of America* 27: 401–93.

51. Interbreeding with archaic *Homo* probably also happened in Africa. See Hammer, M. F., et al. (2011). Genetic evidence for archaic admixture in Africa. *Proceedings of the National Academy of Sciences USA* 108: 15123–28; Harvarti, K., et al. (2011). The Later Stone Age calvaria from Iwo Eleru, Nigeria: Morphology and chronology. *PlosOne* 6: e24024.

52. If hunter-gatherers in Ice Age Europe were like recent subarctic hunter-gatherers who live in territories of 100 square kilometers (38 square miles) per person, then a maximum of three thousand people would have lived in regions such as Italy at any time. See Zubrow, E. (1989). The demographic modeling of Neanderthal extinction. In *The Human Revolution*, ed. P. Mellars and C. B. Stringer. Edinburgh: Edinburgh University Press, 212–31.

53. Caspari, R., and S. H. Lee (2004). Older age becomes common late in human

evolution. *Proceedings of the National Academy of Sciences USA* 101(30): 10895–900.

54. For summaries of these theories, see Stringer, C. (2012). *Lone Survivor: How We Came to Be the Only Humans on Earth.* New York: Times Books; Klein, R. J., and B. Edgar (2002). *The Dawn of Human Culture.* New York: Wiley. You might also enjoy Kuhn, S. L., and M. C. Stiner (2006). What's a mother to do? The division of labor among Neandertals and modern humans in Eurasia. *Current Anthropology* 47: 953–81.

55. Shea, J. J. (2011). Stone tool analysis and human origins research: Some advice from Uncle Screwtape. *Evolutionary Anthropology* 20: 48–53.

56. The basic unit of transmitted biological information is a gene, and the equivalent cultural unit is a meme, usually an idea, such as a symbol, a habit, a practice, or a belief. The word "meme" comes from the Greek word "to imitate." Like genes, memes are passed from one individual to the next, but unlike genes, they are not solely passed from parent to offspring. See Dawkins, R. (1976). *The Selfish Gene.* Oxford: Oxford University Press.

57. Many excellent analyses have been written on cultural evolution and selection, from which I have drawn heavily. For more, see Cavalli-Sforza, L. L., and M. W. Feldman (1981). *Cultural Transmission and Evolution: A Quantitative Approach.* Princeton: Princeton University Press; Boyd, R., and P. J. Richerson (1985). *Culture and the Evolutionary Process.* Chicago: University of Chicago Press; Durham, W. H. (1991). *Co-evolution: Genes, Culture and Human Diversity.* Stanford, CA: Stanford University Press. For more popular accounts, I recommend Richerson, P. J., and R. Boyd (1995). *Not by Genes Alone: How Culture Transformed Human Evolution.* Chicago: University of Chicago Press; and Ehrlich, P. R. (2000). *Human Natures: Genes, Cultures and the Human Prospect.* Washington, DC: Island Press.

58. Lactase is the enzyme that enables you to digest lactose, the sugar present in milk. Until recently, humans, like other mammals, lost the ability to produce lactase after weaning, but mutations that evolved in the LCT gene permit some humans to continue synthesizing the enzyme as adults. Tishkoff, S. A., et al. (2007). Convergent adaptation of human lactase persistence in Africa and Europe. *Nature Genetics* 39: 31–40; Enattah, N. S., et al. (2008). Independent introduction of two lactase-persistence alleles into human populations reflects different history of adaptation to milk culture. *American Journal of Human Genetics* 82: 57–72.

59. Wrangham, R. W. (2009). *Catching Fire: How Cooking Made Us Human.* New York: Basic Books.

60. There are two general principles. The first (known as Bergmann's rule) is that body mass scales to the power of three but surface area scales to the power of two, so bigger individuals have relatively less surface area. Therefore animals in colder climates tend to be bigger. The second principle (known as Allen's rule) is that longer limbs help increase surface area, so in cold climates it is useful to have short limbs.

61. Holliday, T. W. (1997). Body proportions in Late Pleistocene Europe and modern human origins. *Journal of Human Evolution* 32: 423–48; Trinkaus, E. (1981). Neandertal limb proportions and cold adaptation. In *Aspects of Human Evolution*, ed. C. B. Stringer. London: Taylor and Francis, 187–224.

62. Jablonski, N. (2008). *Skin*. Berkeley: University of California Press; Sturm, R. A. (2009). Molecular genetics of human pigmentation diversity. *Human Molecular Genetics* 18: R9–17.

63. Landau, M. (1991) *Narratives of Human Evolution*. New Haven, CT: Yale University Press.

64. Pontzer, H., et al. (2012). Hunter-gatherer energetics and human obesity. *PLoS ONE* 7(7): e40503, doi: 10.1371; Marlowe, F. (2005). Hunter-gatherers and human evolution. *Evolutionary Anthropology* 14: 54–67.

65. There is a slight complication with this analysis, since I have not corrected for the effects of scaling. As animals, including humans, get bigger they spend relatively less energy doing work. Even so, the point is that sedentary Westerners spend less energy per unit of body mass to do work than hunter-gatherers.

66. Lee, R. B. (1979). *The !Kung San: Men, Women and Work in a Foraging Society*. Cambridge: Cambridge University Press.

67. For a summary of hunter-gatherer variation, see Kelly, R. L. (2007). *The Foraging Spectrum: Diversity in Hunter-Gatherer Lifeways*. Clinton Corners, NY: Percheron Press; Lee, R. B., and R. Daly (1999). *The Cambridge Encyclopedia of Hunters and Gatherers*. Cambridge: Cambridge University Press.

7 Progress, Mismatch, and Dysevolution— *The Consequences—Good and Bad—of Having Paleolithic Bodies in a Post-Paleolithic World*

1. Floud R., et al. (2011). *The Changing Body: Health Nutrition and Human Development in the Western Hemisphere Since 1700*. Cambridge: Cambridge University Press.

2. McGuire, M. T., and A. Troisi (1998). *Darwinian Psychiatry*. Oxford: Oxford University Press; see also Baron-Cohen, S., ed. (2012). *The Maladapted Mind: Classic Readings in Evolutionary Psychopathology*. Hove, Sussex: Psychology Press; Mattson, M. P. (2012). Energy intake and exercise as determinants of brain health and vulnerability to injury and disease. *Cell Metabolism* 16: 706–22.

3. There are many excellent books on this topic. Several important ones to consult are Odling-Smee, F. J., K. N. Laland, and M. W. Feldman (2003). *Niche Construction: The Neglected Process in Evolution*. Princeton: Princeton University Press; Richerson, P. J., and R. Boyd (2005). *Not By Genes Alone: How Culture Transformed Human Evolution*. Chicago: University of Chicago Press; Ehrlich, P. R. (2000). *Human Natures: Genes, Cultures and the Human Prospect*. Washington, DC: Island Press; Cochran, G., and H. Harpending (2009). *The 10,000 Year Explosion*. New York: Basic Books.

4. Weeden, J., et al. (2006). Do high-status people really have fewer children? Education, income, and fertility in the contemporary US. *Human Nature* 17: 377–92; Byars, S. G., et al. (2010). Natural selection in a contemporary

human population. *Proceedings of the National Academy of Sciences USA* 107: 1787–92.

5. Williamson, S. H., et al. (2007). Localizing recent adaptive evolution in the human genome. *PLoS Genetics* 3: e90; Sabeti, P. C., et al. (2007). Genome-wide detection and characterization of positive selection in human populations. *Nature* 449: 913–18; Kelley, J. L., and W. J. Swanson (2008). Positive selection in the human genome: From genome scans to biological significance. *Annual Review of Genomics and Human Genetics* 9: 143–60; Laland, K. N., J. Odling-Smee, and S. Myles (2010). How culture shaped the human genome: Bringing genetics and the human sciences together. *Nature Reviews Genetics* 11: 137–48.

6. Brown, E. A., M. Ruvolo, and P. C. Sabeti (2013). Many ways to die, one way to arrive: How selection acts through pregnancy. *Trends in Genetics* S0168-9525.

7. Kamberov, Y. G., et al. (2013). Modeling recent human evolution in mice by expression of a selected EDAR variant. *Cell* 152: 691–702. The gene variant has other effects, including smaller breasts and upper incisors that have a slightly shoveled shape.

8. You can calculate how many generations it takes for gene frequencies to change as $\Delta p = (spq2)/1 - sq2$, where p and q are the frequency of two alleles of the same gene, Δp is the change in the frequency of the allele (p) per generation, and s is the coefficient of selection (0.0 being none and 1.0 being 100 percent).

9. For a review, see Tattersall, I., and R. DeSalle (2011). *Race? Debunking a Scientific Myth.* College Station: Texas A & M Press.

10. Corruccini, R. S. (1999). *How Anthropology Informs the Orthodontic Diagnosis of Malocclusion's Causes.* Lewiston, NY: Edwin Mellen Press; Lieberman, D. E., et al. (2004). Effects of food processing on masticatory strain and craniofacial growth in a retrognathic face. *Journal of Human Evolution* 46: 655–77.

11. Kuno, Y. (1956). *Human Perspiration.* Springfield, IL: Charles C. Thomas.

12. For data on these changes, see Bogin, B. (2001). *The Growth of Humanity.* New York: Wiley; Brace, C. L., K. R. Rosenberg, and K. D. Hunt (1987). Gradual change in human tooth size in the Late Pleistocene and Post-Pleistocene. *Evolution* 41: 705–20; Ruff, C. B., et al. (1993). Postcranial robusticity in *Homo*. I: Temporal trends and mechanical interpretation. *American Journal of Physical Anthropology* 91: 21–53; Lieberman, D. E. (1996). How and why humans grow thin skulls. *American Journal of Physical Anthropology* 101: 217–36; Sachithanandam, V., and B. Joseph (1995). The influence of footwear on the prevalence of flat foot: A survey of 1846 skeletally mature persons. *Journal of Bone and Joint Surgery* 77: 254–57; Hillson, S. (1996). *Dental Anthropology.* Cambridge: Cambridge University Press.

13. Wild, S., et al. (2004). Global prevalence of diabetes. *Diabetes Care* 27: 1047–53.

14. There are several good books on evolutionary medicine. The first major treatment of the topic, still worth reading, is Nesse, R., and G. C. Williams (1994). *Why We Get Sick: The New Science of Darwinian Medicine.* New

York: New York Times Books. Other great books to read are Ewald, P. (1994). *Evolution of Infectious Diseases.* Oxford: Oxford University Press; Stearns, S. C., and J. C. Koella (2008). *Evolution in Health and Disease,* 2nd ed. Oxford: Oxford University Press; Trevathan, W. R., E. O. Smith, and J. J. McKenna (2008). *Evolutionary Medicine and Health.* Oxford: Oxford University Press; Gluckman, P., A. Beedle, and M. Hanson (2009). *Principles of Evolutionary Medicine.* Oxford: Oxford University Press; Trevathan, W. R. (2010). *Ancient Bodies, Modern Lives: How Evolution Has Shaped Women's Health.* Oxford: Oxford University Press.

15. Greaves, M. (2000). *Cancer: The Evolutionary Legacy.* Oxford: Oxford University Press.

16. For a review of this complex topic, see Dunn, R. (2011). *The Wild Life of Our Bodies.* New York: HarperCollins.

17. This is a contentious topic for many kinds of cancer, including prostate cancer. For two different studies published within a year of each other in the same journal but with different conclusions, see Wilt, T. J., et al. (2012). Radical prostatectomy versus observation for localized prostate cancer. *New England Journal of Medicine* 367: 203–13; Bill-Axelson, A., et al. (2011). Radical prostatectomy versus watchful waiting in early prostate cancer. *New England Journal of Medicine* 364: 1708–17.

18. For an entertaining review of the history of dieting, see Foxcroft, L. (2012). *Calories and Corsets: A History of Dieting over Two Thousand Years.* London: Profile Books.

19. See Gluckman, P., and M. Hanson (2006). *Mismatch: The Lifestyle Diseases Timebomb.* Oxford: Oxford University Press.

20. Nesse, R. M. (2005). Maladaptation and natural selection. *The Quarterly Review of Biology* 80: 62–70.

21. This is well studied, but for a key early paper that showed this effect, see Colditz, G. A. (1993). Epidemiology of breast cancer: Findings from the Nurses' Health Study. *Cancer* 71: 1480–89.

22. Baron-Cohen, S. (2008). *Autism and Asperger Syndrome: The Facts.* Oxford: Oxford University Press.

23. Price, W. A. (1939). *Nutrition and Physical Degeneration: A Comparison of Primitive and Modern Diets and Their Effects.* Redlands, CA: Paul B. Hoeber, Inc.

24. See, for example, Mann, G. V., et al. (1962). Cardiovascular disease in African Pygmies: A survey of the health status, serum lipids and diet of Pygmies in Congo. *Journal of Chronic Disease* 15: 341– 71; Mann, G. V., et al. (1962). The health and nutritional status of Alaskan Eskimos. *American Journal of Clinical Nutrition* 11: 31–76; Truswell, A. S., and J. D. L. Hansen (1976). Medical research among the !Kung. In *Kalahari Hunter-Gatherers: Studies of the !Kung San and Their Neighbors,* ed. R. B. Lee and I. DeVore. Cambridge: Harvard University Press, 167–94; Truswell, A. S. (1977). Diet and nutrition of hunter-gatherers. In *Health and Disease in Tribal Societies.* New York: Elsevier, 213–21; Howell, N. (1979). *Demography of the Dobe !Kung.* New York: Academic Press; Kronman, N., and A. Green (1980). Epidemiological studies in the Upernavik District, Greenland. *Acta Medica Scandinavica* 208: 401–6; Trowell, H. C., and D. P. Burkitt (1981). *West-*

ern Diseases: Their Emergence and Prevention. Cambridge, MA: Harvard University Press; Rode, A., and R. J. Shephard (1994). Physiological consequences of acculturation: A 20-year study of fitness in an Inuit community. *European Journal of Applied Physiology and Occupational Physiology* 69: 516–24.

25. See, for example, Wilmsen, E. (1989). *Land Filled with Flies: A Political Economy of the Kalahari.* Chicago: University of Chicago Press.

26. Many animals synthesize vitamin C, but fruit-eating monkeys and apes lost this ability millions of years ago. Modest amounts of vitamin C can therefore be found in the organs of certain animals.

27. Carpenter, K, J. (1988). *The History of Scurvy and Vitamin C.* Cambridge: Cambridge University Press.

28. For more information on the human oral microbiome, check out the website maintained by the Forsyth Dental Institute: http://www.homd.org.

29. For a review of the history and evolution of cavities, see Hillson, S. (2008). The current state of dental decay. In *Technique and Application in Dental Anthropology,* ed. J. D. Irish and G. C. Nelson. Cambridge: Cambridge University Press, 111–35. For data on chimpanzee cavities, see Lovell, N. C. (1990). *Patterns of Injury and Illness in Great Apes: A Skeletal Analysis.* Washington, DC: Smithsonian Press.

30. Vos, T., et al. (2012). Years lived with disability (YLDs) for 1160 sequelae of 289 diseases and injuries 1990–2010: A systematic analysis for the Global Burden of Disease Study 2010. *Lancet* 380: 2163–96.

31. *Oxford English Dictionary,* 3rd ed. (2005). Oxford: Oxford University Press. The most common, contemporary meaning of "palliative" is the alleviation of suffering of terminally ill patients.

32. Boyd, R., and P. J. Richerson (1985). *Culture and the Evolutionary Process.* Chicago: University of Chicago Press; Durham, W. H. (1991). *Co-evolution: Genes, Culture and Human Diversity.* Stanford: Stanford University Press; Ehrlich, P. R. (2000). *Human Natures: Genes, Cultures and the Human Prospect.* Washington, DC: Island Press; Odling-Smee, F. J., K. N. Laland, and M. W. Feldman (2003). *Niche Construction: The Neglected Process in Evolution.* Princeton: Princeton University Press; Richerson, P. J., and R. Boyd (2005). *Not by Genes Alone: How Culture Transformed Human Evolution.* Chicago: University of Chicago Press.

33. Kearney, P. M., et al. (2005). Global burden of hypertension: Analysis of worldwide data. *Lancet* 365: 217–23.

34. Dickinson, H. O., et al. (2006). Lifestyle interventions to reduce raised blood pressure: A systematic review of randomized controlled trials. *Journal of Hypertension* 24: 215–33.

35. Hawkes, K. (2003). Grandmothers and the evolution of human longevity. *American Journal of Human Biology* 15: 380–400.

8 Paradise Lost?—*The Fruits and Follies of Becoming Farmers*

1. Diamond, J. (1987). The worst mistake in the history of the human race. *Discover* 5: 64–66.
2. Ditlevsen, P. D., H. Svensmark, and S. Johnsen (1996). Contrasting atmospheric and climate dynamics of the last-glacial and Holocene periods. *Nature* 379: 810–12.
3. Cohen, M. N. (1977). *The Food Crisis in Prehistory*. New Haven, CT: Yale University Press. See also Cohen, M. N., and G. J. Armelagos (1984). *Paleopathology at the Origins of Agriculture*. Orlando: Academic Press.
4. For a global review of the evidence, see Mithen, S. (2003). *After the Ice: A Global Human History*. Cambridge, MA: Harvard University Press.
5. Doebley, J. F. (2004). The genetics of maize evolution. *Annual Review of Genetics* 38: 37–59.
6. Nadel, D., ed. (2002). *Ohalo II—A 23,000-Year-Old Fisher-Hunter-Gatherers' Camp on the Shore of the Sea of Galilee*. Haifa: Hecht Museum.
7. Bar-Yosef, O. (1998). The Natufian culture of the southern Levant. *Evolutionary Anthropology* 6: 159–77.
8. Alley, R. B., et al. (1993). Abrupt accumulation increase at the Younger Dryas termination in the GISP2 ice core. *Nature* 362: 527–29.
9. The Younger Dryas was a big freeze but is named after a lovely alpine wildflower, *Dryas octopetala*, which became much more abundant at the time.
10. These people are known as the Harifians. Goring-Morris, A. N. (1991). The Harifian of the southern Levant. In *The Natufian Culture in the Levant,* ed. O. Bar-Yosef and F. R. Valla. Ann Arbor, MI: International Monographs in Prehistory, 173–216.
11. See Zeder, M. A. (2011). The origins of agriculture in the Near East. *Current Anthropology* 52(S4): S221–35; Goring-Morris, N., and A. Belfer-Cohen (2011). Neolithisation processes in the Levant. *Current Anthropology* 52(S4): S195–208.
12. For reviews, see Smith, B. D. (2001). *The Emergence of Agriculture*. New York: Scientific American Press; Bellwood, P. (2005). *First Farmers: The Origins of Agricultural Societies*. Oxford: Blackwell Publishing.
13. Wu, X., et al. (2012). Early pottery at 20,000 years ago in Xianrendong Cave, China. *Science* 336: 1696–700.
14. Clutton-Brock, J. (1999). *A Natural History of Domesticated Mammals*, 2nd ed. Cambridge: Cambridge University Press. Also see Connelly, J., et al. (2011). Meta-analysis of zooarchaeological data from SW Asia and SE Europe provides insight into the origins and spread of animal husbandry. *Journal of Archaeological Science* 38: 538–45.
15. Pennington, R. (2001). Hunter-gatherer demography. In *Hunter-Gatherers: An Interdisciplinary Perspective*, ed. C. Panter-Brick, R. Layton, and P. Rowley-Conwy. Cambridge: Cambridge University Press, 170–204.
16. To calculate a population's growth rate, the equation is: $N_t = N_o{}^* e^{rt}$, where N_t is the population size at year t, N_o is the population size at year o, r is the

rate of growth (1% is 0.01), t is the number of years, and e is the base of the natural logarithm (2.718281828).

17. Bocquet-Appel, J. P. (2011). When the world's population took off: The springboard of the Neolithic demographic transition. *Science* 333: 560–61.

18. Price, T. D., and A. B. Gebauer (1996). *Last Hunters, First Farmers: New Perspectives on the Prehistoric Transition to Agriculture*. Santa Fe, NM: School of American Research.

19. It is not easy to define what constitutes a separate language, but for a comprehensive listing, see Lewis, M. P., ed. (2009). *Ethnologue: Languages of the World*, 16th ed. Dallas, TX: SIL International; http://www.ethnologue.com.

20. Kramer, K. L., and P. T. Ellison (2010). Pooled energy budgets: Resituating human energy allocation trade-offs. *Evolutionary Anthropology* 19: 136–47.

21. For an interesting account, see Anderson, A. (1989). *Prodigious Birds*. Cambridge: Cambridge University Press.

22. For a review, see Sée, H. (2004). *Economic and Social Conditions During Eighteenth Century France*. Kitchener, Ontario: Batoche Books. For contemporary accounts, check out Arthur Young's *Travels in France* (1792), available online at http://www.econlib.org/library/YPDBooks/Young/yngTF0.html.

Here is one of Young's many descriptions of the poverty he saw, often compounded by a brutal system of taxation: "Walking up a long hill to ease my mare, I was joined by a poor woman, who complained of the times, and that it was a sad country. Demanding her reasons, she said her husband had but a morsel of land, one cow, and a poor little horse, yet they had [heavy] taxes. She had seven children. . . . This woman, at no great distance, might have been taken for sixty or seventy, her figure was so bent and her face so furrowed and hardened by labor, but she said she was only twenty-eight."

23. See Bogaard, A. (2004). *Neolithic Farming in Central Europe*. London: Routledge.

24. Marlowe, F. W. (2005). Hunter-gatherers and human evolution. *Evolutionary Anthropology* 14: 54–67.

25. Gregg, S. A. (1988). *Foragers and Farmers: Population Interaction and Agricultural Expansion in Prehistoric Europe*. Chicago: University of Chicago Press.

26. Ethnographic studies, which provide only minimum estimates, indicate that the Bushmen of southern Africa regularly eat at least sixty-nine different plant species, the Aché of Paraguay eat at least forty-four plant species, the Efe of the Congo eat at least twenty-eight species, and the Hadza of Tanzania eat at least sixty-two species. For data, see Lee, R. B. (1979). The *!Kung San: Men, Women and Work in a Foraging Society*. Cambridge and New York: Cambridge University Press; Hill, K., et al. (1984). Seasonal variance in the diet of Aché hunter-gatherers of eastern Paraguay. *Human Ecology* 12: 145–80; Bailey, R. C., and N. R. Peacock (1988). Efe Pygmies of northeast Zaire: Subsistence strategies in the Ituri Forest. In *Coping with Uncertainty in Food Supply*, ed. I. de Garine and G. A. Harrison. Oxford: Oxford University Press, 88–117; Marlowe, F. W. (2010). *The Hadza Hunter-Gatherers of Tanzania*. Berkeley: University of California Press.

27. Milton, K. (1999). Nutritional characteristics of wild primate foods: Do the

diets of our closest living relatives have lessons for us? *Nutrition* 15: 488–98; Eaton, S. B., S. B. Eaton III, and M. J. Konner (1997). Paleolithic nutrition revisited: A twelve-year retrospective on its nature and implications. *European Journal of Clinical Nutrition* 51: 207–16.

28. Froment, A. (2001). Evolutionary biology and health of hunter-gatherer populations. In *Hunter-Gatherers: An Interdisciplinary Perspective,* ed. C. Panter-Brick, R. H. Layton, and P. Rowley-Conwy. Cambridge: Cambridge University Press, 239–66.

29. Prentice, A. M., et al. (1981). Long-term energy balance in child-bearing Gambian women. *American Journal of Clinical Nutrition* 34: 279–99; Singh, J., et al. (1989). Energy expenditure of Gambian women. *British Journal of Nutrition* 62: 315–19.

30. Donnelly, J. S. (2001). *The Great Irish Potato Famine.* Norwich, VT: Sutton Books.

31. For a superb summary of the history and causes of famine, see Gráda, C. Ó. (2009). *Famine: A Short History.* Princeton: Princeton University Press.

32. See Hudler, G. (1998). *Magical Mushrooms, Mischievous Molds.* Princeton: Princeton University Press.

33. Hillson, S. (2008). The current state of dental decay. In *Technique and Application in Dental Anthropology,* ed. J. D. Irish and G. C. Nelson. Cambridge: Cambridge University Press, 111–35.

34. Smith, P., O. Bar-Yosef, and A. Sillen (1984). Archaeological and skeletal evidence for dietary change during the late Pleistocene/early Holocene in the Levant. In *Palaeopathology at the Origins of Agriculture,* ed. M. N. Cohen and G. J. Armelagos. New York: Academic Press, 101–36.

35. Chang, C. L., et al. (2011). Identification of metabolic modifiers that underlie phenotypic variations in energy-balance regulation. *Diabetes* 60: 726–34.

36. Lee, R. B. (1979). *The !Kung San: Men, Women and Work in a Foraging Society.* Cambridge: Cambridge University Press; Marlowe, F. W. (2010). *The Hadza Hunter-Gatherers of Tanzania.* Berkeley: University of California Press.

37. Sand, G. (1895). *The Haunted Pool,* trans. F. H. Potter. New York: Dodd, Mead and Co., chapter 2.

38. Leonard, W. R. (2008). Lifestyle, diet, and disease: Comparative perspectives on the determinants of chronic health risks. In *Evolution in Health and Disease,* ed. S. C. Stearns and J. C. Koella. Oxford: Oxford University Press, 265–76.

39. Kramer, K. (2011). The evolution of human parental care and recruitment of juvenile help. *Trends in Ecology and Evolution* 26: 533–40; Kramer, K. (2005). Children's help and the pace of reproduction: Cooperative breeding in humans. *Evolutionary Anthropology* 14: 224–37. Of the populations included in Kramer's study, only one hunter-gatherer group, the Hadza, make children work five or six hours a day.

40. Malthus, T. R. (1798). *An Essay on the Principle of Population.* London: J. Johnson.

41. For crude estimates, see Haub, C. (1995). How many people have ever lived on the Earth? *Population Today* 23: 4–5; Cochran, G., and H. Harpending (2009). *The 10,000 Year Explosion.* New York: Basic Books.

42. Zimmermann, A., J. Hilpert, and K. P. Wendt (2009). Estimations of population density for selected periods between the Neolithic and AD 1800. *Human Biology* 81: 357–80.

43. For a review, see Ewald, P. (1994). *The Evolution of Infectious Disease.* Oxford: Oxford University Press.

44. For a summary of these and other diseases, see Barnes, E. (2005). *Diseases and Human Evolution.* Albuquerque: University of New Mexico Press.

45. Armelagos, G. J., A. H. Goodman, and K. Jacobs (1991). The origins of agriculture: Population growth during a period of declining health. In *Cultural Change and Population Growth: An Evolutionary Perspective*, ed. W. Hern. *Population and Environment* 13: 9–22.

46. Li, Y., et al. (2003). On the origin of smallpox: Correlating variola phylogenics with historical smallpox records. *Proceedings of the National Academy of Sciences USA* 104: 15787–92.

47. Boursot, P., et al. (1993). The evolution of house mice. *Annual Review of Ecology and Systematics* 24: 119–52; Sullivan, R. A. (2004). *Rats: Observations on the History and Habitat of the City's Most Unwanted Inhabitants.* New York: Bloomsbury.

48. Ayala, F. J., A. A. Escalante, and S. M. Rich (1999). Evolution of *Plasmodium* and the recent origin of the world populations of *Plasmodium falciparum. Parassitologia* 41: 55–68.

49. For two lively reviews, see Ewald, P. (1993). *The Evolution of Infectious Disease.* Oxford: Oxford University Press; Diamond, J. (1997). *Guns, Germs, and Steel.* New York: W. W. Norton.

50. The flu spreads more rapidly in the winter, not because people spend more time indoors but because the virus survives longer in cold and dry air after being sneezed or coughed up. See Lowen, A. C., et al. (2007). Influenza virus transmission is dependent on relative humidity and temperature. *PLoS Pathogens* 3: e151.

51. Potter, C. W. (1998). Chronicle of influenza pandemics. In *Textbook of Influenza,* ed. K. G. Nicholson, R. G. Webster, and A. J. Hay. Oxford: Blackwell Science, 395–412.

52. One scary example is smallpox, which has been eradicated by vaccinations, so no one gets the vaccine anymore. Should it reemerge, the consequences could be disastrous in a world where few have any immunity. When Europeans brought smallpox to the New World, where people had never before encountered the virus, the contagion wiped out about 90 percent of native Americans.

53. For the data, see Smith, P. H., and L. K. Horwitz (2007). Ancestors and inheritors: A bio-cultural perspective of the transition to agro-pastoralism in the Southern Levant. In *Ancient Health: Skeletal Indicators of Agricultural and Economic Intensification,* ed. M. N. Cohen and G. M. M. Crane-Kramer. Gainesville: University Press of Florida, 207–22; Eshed, V., et al. (2010). Paleopathology and the origin of agriculture in the Levant. *American Journal of Physical Anthropology* 143: 121–33.

54. Danforth, M. E., et al. (2007). Health and the transition to horticulture in the South-Central U.S. In *Ancient Health: Skeletal Indicators of Agricul-*

tural and Economic Intensification, ed. M. N. Cohen and G. M. M. Crane-Kramer. Gainesville: University Press of Florida: 65–79.

55. Mummert, A., et al. (2011). Stature and robusticity during the agricultural transition: Evidence from the bioarchaeological record. *Economics and Human Biology* 9: 284–301.

56. Pechenkina, E. A., R. A. Benfer, Jr., and Ma Xiaolin (2007). Diet and health in the Neolithic of the Wei and Yellow River Basins, Northern China. In *Ancient Health: Skeletal Indicators of Agricultural and Economic Intensification,* ed. M. N. Cohen and G. M. M. Crane-Kramer. Gainesville: University Press of Florida, 255–72; Temple, D. H., et al. (2008). Variation in limb proportions between Jomon foragers and Yayoi agriculturalists from prehistoric Japan. *American Journal of Physical Anthropology* 137: 164–74.

57. Marquez, M. L., et al. (2002). Health and nutrition in some prehispanic Mesoamerican populations related with their way of life. In *The Backbone of History: Health and Nutrition in the Western Hemisphere,* ed. R. Steckel and J. Rose. Cambridge: Cambridge University Press, 307–38.

58. See Cohen, M. N., and G. J. Armelagos (1984). *Paleopathology at the Origins of Agriculture.* Orlando, FL: Academic Press; Seckel, R. H., and J. C. Rose (2002). *The Backbone of History: Health and Nutrition in the Western Hemisphere.* Cambridge: Cambridge University Press; Cohen, M. N., and G. M. M. Crane-Kramer (2007). *Ancient Health: Skeletal Indicators of Agricultural and Economic Intensification.* Gainesville: University Press of Florida.

59. For reviews of this argument, see Laland, K. N., J. Odling-Smee, and S. Myles (2010). How culture shaped the human genome: Bringing genetics and the human sciences together. *Nature Reviews Genetics* 11: 137–48; Cochran, G., and H. Harpending (2009). *The 10,000 Year Explosion.* New York: Basic Books.

60. Hawks, J., et al. (2007). Recent acceleration of human adaptive evolution. *Proceedings of the National Academy of Sciences USA* 104: 20753–88; Nelson, M. R., et al. (2012). An abundance of rare functional variants in 202 drug target genes sequenced in 14,002 people. *Science* 337: 100–104; Kienan, A., and A. G. Clark (2012). Recent explosive human population growth has resulted in an excess of rare genetic variants. *Science* 336: 740–43; Tennessen, J. A., et al. (2012). Evolution and functional impact of rare coding variation from deep sequencing of human exomes. *Science* 337: 64–69.

61. Fu, W., et al. (2013). Analysis of 6,515 exomes reveals the recent origin of most human protein-coding variants. *Nature* 493: 216–20.

62. Akey, J. M. (2009). Constructing genomic maps of positive selection in humans: Where do we go from here? *Genome Research* 19: 711–22; Bustamante, C. D., et al. (2005). Natural selection on protein-coding genes in the human genome. *Nature* 437: 1153–57; Frazer, K. A., et al. (2007). A second generation human haplotype map of over 3.1 million SNPs. *Nature* 449: 851–61; Sabeti, P. C., et al. (2007). Genome-wide detection and characterization of positive selection in human populations. *Nature* 449, 913–18; Voight, B. F., et al. (2006). A map of recent positive selection in the human genome. *PLoS Biology* 4: e72; Williamson, S. H., et al. (2007). Localizing recent

408 Notes to Pages 206–215

adaptive evolution in the human genome. *PLoS Genetics* 3: e90; Grossman S. R., et al. (2013). Identifying recent adaptations in large-scale genomic data. *Cell* 152: 703–13.

63. López, C., et al. (2010). Mechanisms of genetically-based resistance to malaria. *Gene* 467: 1–12.

64. This response, known as G6PD (glucose-6-phosphate dehydrogenase) deficiency, also occurs after someone with the mutation eats fava beans.

65. Tishkoff, S. A., et al. (2007). Convergent adaptation of human lactase persistence in Africa and Europe. *Nature Genetics* 39: 31–40; Enattah, N. S., et al. (2008). Independent introduction of two lactase-persistence alleles into human populations reflects different history of adaptation to milk culture. *American Journal of Human Genetics* 82: 57–72.

66. Helgason, A., et al. (2007). Refining the impact of *TCF7L2* gene variants on type 2 diabetes and adaptive evolution. *Nature Genetics* 39: 218–25.

67. McGee, H. (2004). *On Food and Cooking,* 2nd ed. New York: Scribner.

9 Modern Times, Modern Bodies—*The Paradox of Human Health in the Industrial Era*

1. The Luddites were opponents of the early stages of the Industrial Revolution in England. They named themselves after a character from folklore, Ned Ludd, a sort of modern Robin Hood.

2. Wegman, M. (2001). Infant mortality in the 20th century: Dramatic but uneven progress. *Journal of Nutrition* 131: 401–8.

3. http://www.cdc.gov/nchs/data/nvsr/nvsr59/nvsr59_01.pdf.

4. Komlos, J., and B. E. Lauderdale (2007). The mysterious trend in American heights in the 20th century. *Annals of Human Biology* 34: 206–15.

5. Ogden, C., and M. Carroll (2010). *Prevalence of Obesity Among Children and Adolescents: United States, Trends 1963–1965 through 2007–2008*; http://www.cdc.gov/nchs/data/hestat/obesity_child_07_08/obesity_child_07_08.

6. Mass spectator sports were another industrial-era invention. According to FIFA, the organization that governs international soccer, billions of people watch soccer (the world's most popular sport), but only about 2.5 million people actually play the sport; www.fifa.com/mm/document/fifafacts/ . . ./emaga_9384_10704.pdf.

7. For a delightful account of the first industrial brewery, see Corcoran, T. (2009). *The Goodness of Guinness: The 250-Year Quest for the Perfect Pint*. New York: Skyhorse Publishing.

8. For a masterful and engaging biography of young Charles Darwin and Victorian science, see Brown, J. (2003). *Charles Darwin: Voyaging*. Princeton: Princeton University Press.

9. For a general history of the Industrial Revolution, see Stearns, P. N. (2007). *The Industrial Revolution in World History*, 3rd ed. Boulder, CO: Westview Press.

10. http://eh.net/encyclopedia/article/whaples.work.hours.us.

11. http://www.globallabourrights.org/reports?id=0034.

12. James, W. P. T., and E. C. Schofield (1990). *Human Energy Requirements: A Manual for Planners and Nutritionists*. Oxford: Oxford University Press.

13. I am assuming eight-hour work days and 260 work days per year. For comparison, an average-sized marathoner expends about 2,800 calories to complete 26.2 miles.

14. Bassett, Jr., D. R., et al. (2008). Walking, cycling, and obesity rates in Europe, North America, and Australia. *Journal of Physical Activity and Health* 5: 795–814.

15. Kerr, J., F. Eves, and D. Carroll (2001). Encouraging stair use: Stair-riser banners are better than posters. *American Journal of Public Health* 91: 1192–93.

16. Archrer, E., et al. (2013). 45-year trends in women's use of time and household management energy expenditure. *PLoS One* 8: e56620.

17. James, W. P. T., and E. C. Schofield (1990). *Human Energy Requirements: A Manual for Planners and Nutritionists*. Oxford: Oxford University Press.

18. Leonard, W. R. (2008). Lifestyle, diet, and disease: Comparative perspectives on the determinants of chronic health risks. In *Evolution in Health and Disease*, ed. S. C. Stearns and J. C. Koella. Oxford: Oxford University Press, 265–76; Pontzer, H., et al. (2012). Hunter-gatherer energetics and human obesity. *PLoS ONE* 7: e40503.

19. For an excellent history of these changes, see Hurt, R. D. (2002). *American Agriculture: A Brief History*, 2nd ed. West Lafayette, IN: Purdue University Press.

20. Abbott, E. (2009). *Sugar: A Bittersweet History*. London: Duckworth.

21. In a market, sugar cost 12 cents a pound in 1913 and 53 cents per pound in 2010. Adjusted for inflation, 12 cents in 1913 is $2.74 in 2010.

22. Haley, S., et al. (2005). Sweetener Consumption in the United States. U.S. Department of Agriculture Electronic Outlook Report from the Economic Research Service; http://www.ers.usda.gov/media/326278/sss24301_002 .pdf.

23. Finkelstein, E. A., C. J. Ruhm, and K. M. Kosa (2005). Economic causes and consequences of obesity. *Annual Review of Public Health* 26: 239–57.

24. Newman, C. (2004). Why are we so fat? The heavy cost of fat. *National Geographic* 206: 46–61.

25. Bray, G. A. (2007). *The Metabolic Syndrome and Obesity*. Totowa, NJ: Humana Press.

26. http://www.cdc.gov/mmwr/preview/mmwrhtml/mm5304a3.htm.

27. Pimentel, D., and M. H. Pimentel (2008). *Food, Energy and Society*, 3rd ed. Boca Raton, FL: CRC Press.

28. L. L. Birch (1999). Development of food preferences. *Annual Review of Nutrition* 19: 41–62.

29. Moss, M. (2013) *Salt Sugar Fat: How the Food Giants Hooked Us*. New York: Random House.

30. Boback, S. M., et al. (2007). Cooking and grinding reduces the cost of meat digestion. *Comparative Biochemistry and Physiology Part A: Molecular and Integrative Physiology* 148: 651–56.

31. For an excellent overview, see Siraisi, N. G. (1990). *Medieval and Early*

Renaissance Medicine: An Introduction to Knowledge and Practice. Chicago: University of Chicago Press.

32. Szreter, S. R. S., and G. Mooney (1998). Urbanisation, mortality and the standard of living debate: New estimates of the expectation of life at birth in nineteenth-century British cities. *Economic History Review* 51: 84–112.

33. Leviticus 13:45 (King James Version).

34. Pasteur has many biographers, but none rise to the level of Paul de Kruif in his 1926 classic, *The Microbe Hunters*, which has recently been updated and edited: De Kruif, P., and F. Gonzalez-Crussi (2002). *The Microbe Hunters*. New York: Houghton Mifflin Harcourt.

35. Snow, S. J. (2008). *Blessed Days of Anaesthesia: How Anaesthetics Changed the World*. Oxford: Oxford University Press.

36. For an amusing, fictionalized description of Kellogg's sanitarium, see Boyle, T. C. (1993). *The Road to Wellville*. New York: Viking Press.

37. Ackroyd, P. (2011). *London Under*. London: Chatto and Windus.

38. Chernow, R. (1998). *Titan: The Life of John D. Rockefeller, Sr.* New York: Warner Books.

39. Many of these details come from Gordon, R. (1993). *The Alarming History of Medicine*. New York: St. Martin's Press.

40. Lauderdale, D. S., et al. (2006). Objectively measured sleep characteristics among early-middle-aged adults: The CARDIA study. *American Journal of Epidemiology* 164: 5–16. See also *Sleep in America Poll, 2001–2002*. Washington, DC: National Sleep Foundation.

41. Worthman, C. M., and M. Melby (2002). Toward a comparative developmental ecology of human sleep. In *Adolescent Sleep Patterns: Biological, Social, and Psychological Influences*, ed. M. S. Carskadon. New York: Cambridge University Press, 69–117.

42. Marlowe, F. (2010). *The Hadza Hunter-Gatherers of Tanzania*. Berkeley: University of California Press.

43. Ekirch, R. A. (2005). *At Day's Close: Night in Times Past*. New York: Norton.

44. The modernization of time is a rich topic, elegantly and comprehensively covered by Landes, D. S. (2000). *Revolution in Time: Clocks and the Making of the Modern Era*, 2nd ed. Cambridge, MA: Harvard University Press.

45. Silber, M. H. (2005). Chronic insomnia. *New England Journal of Medicine* 353: 803–10.

46. Worthman, C. M. (2008). After dark: The evolutionary ecology of human sleep. In *Evolutionary Medicine and Health*, ed. W. R. Trevathan, E. O. Smith, and J. J. McKenna. Oxford: Oxford University Press, 291–313.

47. Roth, T., and T. Roehrs (2003). Insomnia: Epidemiology, characteristics, and consequences. *Clinical Cornerstone* 5: 5–15.

48. Spiegel, K., R. Leproult, and E. Van Cauter (1999). Impact of sleep debt on metabolic and endocrine function. *Lancet* 354: 1435–39.

49. Taheri, S., et al. (2004). Short sleep duration is associated with reduced leptin, elevated ghrelin, and increased body mass index (BMI). *Sleep* 27: A146–47.

50. Lauderdale, D. S., et al. (2006). Objectively measured sleep characteristics among early-middle-aged adults: The CARDIA study. *American Journal of Epidemiology* 164: 5–16.

51. Note that this doesn't mean that no selection has occurred. For a good review, see Stearns, S. C., et al. (2010). Measuring selection in contemporary human populations. *Nature Reviews Genetics* 11: 611–22.

52. Hatton, T. J., and B. E. Bray (2010). Long run trends in the heights of European men, 19th–20th centuries. *Economics and Human Biology* 8: 405–13.

53. Formicola, V., and M. Giannecchini (1999). Evolutionary trends of stature in upper Paleolithic and Mesolithic Europe. *Journal of Human Evolution* 36: 319–33.

54. Bogin, B. (2001). *The Growth of Humanity.* New York: Wiley.

55. Floud, R., et al. (2011). *The Changing Body: Health, Nutrition, and Human Development in the Western World Since 1700.* Cambridge: Cambridge University Press.

56. Villar, J., et al. (1992). Effect of fat and fat-free mass deposition during pregnancy on birth weight. *American Journal of Obstetrics and Gynecology* 167: 1344–52.

57. Floud, R., et al. (2011). *The Changing Body: Health, Nutrition, and Human Development in the Western World Since 1700.* Cambridge: Cambridge University Press.

58. Wang, H., et al. (2012). Age-specific and sex-specific mortality in 187 countries, 1970–2010: A systematic analysis for the Global Burden of Disease Study 2010. *Lancet* 380: 2071–94.

59. Friedlander, D., B. S. Okun, and S. Segal (1999). The demographic transition then and now: Process, perspectives, and analyses. *Journal of Family History* 24: 493–533.

60. http://www.census.gov/population/international/data/idb/worldpopinfo .php.

61. For long-term data on this trend in England, Europe, and America, see Floud, R., et al. (2011). *The Changing Body: Health, Nutrition, and Human Development in the Western World Since 1700.* Cambridge: Cambridge University Press. For mortality data from 1970 to 2010, see Lozano, R., et al. (2012). Global and regional mortality from 235 causes of death for 20 age groups in 1990 and 2010: A systematic analysis for the Global Burden of Disease Study 2010. *Lancet* 380: 2095–128.

62. Aria, E. (2004). United States Life Tables. *National Vital Statistics Reports* 52 (14): 1–40; http://www.cdc.gov/nchs/data/nvsr/nvsr52/nvsr52_14.pdf.

63. For details, see http://www.cdc.gov/nchs/data/nvsr/nvsr59/nvsr59_08.pdf; Vos, T., et al. (2012). Years lived with disability (YLDs) for 1160 sequelae of 289 diseases and injuries 1990–2010: A systematic analysis for the Global Burden of Disease Study 2010. *Lancet* 380: 2163–96.

64. The term comes from a classic 1980 paper by James Fries, who coined the term "compression of morbidity." Fries's hypothesis is that a person's burden of lifetime illness is compressed into a shorter period prior to death if the age of onset of first chronic infirmity is postponed, but that morbidity extends for a longer period if people acquire chronic infirmities at a younger age. See Fries, J. H. (1980). Aging, natural death, and the compression of morbidity. *New England Journal of Medicine* 303: 130–35.

65. Technically, a DALYs score sums the number of years people live with a disability plus the number of years of life they lose to that disability.

66. Murray, C. J. L., et al. (2012). Disability-adjusted life years (DALYs) for 291 diseases and injuries in 21 regions, 1990–2010: A systematic analysis for the Global Burden of Disease Study 2010. *Lancet* 380: 2197–223.

67. Vos, T., et al. (2012). Years lived with disability (YLDs) for 1160 sequelae of 289 diseases and injuries 1990–2010: A systematic analysis for the Global Burden of Disease Study 2010. *Lancet* 380: 2163–96.

68. Vos, T., et al. (2012). Years lived with disability (YLDs) for 1160 sequelae of 289 diseases and injuries 1990–2010: A systematic analysis for the Global Burden of Disease Study 2010. *Lancet* 380: 2163–96.

69. Salomon, J. A., et al. (2012). Healthy life expectancy for 187 countries, 1990–2010: A systematic analysis for the Global Burden Disease Study 2010. *Lancet* 380: 2144–62.

70. Gurven, M., and H. Kaplan (2007). Longevity among hunter-gatherers: A cross-cultural examination. *Population and Development Review* 33: 321–65.

71. Howell, N. (1979). *Demography of the Dobe !Kung.* New York: Academic Press; Hill, K., A. M. Hurtado, and R. Walker (2007). High adult mortality among Hiwi hunter-gatherers: Implications for human evolution. *Journal of Human Evolution* 52: 443–54; Sugiyama, L. S. (2004). Illness, injury, and disability among Shiwiar forager-horticulturalists: Implications of health-risk buffering for the evolution of human life history. *American Journal of Physical Anthropology* 123: 371–89.

72. Mann, G. V., et al. (1962). Cardiovascular disease in African Pygmies: A survey of the health status, serum lipids and diet of Pygmies in Congo. *Journal of Chronic Disease* 15: 341–71; Truswell, A. S., and J. D. L. Hansen (1976). Medical research among the !Kung. In *Kalahari Hunter-Gatherers: Studies of the !Kung San and Their Neighbors,* ed. R. B. Lee and I. DeVore. Cambridge, MA: Harvard University Press, 167–94; Howell, N. (1979). *Demography of the Dobe !Kung.* New York: Academic Press; Kronman, N., and A. Green (1980). Epidemiological studies in the Upernavik District, Greenland. *Acta Medica Scandinavica* 208: 401–6; Rode, A., and R. J. Shephard (1994). Physiological consequences of acculturation: A 20-year study of fitness in an Inuit community. *European Journal of Applied Physiology and Occupational Physiology* 69: 516–24.

73. Data on cancer: Cancer Incidence Data, Office for National Statistics and Welsh Cancer Incidence and Surveillance Unit (WCISU). Available at www.statistics.gov.uk and www.wcisu.wales.nhs.uk. Data on life expectancy: http://www.parliament.uk/documents/commons/lib/research/rp99/rp99-111.pdf.

74. Ford, E. S. (2004). Increasing prevalence of metabolic syndrome among U.S. adults. *Diabetes Care* 27: 2444–49.

75. Talley, N. J., et al. (2011). An evidence-based systematic review on medical therapies for inflammatory bowel disease. *American Journal of Gastroenterology* 106: 2–25.

76. Lim, S. S., et al. (2012). A comparative risk assessment of burden of disease and injury attributable to 67 risk factors and risk factor clusters in 21 regions, 1990–2010: A systematic analysis for the Global Burden of Disease

Study 2010. *Lancet* 380: 2224–60; Ezzati, M., et al. (2004). *Comparative Quantification of Health Risks: Global and Regional Burden of Diseases Attributable to Selected Major Risk Factors*. Geneva: World Health Organization; Mokdad, A. H., et al. (2004). Actual causes of death in the United States, 2000. *Journal of the American Medical Association* 291: 1238–45.

77. Vita, A. J., et al. (1998). Aging, health risks, and cumulative disability. *New England Journal of Medicine* 338: 1035–41.

10 The Vicious Circle of Too Much—*Why Too Much Energy Can Make Us Sick*

1. The oldest of these figurines are dated to about 35,000 years old from Germany. See Conard, N. J. (2009). A female figurine from the basal Aurignacian of Hohle Fels Cave in southwestern Germany. *Nature* 459: 248–52.

2. Johnstone, A. M., et al. (2005). Factors influencing variation in basal metabolic rate include fat-free mass, fat mass, age, and circulating thyroxine but not sex, circulating leptin, or triiodothyronine. *American Journal of Clinical Nutrition* 82: 941–48.

3. Spalding, K. L., et al. (2008). Dynamics of fat cell turnover in humans. *Nature* 453: 783–87.

4. The other basic simple sugar is galactose, which is found in milk and always paired with a glucose.

5. A fraction of the glucose, moreover, binds to proteins throughout the body, where it damages tissue by causing oxidation.

6. I have simplified here. Other hormones, including growth hormone (GH) and epinephrine (also known as adrenaline), also recruit energy in similar fashion.

7. Bray, G. A. (2007). *The Metabolic Syndrome and Obesity*. Totowa, NJ: Humana Press.

8. Technically, BMI is calculated as your mass in kilograms divided by the square of your weight in meters (kg/m2). In hindsight, this was an ill-conceived way to quantify obesity because weight is a cubic parameter (it scales to the power of 3) whereas height is linear (it scales to the power of 1). As a result, millions of tall people think they are fatter than they are, and millions of short people think they are thinner. What's more, BMI correlates weakly with percentage of body fat, and it doesn't account for how much of one's fat is visceral or subcutaneous. Because BMI is so frequently measured it remains widely used.

9. Colditz, G. A., et al. (1995). Weight gain as a risk factor for clinical diabetes mellitus in women. *Annals of Internal Medicine* 122: 481–86; Emberson, J. R., et al. (2005). Lifestyle and cardiovascular disease in middle-aged British men: The effect of adjusting for within-person variation. *European Heart Journal* 26: 1774–82.

10. Pond, C. M., and C. A. Mattacks (1987). The anatomy of adipose tissue in captive *Macaca* monkeys and its implications for human biology. *Folia*

Primatologica 48: 164–85; Kuzawa, C. W. (1998). Adipose tissue in human infancy and childhood: An evolutionary perspective. *Yearbook of Physical Anthropology* 41: 177–209; Eaton, S. B., M. Shostak, and M. Konner (1988). *The Paleolithic Prescription: A Program of Diet and Exercise and a Design for Living.* New York: Harper and Row.

11. Dufour, D. L., and M. L. Sauther (2002). Comparative and evolutionary dimensions of the energetics of human pregnancy and lactation. *American Journal of Human Biology* 14: 584–602; Hinde, K., and L. A. Milligan (2011). Primate milk: Proximate mechanisms and ultimate perspectives. *Evolutionary Anthropology* 20: 9–23.

12. Ellison, P. T. (2001). *On Fertile Ground: A Natural History of Human Reproduction.* Cambridge, MA: Harvard University Press.

13. Fat influences various metabolic functions by producing a hormone known as leptin. The more fat you have, the higher your leptin levels, and vice versa. Leptin has several effects including appetite regulation. Under normal conditions, when the body has lots of fat, leptin levels go up and the brain suppresses appetite; appetite resumes when leptin levels fall from lack of fat. Leptin levels also help regulate when women ovulate. Falling levels of body fat thus lessen a woman's ability to conceive. For more details, see Donato, J., et al. (2011). Hypothalamic sites of leptin action linking metabolism and reproduction. *Neuroendocrinology* 93: 9–18.

14. Neel, J. V. (1962). Diabetes mellitus: A "thrifty" genotype rendered detrimental by "progress"? *American Journal of Human Genetics* 14: 353–62.

15. Knowler, W. C., et al. (1990). Diabetes mellitus in the Pima Indians: Incidence, risk factors, and pathogenesis. *Diabetes Metabolism Review* 6: 1–27.

16. Gluckman, M., A. Beedle, and M. Hanson (2009). *Principles of Evolutionary Medicine.* Oxford: Oxford University Press.

17. Speakman, J. R. (2007). A nonadaptive scenario explaining the genetic predisposition to obesity: The "predation release" hypothesis. *Cell Metabolism* 6: 5–12.

18. Yu, C. H. Y., and B. Zinman (2007). Type 2 diabetes and impaired glucose tolerance in aboriginal populations: A global perspective. *Diabetes Research and Clinical Practice* 78: 159–70.

19. Hales, C. N., and D. J. Barker (1992). Type 2 (non-insulin-dependent) diabetes mellitus: The thrifty phenotype hypothesis. *Diabetologia* 35: 595–601.

20. Painter, R. C., T. J. Rosebloom, and O. P. Bleker (2005). Prenatal exposure to the Dutch famine and disease in later life: An overview. *Reproductive Toxicology* 20: 345–52.

21. Kuzawa, C. W., et al. (2008). Evolution, developmental plasticity, and metabolic disease. In *Evolution in Health and Disease,* 2nd ed., ed. S. C. Stearns and J. C. Koella. Oxford: Oxford University Press, 253–64.

22. Wells, J. C. K. (2011). The thrifty phenotype: An adaptation in growth or metabolism. *American Journal of Human Biology* 23: 65–75.

23. Eriksson, J. G. (2007). Epidemiology, genes and the environment: Lessons learned from the Helsinki Birth Cohort Study. *Journal of Internal Medicine* 261: 418–25.

24. Eriksson, J. G., et al. (2003). Pathways of infant and childhood growth that lead to type 2 diabetes. *Diabetes Care* 26: 3006–10.

25. Ibrahim, M. (2010). Subcutaneous and visceral adipose tissue: Structural and functional differences. *Obesity Reviews* 11: 11–18.

26. Coutinho, T., et al. (2011). Central obesity and survival in subjects with coronary artery disease: A systematic review of the literature and collaborative analysis with individual subject data. *Journal of the American College of Cardiology* 57: 1877–86.

27. For a good review of these and other details, see Wood, P. A. (2009). *How Fat Works.* Cambridge, MA: Harvard University Press.

28. Rosenblum, A. L. (1975). Age-adjusted analysis of insulin responses during normal and abnormal glucose tolerance tests in children and adolescents. *Diabetes* 24: 820–28; Lustig, R. H. (2013). *Fat Chance: Beating the Odds Against Sugar, Processed Food, Obesity, and Disease.* New York: Penguin.

29. There are two common ways to measure this property. The first is the *glycemic index* (GI), which measures how rapidly 100 grams of a food raise blood sugar levels relative to 100 grams of pure glucose. The *glycemic load* (GL) measures how much a portion of food increases blood glucose levels (the GI times the available carbohydrates). For a typical apple, the GI is 39 and the GL is 6; for a fruit roll the GI is 99 and the GL is 24.

30. Weigle, D. S., et al. (2005). A high-protein diet induces sustained reductions in appetite, ad libitum caloric intake, and body weight despite compensatory changes in diurnal plasma leptin and ghrelin concentrations. *American Journal of Clinical Nutrition* 82: 41–8.

31. Small, C. J., et al. (2004). Gut hormones and the control of appetite. *Trends in Endocrinology and Metabolism* 15: 259–63.

32. Samuel, V. T. (2011) Fructose-induced lipogenesis: From sugar to fat to insulin resistance. *Trends in Endocrinology and Metabolism* 22: 60–65.

33. Vos, M. B., et al. (2008). Dietary fructose consumption among U.S. children and adults: The Third National Health and Nutrition Examination Survey. *Medscape Journal of Medicine* 10: 160.

34. A study testing this hypothesis was recently published. The study had twenty-one people (aged 18–40) lose 10 to 15 percent of their body weight on a diet and then randomized them into three groups that ate one of three diets with equal numbers of calories for three months: (1) a low fat diet, (2) a low carb diet, and (3) a low glycemic diet. The low-fat dieters fared the worst; the low-carb dieters burned 300 calories more per day than the low fat dieters, but they showed elevated levels of cortisol and inflammatory markers; the low glycemic dieters burned 150 calories more per day than the low fat dieters but showed none of the negative effects of the low carbohydrate diet. See Ebbeling, C. B., et al. (2012). Effects of dietary composition on energy expenditure during weight-loss maintenance. *Journal of the American Medical Association* 307: 2627–34.

35. This is a big, fast-changing topic. For a good review, see Walley, A. J., J. E. Asher, and P. Froguel (2009). The genetic contribution to non-syndromic human obesity. *Nature Reviews Genetics* 10: 431–42.

36. Frayling, T. M., et al. (2007). A common variant in the *FTO* gene is associated with body mass index and predisposes to childhood and adult obesity. *Science* 316: 889–94; Povel, C. M., et al. (2011). Genetic variants and the metabolic syndrome: A systematic review. *Obesity Reviews* 12: 952–67.

37. Rampersaud, E., et al. (2008). Physical activity and the association of common *FTO* gene variants with body mass index and obesity. *Archives of Internal Medicine* 168: 1791–97.

38. Adam, T. C., and Epel, E. S. (2007). Stress, eating and the reward system. *Physiology and Behavior* 91: 449–58.

39. Epel, E. S., et al. (2000). Stress and body shape: Stress-induced cortisol secretion is consistently greater among women with central fat. *Psychosomatic Medicine* 62: 623–32; Vicennati, V., et al. (2002). Response of the hypothalamic-pituitary-adrenocortical axis to high-protein/fat and high carbohydrate meals in women with different obesity phenotypes. *Journal of Clinical Endocrinology and Metabolism* 87: 3984–88; Anagnostis, P. (2009). Clinical review: The pathogenetic role of cortisol in the metabolic syndrome: A hypothesis. *Journal of Clinical Endocrinology and Metabolism* 94: 2692–701.

40. Mietus-Snyder, M. L., et al. (2008). Childhood obesity: Adrift in the "Limbic Triangle." *Annual Review of Medicine* 59: 119–34.

41. Beccuti, G., and S. Pannain (2011). Sleep and obesity. *Current Opinions in Clinical Nutrition and Metabolic Care* 14: 402–12.

42. Shaw, K., et al. (2006). Exercise for overweight and obesity. Cochrane Database of Systematic Reviews. CD003817.

43. Cook, C. M., and D. A. Schoeller (2011). Physical activity and weight control: Conflicting findings. *Current Opinions in Clinical Nutrition and Metabolic Care* 14: 419–24.

44. Blundell, J. E., and N. A. King (1999). Physical activity and regulation of food intake: Current evidence. *Medicine and Science in Sports and Exercise* 31: S573–83.

45. Poirier, P., and J. P. Després (2001). Exercise in weight management of obesity. *Cardiology Clinics* 19: 459–70.

46. Turnbaugh, P. J., and J. I. Gordon (2009). The core gut microbiome, energy balance and obesity. *Journal of Physiology* 587: 4153–58.

47. Smyth, S., and A. Heron (2006). Diabetes and obesity: The twin epidemics. *Nature Medicine* 12: 75–80.

48. Koyama, K., et al. (1997). Tissue triglycerides, insulin resistance, and insulin production: Implications for hyperinsulinemia of obesity. *American Journal of Physiology* 273: E708–13; Samaha, F. F., G. D. Foster, and A. P. Makris (2007). Low-carbohydrate diets, obesity, and metabolic risk factors for cardiovascular disease. *Current Atherosclerosis Reports* 9: 441–47; Kumashiro, N., et al. (2011). Cellular mechanism of insulin resistance in nonalcoholic fatty liver disease. *Proceedings of the National Academy of Sciences USA* 108: 16381–85.

49. Thomas, E. L., et al. (2012). The missing risk: MRI and MRS phenotyping of abdominal adiposity and ectopic fat. *Obesity* 20: 76–87.

50. Bray, G. A., S. J. Nielsen, and B. M. Popkin (2004). Consumption of high-fructose corn syrup in beverages may play a role in the epidemic of obesity. *American Journal of Clinical Nutrition* 79: 537–43.

51. Lim, E. L., et al. (2011). Reversal of type 2 diabetes: Normalisation of beta cell function in association with decreased pancreas and liver triacylglycerol. *Diabetologia* 54: 2506–14.

52. Borghouts, L. B., and H. A. Keizer (2000). Exercise and insulin sensitivity: A review. *International Journal of Sports Medicine* 21: 1–12.

53. van der Heijden, G. J., et al. (2009). Aerobic exercise increases peripheral and hepatic insulin sensitivity in sedentary adolescents. *Journal of Clinical Endocrinology and Metabolism* 94: 4292–99.

54. O'Dea, K. (1984). Marked improvement in carbohydrate and lipid metabolism in diabetic Australian aborigines after temporary reversion to traditional lifestyle. *Diabetes* 33: 596–603.

55. Basu, S., et al. (2013). The relationship of sugar to population-level diabetes prevalence: An econometric analysis of repeated cross-sectional data. *PLoS One.* 8: e57873.

56. Knowler, W. C., et al. (2002). Reduction in the incidence of Type 2 diabetes with lifestyle intervention or metformin. *New England Journal of Medicine* 346: 393–403.

57. HDLs also shuttle cholesterol to the testes, ovaries, and the adrenal glands of the kidneys, where the cholesterol is transformed into hormones, such as estrogen, testosterone, and cortisol. Note also that neither HDL nor LDL are cholesterol molecules (although they contain them), making the popular terms "good cholesterol" and "bad cholesterol" misleading. I use the terms because they are so well known and commonly used.

58. Thompson, R. C., et al. (2013). Atherosclerosis across 4000 years of human history: The Horus study of four ancient populations. *Lancet* 381: 1211–22.

59. Mann, G. V., et al. (1962). Cardiovascular disease in African Pygmies: A survey of the health status, serum lipids, and diet of Pygmies in Congo. *Journal of Chronic Disease* 15: 341–71; Mann, G. V., et al. (1962). The health and nutritional status of Alaskan Eskimos. *American Journal of Clinical Nutrition* 11: 31–76; Lee, K. T., et al. (1964). Geographic pathology of myocardial infarction. *American Journal of Cardiology* 13: 30–40; Meyer, B. J. (1964). Atherosclerosis in Europeans and Bantu. *Circulation* 29: 415–21; Woods, J. D. (1966). The electrocardiogram of the Australian aboriginal. *Medical Journal of Australia* 1: 238–41; Magarey, F. R., J. Kariks, and L. Arnold (1969). Aortic atherosclerosis in Papua and New Guinea compared with Sydney. *Pathology* 1: 185–91; Mann, G. V., et al. (1972). Atherosclerosis in the Masai. *American Journal of Epidemiology* 95: 26–37; Truswell, A. S., and J. D. L. Hansen (1976). Medical research among the !Kung. In *Kalahari Hunter-Gatherers: Studies of the !Kung San and Their Neighbors,* ed. R. B. Lee and I. DeVore. Cambridge: Harvard University Press, 167–94; Kronman, N., and A. Green (1980). Epidemiological studies in the Upernavik District, Greenland. *Acta Medica Scandinavica* 208: 401–6; Trowell, H. C., and D. P. Burkitt (1981). *Western Diseases: Their Emergence and Prevention.* Cambridge, MA: Harvard University Press; Blackburn, H., and R. Prineas (1983). Diet and hypertension: Anthropology, epidemiology, and public health implications. *Progress in Biochemical Pharmacology* 19: 31–79; Rode, A., and R. J. Shephard (1994). Physiological consequences of acculturation: A 20-year study of fitness in an Inuit community. *European Journal of Applied Physiology and Occupational Physiology* 69: 516–24.

60. Durstine, J. L., et al. (2001). Blood lipid and lipoprotein adaptations to exercise: A quantitative analysis. *Sports Medicine* 31: 1033–62. Note that

exercise does not lower LDLs but instead reduces the percentage of smaller, denser, triglyceride-rich LDLs by burning triglycerides.

61. Ford, E. S. (2002) Does exercise reduce inflammation? Physical activity and C-reactive protein among U.S. adults. *Epidemiology* 13: 561–68.
62. Tanasescu, M., et al. (2002). Exercise type and intensity in relation to coronary heart disease in men. *Journal of the American Medical Association* 288: 1994–2000.
63. Cater, N. B., and A. Garg (1997). Serum low-density lipoprotein response to modification of saturated fat intake: Recent insights. *Current Opinion in Lipidology* 8: 332–36.
64. For reviews, see Willett, W. (1998). *Nutritional Epidemiology*, 2nd ed. Oxford: Oxford University Press; Hu, F. B. (2008). *Obesity Epidemiology*. Oxford: Oxford University Press.
65. These fatty acids are termed N-3 or omega-3 fatty acids because their carbon double bond is at the third to last carbon in the fatty acid chain. For a good summary of the evidence for their health benefits, see McKenney, J. M., and D. Sica (2007). Prescription of omega-3 fatty acids for the treatment of hypertriglyceridemia. *American Journal of Health Systems Pharmacists* 64: 595–605.
66. Mozaffarian, D., A. Aro, and W. C. Willett (2009). Health effects of trans-fatty acids: Experimental and observational evidence. *European Journal of Clinical Nutrition* 63 (suppl. 2): S5–21.
67. Cordain, L., et al. (2002). Fatty acid analysis of wild ruminant tissues: Evolutionary implications for reducing diet-related chronic disease. *European Journal of Clinical Nutrition* 56: 181–91; Leheska, J. M., et al. (2008). Effects of conventional and grass-feeding systems on the nutrient composition of beef. *Journal of Animal Science* 86: 3575–85.
68. Bjerregaard, P., M. E. Jørgensen, and K. Borch-Johnsen (2004). Serum lipids of Greenland Inuit in relation to Inuit genetic heritage, westernisation and migration. *Atherosclerosis* 174: 391–98.
69. Castelli, W. P., et al. (1977). HDL cholesterol and other lipids in coronary heart disease: The cooperative lipoprotein phenotyping study. *Circulation* 55: 767–72; Castelli, W. P., et al. (1992). Lipids and risk of coronary heart disease: The Framingham Study. *Annals of Epidemiology* 2: 23–28; Jeppesen, J., et al. (1998). Triglycerides concentration and ischemic heart disease: An eight-year follow-up in the Copenhagen Male Study. *Circulation* 97: 1029–36; Da Luz, P. L., et al. (2005). Comparison of serum lipid values in patients with coronary artery disease at <50, 50 to 59, 60 to 69, and >70 years of age. *American Journal of Cardiology* 96: 1640–43.
70. Gardner, C. D., et al. (2007). Comparison of the Atkins, Zone, Ornish, and LEARN diets for change in weight and related risk factors among overweight premenopausal women: The A TO Z Weight Loss Study: A randomized trial. *Journal of the American Medical Association* 297: 969–77; Foster, G. D., et al. (2010). Weight and metabolic outcomes after 2 years on a low-carbohydrate versus low-fat diet: A randomized trial. *Annals of Internal Medicine* 153: 147–57.
71. Stampfer, M. J., et al. (1996). A prospective study of triglyceride level, low-density lipoprotein particle diameter, and risk of myocardial infarction.

Journal of the American Medical Association 276: 882–88; Guay, V., et al. (2012). Effect of short-term low- and high-fat diets on low-density lipoprotein particle size in normolipidemic subjects. *Metabolism* 61(1): 76–83.

72. For a thorough review of the literature, see Hooper, L., et al. (2012). Reduced or modified dietary fat for preventing cardiovascular disease. *Cochrane Database of Systematic Reviews* 5: CD002137; Hooper, L., et al. (2012). Effect of reducing total fat intake on body weight: Systematic review and meta-analysis of randomised controlled trials and cohort studies. *British Medical Journal* 345: e7666.

73. As an example, a randomized control study conducted in Spain put 7,447 people aged fifty-five to eighty who were overweight, smoked, or had heart disease on either a low fat diet or a Mediterranean diet, with lots of olive oil, fresh vegetables, and fish. After five years the study was terminated because the individuals on the Mediterranean diet already had a 30 percent lower rate of death from heart attacks, strokes, and other heart diseases. See Estruch, R., et al. (2013). Primary prevention of cardiovascular disease with a Mediterranean diet. *New England Journal of Medicine* 368: 1279–90.

74. Cordain, L., et al. (2005). Origins and evolution of the Western diet: Health implications for the 21st century. *American Journal of Clinical Nutrition.* 81: 341–54.

75. Tropea, B. I., et al. (2000). Reduction of aortic wall motion inhibits hypertension-mediated experimental atherosclerosis. *Arteriosclerosis, Thrombosis, and Vascular Biology* 20: 2127–33.

76. Note that fiber also helps to control appetite by filling the stomach. For a classic summary of the benefits of fiber, see Anderson, J. W., B. M. Smith, and N. J. Gustafson (1994). Health benefits and practical aspects of high-fiber diets. *American Journal of Clinical Nutrition* 59: 1242S–47S.

77. Eaton, S. B. (1992). Humans, lipids and evolution. *Lipids* 27: 814–20.

78. Allam, A. H., et al. (2009). Computed tomographic assessment of atherosclerosis in ancient Egyptian mummies. *Journal of the American Medical Association* 302: 2091–94.

79. American Cancer Society (2011). *Cancer Facts and Figures.* Atlanta: American Cancer Society.

80. Beniashvili, D. S. (1989). An overview of the world literature on spontaneous tumors in nonhuman primates. *Journal of Medical Primatology* 18: 423–37.

81. Rigoni-Stern, D. A. (1842). Fatti statistici relativi alle mallattie cancrose. *Giovnali per servire ai progressi della Patologia e della Terapeutica* 2: 507–17.

82. Greaves, M. (2001). *Cancer: The Evolutionary Legacy.* Oxford: Oxford University Press.

83. One well-studied example is the *p53* gene, which helps cells initiate DNA repair and stops stressed cells from proliferating. Animals, including humans, with mutations in this gene have higher cancer rates when subjected to mutation-inducing stimuli. For a review, see Lane, D. P. (1992). *p53*, guardian of the genome. *Nature* 358: 15–16.

84. Eaton, S. B., et al. (1994). Women's reproductive cancers in evolutionary context. *Quarterly Review of Biology* 69: 353–36.

85. Biologists used to think that nursing frequency suppressed ovulation, but

recent evidence suggests that the overall energetic cost of nursing is the dominant cause of this effect. See Valeggia, C., and P. T. Ellison (2009). Interactions between metabolic and reproductive functions in the resumption of postpartum fecundity. *American Journal of Human Biology* 21: 559–66.

86. Lipworth, L., L. R. Bailey, and D. Trichopoulos (2000). History of breast-feeding in relation to breast cancer risk: A review of the epidemiologic literature. *Journal of the National Cancer Institute* 92: 302–12.

87. For a comprehensive review of this biology from an evolutionary and anthropological perspective, I recommend Trevathan, W. (2010) *Ancient Bodies, Modern Lives: How Evolution Has Shaped Women's Health*. Oxford: Oxford University Press.

88. Austin, H., et al. (1991). Endometrial cancer, obesity, and body fat distribution. *Cancer Research* 51: 568–72.

89. Morimoto, L. M., et al. (2002). Obesity, body size, and risk of postmenopausal breast cancer: The Women's Health Initiative (United States). *Cancer Causes and Control* 13: 741–51.

90. Calistro Alvarado, L. (2010). Population differences in the testosterone levels of young men are associated with prostate cancer disparities in older men. *American Journal of Human Biology* 22: 449–55; Chu, D. I., and S. J. Freedland (2011). Metabolic risk factors in prostate cancer. *Cancer* 117: 2020–23.

91. Jasienska, G., et al. (2006). Habitual physical activity and estradiol levels in women of reproductive age. *European Journal of Cancer Prevention* 15: 439–45.

92. Thune, I., and A. S. Furberg (2001). Physical activity and cancer risk: Dose-response and cancer, all sites and site-specific. *Medicine and Science in Sports and Exercise* 33: S530–50.

93. Peel, B., et al. (2009). Cardiorespiratory fitness and breast cancer mortality: Findings from the Aerobics Center Longitudinal Study (ACLS). *Medicine and Science in Sports and Exercise* 41: 742–48; Ueji, M., et al. (1988). Physical activity and the risk of breast cancer: A case-control study of Japanese women. *Journal of Epidemiology* 8: 116–22.

94. Ellison, P. T. (1999). Reproductive ecology and reproductive cancers. In *Hormones and Human Health*, ed. C. Panter-Brick and C. Worthman. Cambridge: Cambridge University Press, 184–209.

95. For more, see Merlo, L. M. F., et al. (2006). Cancer as an evolutionary and ecological process. *Nature Reviews Cancer* 6: 924–35; Ewald, P. W. (2008). An evolutionary perspective on parasitism as a cause of cancer. *Advances in Parasitology* 68: 21–43.

96. For analyses comparing worldwide rates of death and disability from these diseases in 2010 compared to 1990, see Lozano, R., et al. (2012). Global and regional mortality from 235 causes of death for 20 age groups in 1990 and 2010: A systematic analysis for the Global Burden of Disease Study 2010. *Lancet* 380: 2095–128; Vos, T., et al. (2012). Years lived with disability (YLDs) for 1160 sequelae of 289 diseases and injuries 1990–2010: A systematic analysis for the Global Burden of Disease Study 2010. *Lancet* 380: 2163–96.

97. http://seer.cancer.gov/csr/1975_2009_pops09/results_single/sect_01_table .11_2pgs.pdf.

98. Sobal, J., and A. J. Stunkard (1989). Socioeconomic status and obesity: A review of the literature. *Psychological Bulletin* 105: 260–75.

99. Campos, P., et al. (2006). The epidemiology of overweight and obesity: Public health crisis or moral panic? *International Journal of Epidemiology* 35: 55–60.

100. Wildman, R. P., et al. (2008). The obese without cardiometabolic risk factor clustering and the normal weight with cardiometabolic risk factor clustering: Prevalence and correlates of 2 phenotypes among the U.S. population (NHANES 1999–2004). *Archives of Internal Medicine* 168: 1617–24.

101. McAuley, P. A., et al. (2010). Obesity paradox and cardiorespiratory fitness in 12,417 male veterans aged 40 to 70 years. *Mayo Clinic Proceedings* 85: 115–21; Habbu, A., N. M. Lakkis, and H. Dokainish (2006). The obesity paradox: Fact or fiction? *American Journal of Cardiology* 98: 944–48; McAuley, P. A., and S. N. Blair (2011). Obesity paradoxes. *Journal of Sports Science* 29: 773–82.

102. Lee, C. D., S. N. Blair, and A. S. Jackson (1999). Cardiorespiratory fitness, body composition, and all-cause and cardiovascular disease mortality in men. *American Journal of Clinical Nutrition* 69: 373–80.

11 Disuse—*Why We Are Losing It by Not Using It*

1. Technically, a safety factor is a structure's maximum strength or capacity divided by its maximum load.

2. See Horstman, J. (2012). *The Scientific American Healthy Aging Brain: The Neuroscience of Making the Most of Your Mature Mind.* San Francisco: Jossey-Bass.

3. When Japanese researchers were trying to understand why some soldiers were better able than others to acclimatize to the hot, humid conditions of the southern Pacific, they discovered that people who experienced more heat stress in the first three years of life developed many more functional sweat glands, which they then maintained as adults. See Kuno, Y. (1956). *Human Perspiration.* Springfield, IL: Charles C. Thomas.

4. Reptiles provide lots of great examples. If you raise lizards on narrower branches, they develop shorter limbs, and in some species changing the temperature of an egg determines whether the hatchling is male or female. See Losos, J. B., et al. (2000). Evolutionary implications of phenotypic plasticity in the hindlimb of the lizard *Anolis sagrei. Evolution* 54: 301–5. Shine, R. (1999). Why is sex determined by nest temperature in many reptiles? *Trends in Ecology and Evolution* 14: 186–89.

5. For a terrific account of this biology, see Jablonski, N. (2007). *Skin: A Natural History.* Berkeley: University of California Press.

6. The notion that bodies adapt their structures to match but not exceed demand is known as the hypothesis of symmorphosis. For more, see Weibel, E. R., C. R. Taylor, and H. Hoppeler (1991). The concept of symmorphosis: A testable hypothesis of structure-function relationship. *Proceedings of the National Academy of Sciences USA* 88: 10357–61.

7. Jones, H. H., et al. (1977). Humeral hypertrophy in response to exercise. *Journal of Bone and Joint Surgery* 59: 204–8.

8. For a review, see Lieberman, D. E. (2011). *The Evolution of the Human Head*. Cambridge, MA: Harvard University Press.

9. For a review, see Carter, D. R., and G. S. Beaupré (2001). *Skeletal Function and Form: Mechanobiology of Skeletal Development, Aging, and Regeneration*. Cambridge: Cambridge University Press.

10. Currey, J. D. (2002). *Bone: Structure and Mechanics*. Princeton: Princeton University Press.

11. Riggs, B. L., and L. J. Melton III (2005). The worldwide problem of osteoporosis: Insights afforded by epidemiology. *Bone* 17 (suppl. 5): 505–11.

12. Roberts, C. A., and K. Manchester (1995). *The Archaeology of Disease*, 2nd ed. Ithaca, NY: Cornell University Press.

13. Martin, R. B., D. B. Burr, and N. A. Sharkey (1998). *Skeletal Tissue Mechanics*. New York: Springer.

14. Guadalupe-Grau, A., et al. (2009). Exercise and bone mass in adults. *Sports Medicine* 39: 439–68.

15. Devlin, M. J. (2011). Estrogen, exercise, and the skeleton. *Evolutionary Anthropology* 20: 54–61.

16. See http://www.ars.usda.gov/foodsurvey; Eaton, S. B., S. B. Eaton III, and M. J. Konner (1997). Paleolithic nutrition revisited: A twelve-year retrospective on its nature and implications. *European Journal of Clinical Nutrition* 51: 207–16.

17. Bonjour, J. P. (2005). Dietary protein: An essential nutrient for bone health. *Journal of the American College of Nutrition* 24: 526S–36S.

18. Corruccini, R. S. (1999). *How Anthropology Informs the Orthodontic Diagnosis of Malocclusion's Causes*. Lewiston, NY: Mellen Press.

19. Hagberg, C. (1987). Assessment of bite force: A review. *Journal of Craniomandibular Disorders: Facial and Oral Pain* 1: 162–69.

20. These forces have been measured in nonhuman primates. For an example, see Hylander, W. L., K. R. Johnson, and A. W. Crompton (1987). Loading patterns and jaw movements during mastication in *Macaca fascicularis*: A bone-strain, electromyographic, and cineradiographic analysis. *American Journal of Physical Anthropology* 72: 287–314.

21. Lieberman, D. E., et al. (2004). Effects of food processing on masticatory strain and craniofacial growth in a retrognathic face. *Journal of Human Evolution* 46: 655–77.

22. Corruccini, R. S., and R. M. Beecher (1982). Occlusal variation related to soft diet in a nonhuman primate. *Science* 218: 74–76; Ciochon, R. L., R. A. Nisbett, and R. S. Corruccini (1997). Dietary consistency and craniofacial development related to masticatory function in minipigs. *Journal of Craniofacial Genetics and Developmental Biology* 17: 96–102.

23. Corruccini, R. S. (1984). An epidemiologic transition in dental occlusion in world populations. *American Journal of Orthodontics and Dentofacial Orthopaedics* 86: 419–26; Lukacs, J. R. (1989). Dental paleopathology: Methods for reconstructing dietary patterns. In *Reconstruction of Life from the Skeleton*, ed. M. R. Iscan and K. A. R. Kennedy. New York: Alan R. Liss, 261–86.

24. For more details, see Lieberman, D. E. (2011). *The Evolution of the Human Head*. Cambridge, MA: Harvard University Press.

25. Twetman, S. (2009). Consistent evidence to support the use of xylitol- and sorbitol-containing chewing gum to prevent dental caries. *Evidence Based Dentistry* 10: 10–11.

26. Ingervall, B., and E. Bitsanis (1987). A pilot study of the effect of masticatory muscle training on facial growth in long-face children. *European Journal of Orthodontics* 9: 15–23.

27. Savage, D. C. (1977). Microbial ecology of the gastrointestinal tract. *Annual Review of Microbiology* 31: 107–33.

28. Dethlefsen, L., M. McFall-Ngai, and D. A. Relman (2007). An ecological and evolutionary perspective on human-microbe mutualism and disease. *Nature* 449: 811–18.

29. Ruebush, M. (2009). *Why Dirt Is Good*. New York: Kaplan.

30. Brantzaeg, P. (2010). The mucosal immune system and its integration with the mammary glands. *Journal of Pediatrics* 156: S8–15.

31. Strachan, D. J. (1989). Hay fever, hygiene, and household size. *British Medical Journal* 299: 1259–60.

32. See Correale, J., and M. Farez (2007). Association between parasite infection and immune responses in multiple sclerosis. *Annals of Neurology* 61: 97–108; Summers, R. W., et al. (2005). *Trichuris suis* therapy in Crohn's disease. *Gut* 54: 87–90; Finegold, S. M., et al. (2010). Pyrosequencing study of fecal microflora of autistic and control children. *Anaerobe* 16: 444–53.

33. Bach, J. F. (2002). The effect of infections on susceptibility to autoimmune and allergic diseases. *New England Journal of Medicine* 347: 911–20.

34. Otsu, K., and S. C. Dreskin (2011). Peanut allergy: An evolving clinical challenge. *Discovery Medicine* 12: 319–28.

35. Prescott, S. L., et al. (1999). Development of allergen-specific T-cell memory in atopic and normal children. *Lancet* 353: 196–200; Sheikh, A., and D. P. Strachan (2004). The hygiene theory: Fact or fiction? *Current Opinions in Otolaryngology and Head and Neck Surgery* 12: 232–36.

36. Hansen, G., et al. (1999). Allergen-specific Th1 cells fail to counterbalance Th2 cell-induced airway hyperreactivity but cause severe airway inflammation. *Journal of Clinical Investigation* 103: 175–83.

37. Benn, C. S., et al. (2004). Cohort study of sibling effect, infectious diseases, and risk of atopic dermatitis during first 18 months of life. *British Medical Journal* 328: 1223–27.

38. Rook, G. A. (2009). Review series on helminths, immune modulation and the hygiene hypothesis: The broader implications of the hygiene hypothesis. *Immunology* 126: 3–11.

39. Braun-Fahrlander, C., et al. (2002). Environmental exposure to endotoxin and its relation to asthma in school-age children. *New England Journal of Medicine* 347: 869–77; Yazdanbakhsh, M., P. G. Kremsner, and R. van Ree (2002). Allergy, parasites, and the hygiene hypothesis. *Science* 296: 490–94.

40. Rook, G. A. (2012). Hygiene hypothesis and autoimmune diseases. *Clinical Reviews in Allergy and Immunology* 42: 5–15.

41. Van Nood, E., et al. (2013). Duodenal infusion of donor feces for recurrent *Clostridium difficile*. *New England Journal of Medicine* 368: 407–15.

42. Feijen, M., J. Gerritsen, and D. S. Postma (2000). Genetics of allergic disease. *British Medical Bulletin* 56: 894–907. One interesting caveat, however, is that twins also tend to share the same microbiome, inflating estimates of the role of genes. For more, see Turnbaugh, P. J., et al. (2009). A core gut microbiome in obese and lean twins. *Nature* 457: 480–84.

43. Among the many studies that have examined this effect, one of my favorites is the Stanford Runners Study, conducted by Dr. James Fries and colleagues. This study has been tracking since 1984 two groups of Americans over the age of fifty: one group included 538 amateur runners; the other group included 423 healthy, nonoverweight but sedentary controls who did not exercise very much. After two decades, the amateur runners had a 20 percent lower chance of dying in a given year than the sedentary controls, and of the 225 participants who had died, only one-third were runners (a twofold difference). In addition, the runners had 50 percent lower levels of disability—the equivalent of bodies that were fourteen years younger. See Chakravarty, E.F., et al. (2008) Reduced disability and mortality among aging runners: a 21-year longitudinal study. *Archives of Internal Medicine* 168: 1638–46.

12 The Hidden Dangers of Novelty and Comfort— *Why Everyday Innovations Can Damage Us*

1. Paik, D. C., et al. (2001). The epidemiological enigma of gastric cancer rates in the U.S.: Was grandmother's sausage the cause? *International Journal of Epidemiology* 30: 181–82; Jakszyn, P., and C. A. Gonzalez (2006). Nitrosamine and related food intake and gastric and oesophageal cancer risk: A systematic review of the epidemiological evidence. *World Journal of Gastroenterology* 12: 4296–303.

2. I suspect that Neanderthals figured out how to wrap skins around their feet during the winter, but such materials don't survive that long in the archaeological record, and the first indirect evidence for shoes comes from studies of toe bone thickness based on observations that shod people have relatively thinner toe bones than barefoot people. See Trinkaus, E., and H. Shang (2008). Anatomical evidence for the antiquity of human footwear: Tianyuan and Sunghir. *Journal of Archaeological Science* 35: 1928–33.

3. Pinhasi, R., et al. (2010). First direct evidence of chalcolithic footwear from the Near Eastern Highlands. *PLoS ONE* 5(6): e10984; Bedwell, S. F., and L. S. Cressman (1971). Fort Rock Report: Prehistory and environment of the pluvial Fort Rock Lake area of South-Central Oregon. In *Great Basin Anthropological Conference*, ed. M. C. Aikens. Eugene: University of Oregon Anthropological Papers, 1–25.

4. The American Podiatric Medical Association's website states that "cushioned-sole shoes that give good support are essential for those who spend most of their working days on their feet." http://www.apma.org/MainMenu/FootHealth/Brochures/Footwear.aspx.

5. McDougall, C. (2009). *Born to Run: A Hidden Tribe, Superathletes, and the Greatest Race the World Has Never Seen.* New York: Knopf.

6. Lieberman, D. E., et al. (2010). Foot strike patterns and collision forces in habitually barefoot versus shod runners. *Nature* 463: 531–35.

7. Kirby, K. A. (2010). Is barefoot running a growing trend or a passing fad? *Podiatry Today* 23: 73.

8. Chi, K. J., and D. Schmitt (2005). Mechanical energy and effective foot mass during impact loading of walking and running. *Journal of Biomechanics* 38: 1387–95.

9. This is also true of walking (tiptoeing), but it is not a common gait because it is so inefficient and usually unnecessary.

10. Nigg, B. M. (2010). *Biomechanics of Sports Shoes*. Calgary: Topline Printing.

11. Variation is always present, and while many experienced barefoot runners prefer to forefoot strike, some habitually barefoot people do sometimes land on their heels. We don't yet know how much this variation is affected by factors such as skill, distance run, surface hardness, speed, and fatigue. Although habitually barefoot runners from the Kalenjin tribe in Kenya, famous long-distance runners, typically forefoot strike when barefoot, one study of barefoot people from northern Kenya, the Daasenach, found they land often on their heels, especially when going slowly. The Daasenach, however, are pastoralists who live in a hot, sandy desert and don't run very much. See Lieberman, D. E., et al. (2010). Foot strike patterns and collision forces in habitually barefoot versus shod runners. *Nature* 463: 531–35; Hatala, K. G., et al. (2013). Variation in foot strike patterns during running among habitually barefoot populations. *PLoS One* 8: e52548.

12. My personal opinion is that running *well* for long distances is a skill, just like other athletic skills such as swimming, throwing, or climbing, and that there is much to learn from the way experienced barefoot runners move. More research is needed, but many coaches and experts believe that good running form generally involves landing gently on a nearly flat foot, having short strides, in which the foot lands below the knee, using a high cadence of about 170–180 steps per minute, and not leaning too much at the hips. An important consideration, however, is that this style of running requires more strength in your feet and calf muscles. In addition, if you haven't run this way before, it is crucial to transition slowly and carefully to build up muscle strength and adapt your tendons, ligaments, and bones. Otherwise, you risk injury.

13. Milner, C. E., et al. (2006). Biomechanical factors associated with tibial stress fracture in female runners. *Medicine and Science in Sports and Exercise* 38: 323–28; Pohl, M. B., J. Hamill, and I. S. Davis (2009). Biomechanical and anatomic factors associated with a history of plantar fasciitis in female runners. *Clinical Journal of Sports Medicine* 19: 372–76. For a contrary hypothesis that impact peaks are not injurious because your body dampens the forces, see Nigg, B. M. (2010). *Biomechanics of Sports Shoes*. Calgary: Topline Printing.

14. Daoud, A. I., et al. (2012). Foot strike and injury rates in endurance runners: A Retrospective Study. *Medicine and Science in Sports and Exercise* 44: 1325–44.

15. Dunn, J. E., et al. (2004). Prevalence of foot and ankle conditions in a multiethnic community sample of older adults. *American Journal of Epidemiology* 159: 491–98.

16. Rao, U. B., and B. Joseph (1992). The influence of footwear on the prevalence of flat foot: A survey of 2300 children. *Journal of Bone and Joint Surgery* 74: 525–27; D'Août, K., et al. (2009). The effects of habitual footwear use: Foot shape and function in native barefoot walkers. *Footwear Science* 1: 81–94.

17. Chandler, T. J., and W. B. Kibler (1993). A biomechanical approach to the prevention, treatment and rehabilitation of plantar fasciitis. *Sports Medicine* 15: 344–52.

18. See Ryan, M. B., et al. (2011). The effect of three different levels of footwear stability on pain outcomes in women runners: A randomised control trial. *British Journal of Sports Medicine* 45: 715–21; Richards, C. E., P. J. Magin, and R. Callister (2009). Is your prescription of distance running shoes evidence-based? *British Journal of Sports Medicine* 43: 159–62; Knapick, J. J., et al. (2010). Injury reduction effectiveness of assigning running shoes based on plantar shape in Marine Corps basic training. *American Journal of Sports Medicine* 36: 1469–75.

19. Marti, B., et al. (1988). On the epidemiology of running injuries: The 1984 Bern Grand-Prix Study. *American Journal of Sports Medicine* 16: 285–94.

20. van Gent, R. M., et al. (2007). Incidence and determinants of lower extremity running injuries in long distance runners: A systematic review. *British Journal of Sports Medicine* 41: 469–80.

21. Nguyen, U. S., et al. (2010). Factors associated with hallux valgus in a population-based study of older women and men: The MOBILIZE Boston Study. *Osteoarthritis Cartilage* 18: 41–46; Goud, A., et al. (2011). Women's musculoskeletal foot conditions exacerbated by shoe wear: An imaging perspective. *American Journal of Orthopedics* 40: 183–91.

22. Kerrigan, D. C., et al. (2005). Moderate-heeled shoes and knee joint torques relevant to the development and progression of knee osteoarthritis. *Archives of Physical Medicine and Rehabilitation* 86: 871–75.

23. Also, where did we get the idea that shoes are less dirty than feet? How often do you clean your shoes compared to your feet? For a review of these and other issues, see Howell, L. D. (2010). *The Barefoot Book*. Alameda, CA: Hunter House.

24. Zierold, N. (1969). *Moguls*. New York: Coward-McCann.

25. Au Eong, K. G., T. H. Tay, and M. K. Lim (1993). Race, culture and myopia in 110,236 young Singaporean males. *Singapore Medical Journal* 34: 29–32; Sperduto, R. D., et al. (1983). Prevalence of myopia in the United States. *Archives of Ophthalmology* 101: 405–7.

26. Holm, S. (1937). The ocular refraction state of the Palaeo-Negroids in Gabon, French Equatorial Africa. *Acta Ophthalmology* 13(suppl.):1–299; Saw, S. M., et al. (1996). Epidemiology of myopia. *Epidemiologic Reviews* 18: 175–87.

27. Ware, J. (1813). Observations relative to the near and distant sight of different persons. *Philosophical Transactions of the Royal Society, London* 103: 31–50.

28. Tscherning, M. (1882). *Studier over Myopiers Aetiologi*. Copenhagen: C. Myhre.

29. Young, F. A., et al. (1969). The transmission of refractive errors within

Eskimo families. *American Journal of Optometry and Archives of the American Academy of Optometry* 46: 676–85.

30. For excellent reviews, see Foulds, W. S., and C. D. Luu (2010). Physical factors in myopia and potential therapies. In *Myopia: Animal Models to Clinical Trials*, ed. R. W. Beuerman, et al. Hackensack, NJ: World Scientific, 361–86; Wojciechowski, R. (2011). Nature and nurture: The complex genetics of myopia and refractive error. *Clinical Genetics* 79: 301–20; Young, T. L. (2009). Molecular genetics of human myopia: An update. *Optometry and Vision Science* 86: E8–E22.

31. Saw, S. M., et al. (2002). Nearwork in early onset myopia. *Investigative Ophthalmology and Vision Science* 43: 332–39.

32. Saw, S.M., et al. (2002). Component dependent risk factors for ocular parameters in Singapore Chinese children. *Ophthalmology* 109: 2065–71.

33. Jones, L. A. (2007). Parental history of myopia, sports and outdoor activities, and future myopia. *Investigative Ophthalmology and Vision Science* 48: 3524–32; Rose, K. A., et al. (2008). Outdoor activity reduces the prevalence of myopia in children. *Ophthalmology* 115: 1279–85; Dirani, M., et al. (2009). Outdoor activity and myopia in Singapore teenage children. *British Journal of Ophthalmology* 93: 997–1000.

34. The proposed dietary mechanism is that starchy meals elevate insulin levels, which then lead to higher levels of a particular growth factor in the blood (known as IGF-1) that acts not only on growth plates in bones but also on the walls of the eyeball. This mechanism, if correct, could help explain evidence that nearsighted people tend to be taller and grow faster than normal-sighted people, and that individuals with type 2 diabetes (who have high insulin levels) also tend to be more likely to be nearsighted. For more information, see Gardiner, P. A. (1954). The relation of myopia to growth. *Lancet* 1: 476–79; Cordain, L., et al. (2002). An evolutionary analysis of the aetiology and pathogenesis of juvenile-onset myopia. *Acta Ophthalmologica Scandinavica* 80: 125–35; Teikari, J. M. (1987). Myopia and stature. *Acta Ophthalmologica Scandinavica* 65: 673–76; Fledelius, H. C., J. Fuchs, and A. Reck (1990). Refraction in diabetics during metabolic dysregulation, acute or chronic with special reference to the diabetic myopia concept. *Acta Ophthalmologica Scandinavica* 68: 275–80.

35. These fibers are often termed the zonular fibers, and the old-fashioned name for them is the zonules of Zinn, named after the German naturalist Johann Gottfried Zinn (who also gives his name to the zinnia).

36. Sorsby, A., et al. (1957). *Emmetropia and Its Aberrations.* London: Her Majesty's Stationery Office.

37. Grosvenor, T. (2002). *Primary Care Optometry,* 4th ed. Boston: Butterworth-Heinemann.

38. McBrien, N. A., A. I. Jobling, and A. Gentle (2009). Biomechanics of the sclera in myopia: Extracellular and cellular factors. *Ophthalmology and Vision Science* 86: E23–30.

39. Young, F. A. (1977). The nature and control of myopia. *Journal of the American Optometric Association* 48: 451–57; Young, F. A. (1981). Primate myopia. *American Journal of Optometry and Physiological Optics* 58: 560–66.

428 Notes to Pages 333–338

40. Woodman, E. C., et al. (2011). Axial elongation following prolonged near work in myopes and emmetropes. *British Journal of Ophthalmology* 5: 652–56; Drexler, W., et al. (1998). Eye elongation during accommodation in humans: Differences between emmetropes and myopes. *Investigative Ophthalmology and Vision Science* 39: 2140–47; Mallen, E. A., P. Kashyap, and K. M. Hampson (2006). Transient axial length change during the accommodation response in young adults. *Investigative Ophthalmology and Vision Science* 47: 1251–54.
41. McBrien, N. A., and D. W. Adams (1997). A longitudinal investigation of adult-onset and adult-progression of myopia in an occupational group: Refractive and biometric findings. *Investigative Ophthalmology and Vision Science* 38: 321–33.
42. Hubel D., T. N. Wiesel, and E. Raviola (1977). Myopia and eye enlargement after neonatal lid fusion in monkeys. *Nature* 266: 485–88.
43. Raviola, E., and T. N. Weisel (1985). An animal model of myopia. *New England Journal of Medicine* 312: 1609–15.
44. Smith III, E. L., G. W. Maguire, and J. T. Watson (1980). Axial lengths and refractive errors in kittens reared with an optically induced anisometropia. *Investigative Ophthalmology and Vision Science* 19: 1250–55; Wallman, J., et al. (1987). Local retinal regions control local eye growth and myopia. *Science* 237: 73–77.
45. Rose, K. A., et al. (2008). Outdoor activity reduces the prevalence of myopia in children. *Ophthalmology* 115: 1279–85.
46. 2 Peter 1:9 (King James Version).
47. Nadell, M. C., and M. J. Hirsch (1958). The relationship between intelligence and the refractive state in a selected high school sample. *American Journal of Optometry and Archives of American Academy of Optometry* 35: 321–26; Czepita, D., E. Lodygowska, and M. Czepita (2008). Are children with myopia more intelligent? A literature review. *Annales Academiae Medicae Stetinensis* 54: 13–16.
48. Miller, E. M. (1992). On the correlation of myopia and intelligence. *Genetic, Social, and General Psychology Monographs* 118: 363–83.
49. Saw, S. M., et al. (2004). IQ and the association with myopia in children. *Investigative Ophthalmology and Vision Science* 45: 2943–48.
50. Rehm, D. (2001). *The Myopia Myth;* http://www.myopia.org/ebook/index.htm.
51. Leung, J. T., and B. Brown (1999). Progression of myopia in Hong Kong Chinese schoolchildren is slowed by wearing progressive lenses. *Optometry and Vision Science* 76: 346–54; Gwiazda, J., et al. (2003). A randomized clinical trial of progressive addition lenses versus single vision lenses on the progression of myopia in children. *Investigative Ophthalmology and Vision Science* 44: 1492–1500.
52. Rieff, C., K. Marlatt, and D. R. Denge (2011). Difference in caloric expenditure in sitting versus standing desks. *Journal of Physical Activity and Health* 9: 1009–11.
53. Convertino, V. A., S. A. Bloomfield, and J. E. Greenleaf (1997). An overview of the issues: Physiological effects of bed rest and restricted physical activity. *Medicine and Science in Sports and Exercise* 29: 187–90.

54. O'Sullivan, P. B., et al. (2006). Effect of different upright sitting postures on spinal-pelvic curvature and trunk muscle activation in a pain-free population. *Spine* 31: E707–12.

55. Lieber, R. L. (2002). *Skeletal Muscle Structure, Function, and Plasticity: The Physiological Basis of Rehabilitation.* Philadelphia: Lippincott, Williams and Wilkins.

56. Nag, P. K., et al. (1996). EMG analysis of sitting work postures in women. *Applied Ergonomics* 17: 195–97.

57. Riley, D. A., and J. M. Van Dyke (2012). The effects of active and passive stretching on muscle length. *Physical Medicine and Rehabilitation Clinics of North America* 23: 51–57.

58. Dunn, K. M., and P. R. Croft (2004). Epidemiology and natural history of lower back pain. *European Journal of Physical and Rehabilitation Medicine* 40: 9–13.

59. For a superb review, I recommend Waddell, G. (2004). *The Back Pain Revolution,* 2nd ed. Edinburgh: Churchill-Livingstone.

60. Violinn, E. (1997). The epidemiology of low back pain in the rest of the world: A review of surveys in low- and middle-income countries. *Spine* 22: 1747–54.

61. Hoy, D., et al. (2003). Low back pain in rural Tibet. *Lancet* 361: 225–26; Nag, A., H. Desai, and P. K. Nag (1992). Work stress of women in sewing machine operation. *Journal of Human Ergonomics* 21: 47–55.

62. The oldest known evidence for a mattress is 77,000 years old from the cave of Sidubu in South Africa. Apparently, the occupants of this cave slept on a bed of grasses and aromatic leaves (which keep off insects). See Wadley, L., et al. (2011). Middle Stone Age bedding construction and settlement patterns at Sibudu, South Africa. *Science* 334: 1388–91.

63. Adams, M. A., et al. (2002). *The Biomechanics of Back Pain.* Edinburgh: Churchill-Livingstone.

64. Mannion, A. F. (1999). Fibre type characteristics and function of the human paraspinal muscles: Normal values and changes in association with low back pain. *Journal of Electromyography and Kinesiology* 9: 363–77; Cassisi, J. E., et al. (1993). Trunk strength and lumbar paraspinal muscle activity during isometric exercise in chronic low-back pain patients and controls. *Spine* 18: 245–51; Marras, W. S., et al. (2005). Functional impairment as a predictor of spine loading. *Spine* 30: 729–37.

65. Mannion, A. F., et al. (2001). Comparison of three active therapies for chronic low back pain: Results of a randomized clinical trial with one-year follow-up. *Rheumatology* 40: 772–78.

13 Survival of the Fitter—*Can Evolutionary Logic Help Cultivate a Better Future for the Human Body?*

1. May, A. L., E. V. Kuklina, and P. W. Yoon (2012). Prevalence of cardiovascular disease risk factors among U.S. adolescents, 1999–2008. *Pediatrics* 129: 1035–41.

2. Olshansky, S. J., et al. (2005). A potential decline in life expectancy in the United States in the 21st century. *New England Journal of Medicine* 352: 1138–45.
3. World Heath Organization (2011). *Global Status Report on Noncommunicable Diseases 2010.* Geneva: WHO Press; http://whqlibdoc.who.int /publications/2011/9789240686458_eng.pdf.
4. Shetty, P. (2012). Public health: India's diabetes time bomb. *Nature* 485: S14–S16.
5. This number is based on a count of 18.8 million Americans with diagnosed type 2 diabetes in 2011 (another 7 million undiagnosed Americans are estimated to have the disease), and a total direct cost of $116 billion in 2007. For more, see http://www.cdc.gov/chronicdisease/resources/publications/AAG/ ddt.htm.
6. Russo, P. (2011). Population health. In *Health Care Delivery in the United States,* ed. A. R. Kovner and J. R. Knickman. New York: Springer, 85–102.
7. Byars, S. G., et al. (2009). Natural selection in a contemporary human population. *Proceedings of the National Academy of Sciences USA* 107 (suppl. 1): 1787–92.
8. Elbers, C. C. (2011). Low fertility and the risk of type 2 diabetes in women. *Human Reproduction* 26: 3472–78.
9. Pettigrew, R., and D. Hamilton-Fairley (1997). Obesity and female reproductive function. *British Medical Bulletin* 53: 341–58.
10. de Condorcet, M. J. A. (1795). *Esquisse d'un Tableau Historique des Progrès de l'Esprit Humain.* Paris: Agasse. For a modern-day futurist, see Ray Kurzweil's predictions: http://www.kurzweilai.net/predictions/download.php.
11. TODAY Study Group (2012). A clinical trial to maintain glycemic control in youth with type 2 diabetes. *New England Journal of Medicine* 366: 2247–56.
12. You can check the data for yourself at http://www.cdc.gov/nchs/. Note that mortality rates are adjusted to account for changes in the size and age of the population, and they are not biased by changes in how many people are diagnosed with a disease, because these are solely death rates.
13. Pritchard, J. K. (2001). Are rare variants responsible for susceptibility to common diseases? *American Journal of Human Genetics* 69: 124–37; Tennessen, J. A. (2012). Evolution and functional impact of rare coding variation from deep sequencing of human exomes. *Science* 337: 64–69; Nelson, M. R. (2012). An abundance of rare functional variants in 202 drug target genes sequenced in 14,002 people. *Science* 337: 100–4.
14. Yusuf, S., et al. (2004). Effect of potentially modifiable risk factors associated with myocardial infarction in 52 countries (the INTERHEART study): Case-control study. *Lancet* 364: 937–52.
15. Blair, S. N., et al. (1995). Changes in physical fitness and all-cause mortality: A prospective study of healthy and unhealthy men. *Journal of the American Medical Association* 273: 1093–98.
16. This number is based only on the thirteen million Americans who suffered from a stroke or a heart attack in 2011, but this is clearly an underestimate as these individuals are just a fraction of the number of people with heart disease. For more data, see Kovner, A. R., and J. R. Knickman (2011). *Health Care Delivery in the United States.* New York: Springer.

17. Russo, P. (2011). Population health. In *Health Care Delivery in the United States,* ed. A. R. Kovner and J. R. Knickman. New York: Springer, 85–102; see also http://report.nih.gov/award/.

18. Trust for America's Health (2008). *Prevention for a Healthier America: Investments in Disease Prevention Yield Significant Savings, Stronger Communities.* Washington, DC: Trust for America's Health. You can read the report at http://healthyamericans.org/reports/prevention08/.

19. Brandt, A. M. (2007). *The Cigarette Century.* New York: Basic Books.

20. McTigue, K. M., et al. (2003). Screening and interventions for obesity in adults: Summary of the evidence for the U.S. Preventive Services Task Force. *Archives of Internal Medicine* 139: 933–49; http://www.cdc.gov/nchs/data/hus/hus11.pdf#073.

21. For a vigorous critique of the way profit motives warp medicine, see Bortz, W. M. (2011). *Next Medicine: The Science and Civics of Health.* Oxford: Oxford University Press.

22. Glanz, K., B. K. Rimer, and K. Viswanath (2008). Theory, research and practice in health behavior and health education. In *Health Behavior in Education: Theory, Research and Practice,* 4th ed. San Francisco: Jossey-Bass, 23–41.

23. Institute of Medicine (2000). *Promoting Health: Intervention Strategies from Social and Behavioral Research.* Washington, DC: National Academy Press.

24. Orleans, C. T., and E. F. Cassidy (2011). Health and behavior. In *Health Care Delivery in the United States,* ed. A. R. Kovner and J. R. Knickman. New York: Springer, 135–49.

25. Gantz, W., et al. (2007). *Food for Thought: Television Food Advertising to Children in the United States.* Menlo Park, CA: Kaiser Family Foundation.

26. Hager, R., et al. (2012). Evaluation of a university general education health and wellness course delivered by lecture or online. *American Journal of Health Promotion* 26: 263–69.

27. Cardinal, B. J., K. M. Jacques, and S. S. Levy (2002). Evaluation of a university course aimed at promoting exercise behavior. *Journal of Sports Medicine and Physical Fitness* 42: 113–19; Wallace, L. S., and J. Buckworth (2003). Longitudinal shifts in exercise stages of change in college students. *Journal of Sports Medicine and Physical Fitness* 43: 209–12; Sallis, J. F., et al. (1999). Evaluation of a university course to promote physical activity: Project GRAD. *Research Quarterly for Exercise and Sport* 70: 1–10.

28. Galef Jr., B. G. (1991). A contrarian view of the wisdom of the body as it relates to dietary self-selection. *Psychology Reviews* 98: 218–23.

29. See Birch, L. L. (1999). Development of food preferences. *Annual Review of Nutrition* 19: 41–62; Popkin, B. M., K. Duffey, and P. Gordon-Larsen (2005). Environmental influences on food choice, physical activity and energy balance. *Physiology and Behavior* 86: 603–13.

30. Webb, O. J., F. F. Eves, and J. Kerr (2011). A statistical summary of mall-based stair-climbing interventions. *Journal of Physical Activity and Health* 8: 558–65.

31. For two popular books on behavioral economics that explain how we make such decisions, I recommend Kahneman, D. (2011). *Thinking Fast and*

Thinking Slow. New York: Farrar, Straus and Giroux; and Ariely, D. (2008). _Predictably Irrational: The Hidden Forces That Shape Our Decisions_. New York: Harper.

32. National child labor laws in the United States, which limit the hours and kinds of work that children may do, were not passed until 1938.

33. See Feinberg, J. (1986). _Harm to Self_. Oxford: Oxford University Press; Sunstein, C., and R. Thaler (2008). _Nudge: Improving Decisions About Health, Wealth, and Happiness_. New Haven, CT: Yale University Press.

34. http://www.surgeongeneral.gov/initiatives/healthy-fit-nation/obesityvision2010.pdf.

35. Johnstone, L. D., J. Delva, and P. M. O'Malley (2007). Sports participation and physical education in American secondary schools. _American Journal of Preventive Medicine_ 33(4S): S195–S208.

36. Avena, N. M., P. Rada, and B. G. Hoebel (2008). Evidence for sugar addiction: Behavioral and neurochemical effects of intermittent, excessive sugar intake. _Neuroscience Biobehavioral Reviews_ 32: 20–39.

37. Garber, A. K., and R. H. Lustig (2011). Is fast food addictive? _Current Drug Abuse Reviews_ 4: 146–62.

Index

Page numbers in *italics* refer to illustrations.

ALLEN LANE
an imprint of
PENGUIN BOOKS

Recently Published

Vincent Deary, *How We Are: Book One of the How to Live Trilogy*

Henry Kissinger, *World Order*

Alexander Watson, *Ring of Steel: Germany and Austria-Hungary at War, 1914-1918*

Richard Vinen, *National Service: Conscription in Britain, 1945-1963*

Paul Dolan, *Happiness by Design: Finding Pleasure and Purpose in Everyday Life*

Mark Greengrass, *Christendom Destroyed: Europe 1517-1650*

Hugh Thomas, *World Without End: The Global Empire of Philip II*

Richard Layard and David M. Clark, *Thrive: The Power of Evidence-Based Psychological Therapies*

Uwe Tellkamp, *The Tower: A Novel*

Zelda la Grange, *Good Morning, Mr Mandela*

Ahron Bregman, *Cursed Victory: A History of Israel and the Occupied Territories*

Tristram Hunt, *Ten Cities that Made an Empire*

Jordan Ellenberg, *How Not to Be Wrong: The Power of Mathematical Thinking*

David Marquand, *Mammon's Kingdom: An Essay on Britain, Now*

Justin Marozzi, *Baghdad: City of Peace, City of Blood*

Adam Tooze, *The Deluge: The Great War and the Remaking of Global Order 1916-1931*

John Micklethwait and Adrian Wooldridge, *The Fourth Revolution: The Global Race to Reinvent the State*

Steven D. Levitt and Stephen J. Dubner, *Think Like a Freak: How to Solve Problems, Win Fights and Be a Slightly Better Person*

Alexander Monro, *The Paper Trail: An Unexpected History of the World's Greatest Invention*

Jacob Soll, *The Reckoning: Financial Accountability and the Making and Breaking of Nations*

Gerd Gigerenzer, *Risk Savvy: How to Make Good Decisions*

James Lovelock, *A Rough Ride to the Future*

Michael Lewis, *Flash Boys*

Hans Ulrich Obrist, *Ways of Curating*

Mai Jia, *Decoded: A Novel*

Richard Mabey, *Dreams of the Good Life: The Life of Flora Thompson and the Creation of* Lark Rise to Candleford

Danny Dorling, *All That Is Solid: The Great Housing Disaster*

Leonard Susskind and Art Friedman, *Quantum Mechanics: The Theoretical Minimum*

Michio Kaku, *The Future of the Mind: The Scientific Quest to Understand, Enhance and Empower the Mind*

Nicholas Epley, *Mindwise: How we Understand what others Think, Believe, Feel and Want*

Geoff Dyer, *Contest of the Century: The New Era of Competition with China*

Yaron Matras, *I Met Lucky People: The Story of the Romani Gypsies*

Larry Siedentop, *Inventing the Individual: The Origins of Western Liberalism*

Dick Swaab, *We Are Our Brains: A Neurobiography of the Brain, from the Womb to Alzheimer's*

Max Tegmark, *Our Mathematical Universe: My Quest for the Ultimate Nature of Reality*

David Pilling, *Bending Adversity: Japan and the Art of Survival*

Hooman Majd, *The Ministry of Guidance Invites You to Not Stay: An American Family in Iran*

Roger Knight, *Britain Against Napoleon: The Organisation of Victory, 1793-1815*

Alan Greenspan, *The Map and the Territory: Risk, Human Nature and the Future of Forecasting*

Daniel Lieberman, *Story of the Human Body: Evolution, Health and Disease*

Malcolm Gladwell, *David and Goliath: Underdogs, Misfits and the Art of Battling Giants*

Paul Collier, *Exodus: Immigration and Multiculturalism in the 21st Century*

John Eliot Gardiner, *Music in the Castle of Heaven: Immigration and Multiculturalism in the 21st Century*

Catherine Merridale, *Red Fortress: The Secret Heart of Russia's History*

Ramachandra Guha, *Gandhi Before India*

Vic Gatrell, *The First Bohemians: Life and Art in London's Golden Age*

Richard Overy, *The Bombing War: Europe 1939-1945*

Charles Townshend, *The Republic: The Fight for Irish Independence, 1918-1923*

Eric Schlosser, *Command and Control*

Sudhir Venkatesh, *Floating City: Hustlers, Strivers, Dealers, Call Girls and Other Lives in Illicit New York*

Sendhil Mullainathan and Eldar Shafir, *Scarcity: Why Having Too Little Means So Much*

John Drury, *Music at Midnight: The Life and Poetry of George Herbert*

Philip Coggan, *The Last Vote: The Threats to Western Democracy*

Richard Barber, *Edward III and the Triumph of England*

Daniel M Davis, *The Compatibility Gene*

John Bradshaw, *Cat Sense: The Feline Enigma Revealed*

Roger Knight, *Britain Against Napoleon: The Organisation of Victory, 1793-1815*

Thurston Clarke, *JFK's Last Hundred Days: An Intimate Portrait of a Great President*

Jean Drèze and Amartya Sen, *An Uncertain Glory: India and its Contradictions*

Rana Mitter, *China's War with Japan, 1937-1945: The Struggle for Survival*

Tom Burns, *Our Necessary Shadow: The Nature and Meaning of Psychiatry*

Sylvain Tesson, *Consolations of the Forest: Alone in a Cabin in the Middle Taiga*

George Monbiot, *Feral: Searching for Enchantment on the Frontiers of Rewilding*

Ken Robinson and Lou Aronica, *Finding Your Element: How to Discover Your Talents and Passions and Transform Your Life*

David Stuckler and Sanjay Basu, *The Body Economic: Why Austerity Kills*

Suzanne Corkin, *Permanent Present Tense: The Man with No Memory, and What He Taught the World*

Daniel C. Dennett, *Intuition Pumps and Other Tools for Thinking*

Adrian Raine, *The Anatomy of Violence: The Biological Roots of Crime*

Eduardo Galeano, *Children of the Days: A Calendar of Human History*